Springer Proceedings in Physics

Volume 349

Indexed by Scopus

The series Springer Proceedings in Physics, founded in 1984, is devoted to timely reports of state-of-the-art developments in physics and related sciences. Typically based on material presented at conferences, workshops and similar scientific meetings, volumes published in this series will constitute a comprehensive up to date source of reference on a field or subfield of relevance in contemporary physics. Proposals must include the following:

- Name, place and date of the scientific meeting
- A link to the committees (local organization, international advisors etc.)
- Scientific description of the meeting
- List of invited/plenary speakers
- An estimate of the planned proceedings book parameters (number of pages/ articles, requested number of bulk copies, submission deadline).

Please contact:

For Americas and Europe: Dr. Zachary Evenson; zachary.evenson@springer.com
For Asia, Australia and New Zealand: Dr. Loyola DSilva; loyola.dsilva@springer.com

Ashok Vaseashta · Mohammed Essaid Achour · Mustapha Mabrouki · Amina Tachafine · Mustapha Aitali
Editors

Proceedings of the Eighth International Symposium on Dielectric Materials and Applications (ISyDMA'8)

Springer

Editors
Ashok Vaseashta
International Clean Water Institute
Manassas, VA, USA

Ghitu Institute of Electronic Engineering and Nanotechnologies
Academy of Sciences of Moldova
Chisinau, Moldova

University of Bucharest
Măgurele, Romania

Mohammed Essaid Achour
Université Ibn-Tofail
Kenitra, Morocco

Amina Tachafine
Unité de Dynamique et Structure des Matériaux Moléculaires
Université du Littoral Côte d'Opale (ULCO)
Calais, France

Mustapha Mabrouki
Laboratoire Génie Industriel Ingénierie des Surfaces (LGIIS)
Industrial Engineering Laboratory
Sultan Moulay Slimane University
Beni Mellal, Morocco

Mustapha Aitali
Faculty of Sciences Semlalia
Cadi Ayyad University
Marrakech, Morocco

ISSN 0930-8989 ISSN 1867-4941 (electronic)
Springer Proceedings in Physics
ISBN 978-3-031-84811-7 ISBN 978-3-031-84809-4 (eBook)
https://doi.org/10.1007/978-3-031-84809-4

This Springer imprint is published by the registered company Springer Nature Switzerland AG
The registered company address is: Gewerbestrasse 11, 6330 Cham, Switzerland

Foreword

I am very pleased to provide a few words for the proceedings of the Eighth International Symposium on Dielectric Materials and Applications (ISyDMA'8), which was held in Orlando, Florida, USA from May 12 to 16, 2024. Professor Ashok Vaseashta served as the host chair of the conference. The conference co-chairs included Prof. Khudaverdyan, Armenia; Prof. Popov, Germany; Prof. Balazsi, Hungary; Prof. Dekhtyar, Latvia; and Profs. Oueriagli, Achour, Aitali, and Mabrouki, Morocco. Over 100 participants from 28 countries participated in the event, in a hybrid modality, consisting of 12 keynote lectures, 28 invited presentations, 15 contributed talks, 10 online presentations, and 15 posters, covering a broad range of topics from basic to applied sciences related to major issues such as energy, environment, biotechnology, electronics, and optoelectronics.

The conference was the eighth in the series of ISyDMA, which are focused on recent developments in dielectric materials and their applications. Due to the further miniaturization of devices around us, the study of dielectrics is significant and vital to the exponential growth of technologies worldwide. Hence, the conference series is aimed at materials specialists, physicists, chemists, biologists, and engineers engaged in fundamental and applied research on dielectric materials and their innovative applications. A few selected publications will be published in conference proceedings to be published by Springer Nature, the Netherlands. I am delighted that Springer Nature has provided us with an opportunity to publish conference proceedings in a series dedicated to ISyDMA. The contents of these proceedings cover basic physics, materials, device processing, and scientific applications in electronics, biomedical, aviation, decentralized power generation, tactile sensing, safety, and security. The contents of the proceedings are very timely, due to the increased use of dielectrics in commercialization, urbanization, and industrialization to support our growing consumption in the era of technological advances. The topics presented in this edition are extremely unique and distinct, and I foresee that this and future editions of the ISyDMA proceedings will cover many more exciting topics to address global challenges, not only in the areas of electronics but also for our future sustainability.

I express my thanks to the chair, co-chairs, and co-editors for their contributions to making this event successful.

Prof. Dr. Luís Cadillon Costa
I3N and Physics Department
University of Aveiro
Aveiro, Portugal

The original version of the book has been revised. A correction to this book can be found at https://doi.org/10.1007/978-3-031-84809-4_27

Preface

The ISyDMA conferences focus on recent developments in dielectric materials and their applications. The conference series is aimed at PhD students, post-docs, researchers, and engineers from public and private research organizations concerned with innovations in advanced dielectric materials for biomedical, biotechnological, and industrial applications. The target audience comprises materials specialists, physicists, chemists, biologists, and engineers engaged in fundamental and applied research on dielectric materials and their innovative applications. Due to broad interest in this topic, the conference series continues to gain momentum in size, publicity, and overall quality of publications.

The first ISyDMA conference was held in Rabat-Kenitra, Morocco, in 2016 under the founding chair Prof. Achour and co-chairs, Profs. Messoussi, Touahni, Oueriagli, Berrada, Outzoughit and Mabrouki. The following editions took place in Bucharest, Romania, in 2017, in Beni Mellal (Morocco) in 2018, in Amman (Jordan) in 2019, and as a virtual event due to the COVID-19 global pandemic, in 2020. The 6th edition of the ISyDMA was held in Calais, France, with Prof. Tachafine, Fasquelle, Singh, Carru, Leroy, Desfeux, Courtois, Gagou, Rguiti, Lorgouilloux, Chambrier, Ferri, Da Costa, Jouiad, Lemee, Achour, Oueriagli, Mabrouki, Ait Ali, Leblanc, and Vaseashta as co-directors. Due to the global COVID pandemic, the conference in 2020 was organized completely online. Although conditions were relaxed somewhat in 2021, the restrictions were reimposed due to the emergence of a couple of new variants, such as Delta and Omicron. Hence, despite shifting the conference schedule, the mobility was limited, so the ISyDMA'6 was organized in a hybrid format. ISyDMA'7 was organized primarily online to accommodate the influx of people from Ukraine due to the regional conflict.

Additionally, due to the success of the ISyDMA'3 pre-conference summer school, viz. First Moroccan Spring School on Advanced Materials (MoSSAM'1), in Marrakech, Morocco, from 15 to 17 April 2018, subsequent summer schools, viz., the ISyDMA'4 Pre-Conference School in Irbid, Jordan, from 30 April to 1 May 2019, the ISyDMA'6 Pre-Conference summer school, viz. International Summer School on Advanced Materials for Energy (ISSAME) was organized from 13 to 14 December 2021 in Saint-Omer, France. The ISSAME was intended for young scientists and

Ph.D. students interested in Materials Science for energy. The scope of the Summer School is to familiarize students with the state of the art in advanced materials for energy, hence the ISSAME was focused on theoretical and experimental aspects of synthesis and characterization of advanced materials for energy, i.e. materials that meet strong societal demand for the conversion and storage of electrical energy as well as energy savings. The school aimed to present an overview of the latest developments in the theory, modeling, processing, characterization, and applications of advanced materials. It covered topics such as Electronic and Optical Materials, Polymers and Composites, oxides, semiconducting, metallic, and organic materials, materials for supercapacitors, fuel cells, and batteries, photovoltaic devices, and related technologies. Lectures in these areas were given by high-level researchers, and the participants had the opportunity to discuss their research work and exchange ideas with the scientists involved in this area. As before, the pre-conference School's target audience was graduate students and junior researchers (Ph.D. students, postdocs).

The Proceedings of ISYDMA'8 result from a collection of papers presented during the conference hosted by the International Clean Water Institute in Orlando, FL, from May 12–16, 2024. The location of the ISYDMA'8 was decided, after much consideration, to be Orlando, Florida, US, since the location is ideal for researchers to spend some time with their friends and families and to enjoy a few days of new ideas while having the opportunity to visit Orlando and its nearby attractions.

The scientific basis of the conference has deep-rooted significance, which stems from history, viz. Faraday first published numerical measurements on dielectrics and found that the capacity of a condenser was dependent on the nature of the material separating the conducting surfaces. This discovery stimulated further empirical understanding of insulating materials aimed at maximizing the amount of charge that can be stored by a capacitor, first constructed by Cunaeus and Musschenbroek in 1745, to be termed the Leyden jar. Throughout much of the 19th century, several comprehensive mechanisms leading to an understanding of the behavior of these materials appeared. Clausius and Mossotti conducted a systematic investigation of the dielectric properties of materials to correlate the specific inductive capacity, introduced by Faraday, and termed as "dielectric constant" realizing the microscopic structure of the material, in terms of the real part of the dielectric constant and the volume fraction occupied by the conducting particles in the dielectric. Debye later succeeded in extending the Clausius-Mossotti theory to consider the permanent moments of the molecules, which allowed him and others to determine the molecular dipole moment from the measurement of the dielectric constant. His theory was later extended by Onsager and Kirkwood and served as an excellent agreement with experimental results for most of the polar liquids. Debye's other significant contributions included applying the molecular permanent dipole moment concept to explain the anomalous dispersion of the dielectric constant observed by Drude. The extension of Debye's theory was pioneered by Cole, Davidson, and Williams, who interpreted the non-Debye relaxation behavior of the material in terms of the superposition of an exponentially relaxing process, which led to a distribution of relaxation times. The second extension was offered by Joncher, who proposed that the relaxation behavior at the molecular level is intrinsically non-Debye-like due to the cooperative molecular motions.

After over ninety years of research, the theory of dielectrics is still very active in understanding the behavior of dielectric material and is critical for present-day high-k electronics. The reduced dimensions of metal-oxide-semiconductor (MOS) devices for ultra-large-scale integration (ULSI) applications have been placing increasing pressure on the performance of gate dielectrics. The durability of gate dielectric material has become an important issue as device gate thickness is significantly reduced, leading to the search for high-k dielectrics, especially for high-frequency wireless communication technologies. The constant need for miniaturization provides a continuing driving force for the discovery of increasingly sophisticated materials to perform the same or improved function with decreased size and weight. The dielectric materials mentioned above are used as the basis for resonators and filters for the microwave carrying the desired information. These materials are presently employed as bulk ceramics in microwave communication devices. The need for better dielectrics with improved properties suitable for modern integrated manufacturing needs is the motivation behind the ISyDMA conferences, open to scientists, physicists, chemists, biologists, and electrical engineers engaged in fundamental and applied research or technical investigations on such materials.

The ISyDMA'8 proceeding consists of 26 articles selected to represent research topics discussed during the conference. The proceeding begins with several keynote speakers who provide an overview of mechanisms, processes, and applications. An article by Tenne et al. provides an update on the synthesis and characterization of new inorganic nanotubes from 2D compounds, such as WS_2, and from misfit layered compounds from theoretical and experimental perspectives. Another article by Fomin et al. describes first-order resonant Raman spectroscopy in an external magnetic field as a method for exploring configurations of energy levels of two-dimensional transition metal dichalcogenide semiconductors for its use in device applications. The behavior of the magneto-polaron resonances as a function of the phonon symmetry inherent in the monolayer transition metal dichalcogenide is described. An article by Sergey et al. describes dielectric barrier discharge and its application in generating and enhancing H_2O_2 action for various applications. The following article by Prof. Mabrouki discusses energy harvesting efficiency analysis using artificial intelligence, notably the optimization of piezoelectric and triboelectric nanogenerators regarding recognition, production, and consumption. It proposes an adaptive approach using machine learning and neural network techniques to collect and optimize vibration energy. The following article by Prof. Kervalishvili on Spintronics and Spinquant 2D Structures was prepared using the laser plasma method. The paper elaborates on using leptons-based spintronic and spinqubit systems obtained by Laser Plasma technology of 2D materials. A chapter by Prof. Khudaverdyan discusses new functionalities of semiconductor photodetectors with a high-resistivity layer, which accounts for the mutual compensation of photocurrents from the potential barriers and, consequently, the short-wave and long-wave spectral maxima. The next chapter by Dr. Boukheir describes superhydrophobic coating based on functionalized SiO_2 nanoparticles for self-cleaning and anti-icing. An article by Prof. Phanjoubam describes magnetoelectric coupling enhancement at room temperature in $Na_{0.5}Bi_{0.5}TiO_3$–$BaFe_{12}O_{19}$ based flexible polymer films. Next, Dr. Le Sage presents a fascinating study of compact microwave absorbers using additive manufacturing. Prof. Hamouda describes, in the following article, the effect of pollution on high-voltage power line insulators. In her article, Dr. Soro presents the impact of the change of chirality of green dielectric gold nanoparticles on antimicrobial activity. Prof. Villagracia, next, presents a meta-GGA calculation of the optical and electronic properties of pristine and N-doped penta-graphene. Proof Neves presents his work on the effects of laser-power dependence on the structural stability of CeZr and ZrMn catalyst oxide studied by Raman Spectroscopy. Prof. Dumitru presents a study of the intermediate phase glasses of the As-S-Ge ternary in terms of their self-organization, properties, and applications in programmable metallization cells. Prof. El-Kissani presents his work on eco-friendly chalcogenide semiconductors in terms of elaboration, characterization, and application in photovoltaics. Next, Prof. Rahmani presents Raman spectroscopy of pristine and filled single-walled carbon nanotubes. Next, a chapter by Noorali, Baldwin, and Vaseashta presents an interesting study on next-generation sensors using artificial intelligence as enabling technologies for biomedical and healthcare applications, more specifically for antibiotic-resistant diseases. An article by Davis presents innovative nanocomposites for energy harvesting and biomedical sensing applications. Dr. Barnoss, next, presents her work on dielectric behavior and AC electrical conductivity of composite materials based on conducting polymers. Dr. El Hamzi presents her work on modeling the variation of time, temperature, and composition in the thermal treatment of Al-Si alloys. There are several student articles on various aspects of dynamic simulation of buildings in Morocco, including one on the machine learning-driven optimization of green hydrogen production in porous electrode alkaline electrolysis cells. Overall, the articles presented here covered several unique aspects of dielectrics and their various applications.

Disclaimers: The opinions, recommendations, and suggestions presented in chapters represent the findings and views of the authors and may not express the position(s) of the editors. The editors do not promote, endorse, or express their dissent against any product and/or commercial entities cited in the articles. Due precautions are exercised before publication, and the editors do not take any responsibility for infringement by authors of any copyrighted material or copied from elsewhere.

Manassas, VA, USA	Prof. Ashok Vaseashta
Kenitra, Morocco	Prof. Mohammed Essaid Achour
Beni Mellal, Morocco	Prof. Mustapha Mabrouki
Calais, France	Prof. Amina Tachafine
Marrakech, Morocco	Prof. Mustapha Aitali

ISYDMA8 Photos

Springer

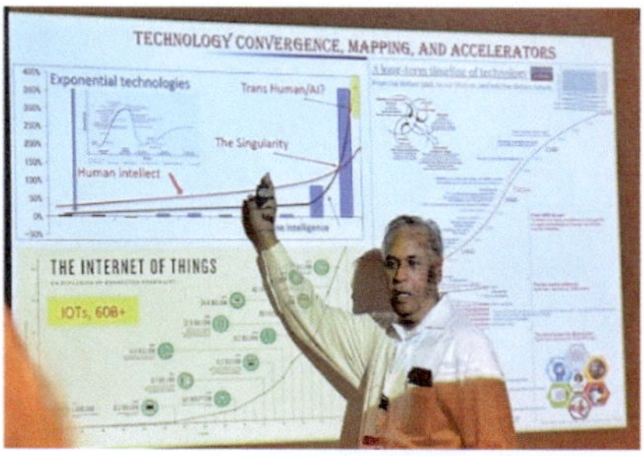
TECHNOLOGY CONVERGENCE, MAPPING, AND ACCELERATORS
Exponential technologies
Trans Human/AI?
The Singularity
Human intellect
THE INTERNET OF THINGS
IOTs, 60B+

Galaxy A15

ISYDMA8

Contents

Contributed Articles

About the Editors

Ashok Vaseashta is CEO/CTO and Executive Director for research with the International Clean Water Institute in Virginia, USA. He also serves in several honorary positions, such as a Professor at the Transylvania University of Brasov; Chaired Professor of Nanotechnology at the Ghitu Institute of Electrical Engineering and Nanotechnologies, Academy of Sciences of Moldova; Academician at the Euro-Mediterranean Academy of Arts and Sciences, and Senior Strategic Research Advisor for several organizations. Inspired by nature and guided by societal necessities, he strives for technological innovations to address the global challenges of the 21st century. He is a scholar, visionary, strategist, and dedicated futurist providing strategic leadership to promote and advance research initiatives and priorities using data-driven decisions. He received Ph.D. from the Virginia Polytechnic Institute and State University, Blacksburg, VA in 1990 followed by Kobe's post-doctoral fellowship. Following his Ph.D., he served as a professor and researcher at Virginia Tech and Marshall University. He also served as the Director of Research at the Institute for Advanced Sciences Convergence and International Clean Water Institute for Norwich University Applied Research Institutes, Vice-Provost (Rector) for Research in South Carolina, and Executive Director and Chair of the Institutional Review Board at a State University in New Jersey. His earlier honorary positions included a Visiting Professor at the Riga Technical University, Latvia, 3 Nano-SAE Research Centre, University of Bucharest, Romania, Transilvania University of Brasov,

Romania, and a Visiting Scientist at the Weizmann Institute of Science, Israel. He served the U.S. Department of State in two rotations, as a strategic S&T advisor in the Bureau of International Security and Nonproliferation, Office of Weapons of Mass Destruction and Terrorism, and U.S. diplomat. His research interests span foresight, nanotechnology, environmental/ ecological science, green chemistry, the environmental impact of micro/nano plastics, and critical infrastructure safety and security- all using the nexus of advanced technological solution platforms. He is the author/editor of 18 books and has published over 350 articles in scientific journals, book chapters, and conferences. He serves on the editorial boards of several international journals and is an active member of various professional organizations.

Mohammed Essaid Achour received his "Thèse de 3ème cycle" from The Bordeaux University (France) and "Thèse d'état" from the Moulay Ismail University of Meknes (Morocco) degrees in the field of Physics in 1983 and 1991, respectively. From 1983 to 1992, he was a "Maitre assistant", at the Sciences Faculty of Meknes (Morocco), "Maitre de Conférence" (1992–1996), and a Professor. Professor Achour joined the Sciences Faculty, Ibn Tofail University at Kenitra in 1999. From 1997 to 2011, he also taught at the Royal Military Academy in Meknes, Morocco. He is an Expert evaluator and, a member of the scientific committee of the National Center for Scientific and Technical Research (CNRST) scientific committee, the Moroccan Center for Innovation (CIM). He is an honorary guest professor at Brest University in France and a visiting scientist/ researcher at different universities and research Centers in several countries worldwide. His research interests include electromagnetic and electrical properties, microwave characterization, and dielectric responses of composite materials: carbon dots, graphene, carbon nanotubes, carbon black, and natural fibers in natural or synthetic polymers. He has co-authored peer-reviewed more than 100 scientific papers published in leading refereed journals, 14 book chapters, and 6 guest editorials. Pr. Achour was the founder and chairs of different editions of the International Symposium on Dielectric Materials and Applications (ISyDMA), the chair of the Moroccan Spring School on Advanced Materials (MoSSAM'1: Marrakech, Morocco April 15–17, 2018 and MoSSAM'2: MAScIR, Rabat, Morocco June 24–25, 2019), the International School on Advanced Materials (InSAM'3; Ouarzazate, Morocco May 2–4, 2023), the Fourth International Meeting on Dielectric Materials (IMDM'4; Marrakech, Morocco May 29–31, 2013), the Co-director of two editions of NATO Advanced Study Institute, Advanced Technologies for Detection and Defence Against CBRN Agents (2019, 2024), Sozopol, Bulgaria and the Co-Chair of the International Symposium on the Advanced Materials for Optics Micro-Electronics and Nanoelectronics (AMOMEN; Kenitra, Morocco, October 27–29, 2011).

Mustapha Mabrouki is a Permanent Professor in Sultan Moulay Slimane University. He obtained a doctorate 3rd cycle graduate and state thesis (Thèse d'Etat) from Cadi Ayyad University in 2004. Followed by a Postdoctoral Fellowship at the University of Miami (USA) with Doctor Roger M. LEBLANC in Supramolecular Research Center. He is a professor at the Faculty of Sciences and Technologies Beni Mellal since 1994. Head and member of the Industrial Engineering Laboratory (LGI). His field of interest is the organic and inorganic materials applied in electronic and optoelectronic areas. His actual work is to understand how the surface is involved in biological adhesion. Professor Mabrouki co-authors more than seventy articles and a hundred papers at national and international conferences. He has participated in several scientific events (conferences, workshops, and meetings) as chairman. He was the chair of the Third International Symposium on Dielectric Materials and Applications "ISyDMA'2018" (Beni Mellal, Morocco April 17–19, 2018), a member and an organizer. He was also the source of several cooperation projects at the national level and international level. He is also a member of many scientific associations like the European Physical Society (EPS), Moroccan Society of Applied Physics (FSSM) Marrakech, and an active member of the Moroccan Society of Nanotechnology (MANAT).

Amina Tachafine is an Associate Professor at the Littoral-Côte d'Opale University in Calais, France. She obtained her engineering degree in electronics at the Polytech' Lille School in 1989 and her Ph.D. in electronics from the Electronics, Microelectronics and Nanotechnology Institute, University of Lille, France in 1994. She is currently part of the Dynamics and Structure of Molecular Materials Unit. She is interested in lead-free ferroelectric and piezoelectric materials for electronic applications. Her research activities concern frequency-tunable systems for radiofrequency and microwave applications, energy harvesting and storage devices for powering autonomous systems, and new Aurivillius materials for information storage. She co-authored some thirty articles and made around forty oral communications at national and international conferences. She has also been in charge of European and international research projects.

Mustapha Aitali is a Professor at the Cadi Ayyad University. Professor Aitali Mustapha is a full professor and Head of the Chemistry Department at the University Cadi Ayyad in Marrakech. He is working in Organic Chemistry, Organometallic Chemistry, and catalysis in the Department of Chemistry, Faculty of Sciences Semlalia, Marrakech-Morocco. His research interests include Coordination Chemistry, asymmetric Catalysis, Green chemistry, and the chemistry of nanostructured materials: graphene; silicene, and phosphorene. Pr. Aitali served as a guest professor at Villeneuve d'Ascq University, France, ENS Chimie de Rennes France and at the University of Cergy Pontoise France. He co-authored more than 150 papers published in leading refereed journals. He participated in more than 70 congresses and supervised more than 12 Ph.D. students. He also participated in several international cooperation projects, and he was an active member in the organization of several international conferences.

Keynote Articles

Inorganic Nanotubes: From WS_2 to "Misfit" Layer Compounds

V. Kundrat, M. B. Sreedhara, and R. Tenne

Abstract Update on the synthesis and characterization of new inorganic nanotubes from 2D compounds, like WS_2, and from misfit layered compounds (MLC) will be reported. The stability of nanotubes from 2D compounds will be discussed from theoretical and experimental perspectives. A few recent experiments encompassing the optical, electrical, electro-optical, and electromechanical effects of WS_2 (MoS_2) nanotubes will be discussed as well. Potential applications are presented.

Keywords 2D materials · Inorganic nanotubes · Misfit layer compounds · WS_2

1 Introduction

Following the invention of ray diffraction (XRD) and the establishment of the Bragg equation over 100 years ago, the structure of anisotropic compounds with layered structures (2D materials) was elucidated.[1–3] In these compounds, the chemical bond between each two atoms within the layer (*a-b* plane) is very strong (> 2 eV) and highly oriented. On the other hand, large interlayer spacing (>4 Å) along the *c*-axis suggested that the layers are staked together via weak van der Waals forces (< 0.1 eV/atom) which lack any angular dependence. This anisotropy explained the large differences in the physio-chemical properties of 2D materials within the molecular slab (the *a-b* plane) and along the *c*-axis. For instance, the electrical conductivity of graphite is about 100 times larger in the *a-b* plane compared with that along the

V. Kundrat · R. Tenne (✉)
Department of Molecular Chemistry and Materials Science, Weizmann Institute, Rehovot, Israel
e-mail: reshef.tenne@weizmann.ac.il

V. Kundrat
Thermo Fisher Scientific, Brno, Czech Republic

M. B. Sreedhara
Solid State and Structural Chemistry Unit, Indian Institute of Science, Bangalore, India

A. Vaseashta et al. (eds.), *Proceedings of the Eighth International Symposium on Dielectric Materials and Applications (ISyDMA'8)*, Springer Proceedings in Physics 349, https://doi.org/10.1007/978-3-031-84809-4_1

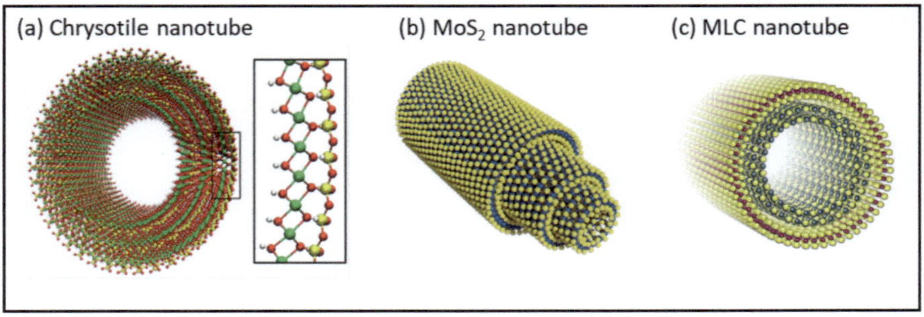

Fig. 1 **a** Schematics of a chrysotile nanotube and magnified portion of its wall: green, red, yellow, and white balls are Mg, O, Si, and H atoms, respectively (graciously provided by Dr. A. N. Enyashin); **b** Schematics of three wall WS_2 nanotubes: yellow and blue balls are S and W atoms, respectively (graciously provided by Dr. Sudarson S. Sinha); **c** schematic drawing of an MLC nanotube

c-axis. Layer compounds can be cleaved using a sharp wedge or by sticking and lifting an adhesive tape.

The propensity of compounds with layered structures to form nanotubes is not new. Already in 1930, Pauling proposed[4] that kaolinite $Al_2Si_2O_5(OH)_4$) and chrysotile $Mg_3(Si_2O_5)(OH)_4$ made of a layer of alumina (magnesia) octahedra connected to silica tetrahedra via a common oxygen atom, possess an asymmetric structure in the *a-b* plane. Hence, he conjectured that the layer would not remain flat and bend (see schematics in **Fig. 1a**). Chrysotile nanotubes were confirmed 20 years later when transmission electron microscopy acquired sufficient resolution. [5]

More recently, nanoparticles (NPs) of layer compounds, like WS_2 and MoS_2, were found to be unstable in the planar form owing to the abundance of rim atoms with extra dangling bonds. [6–8] Therefore, in analogy to carbon fullerenes and nanotubes, inorganic compounds with layered structure were shown to form fullerene-like nanoparticles (IF) and nanotubes (INT) - see schematics in **Fig. 1b**. Following many years of research, IF NPs and nanotubes of both WS_2 [9, 10] and MoS_2 [11, 12] could be synthesized as a pure phase and in substantial amounts. In the two-step synthesis, $W_{18}O_{49}$ nanowhiskers are obtained first and are subsequently converted to WS_2 nanotubes. [10]

More recently, sub-mm long WS_2 nanotubes with an aspect ratio larger than a few thousands were obtained via sulfidation of W_5O_{14} nanowhiskers, which were produced under a pressure of few bars. [13] **Fig. 2a** shows a scanning electron microscopy (SEM) image of the WS_2 nanotubes-mat, while **Fig. 2b** displays a transmission electron microscopy (TEM) image of one such nanotube. The highly uniform and crystalline structure of the nanotubes can be appreciated. A mechanically stable "buckypaper" < 1 mm thick was prepared from the long nanotubes- see photograph in **Fig. 2c**. This buckypaper served as a membrane to filter out colloidal nanoparticles of heavy metals, like gold from aqueous solution. **Figure 2d** shows the spectrum of the colloidal solution of gold nanoparticles before and after filtration. This result demonstrates the significant potential for such nanotubes to serve as filtration membranes for heavy metal recovery.

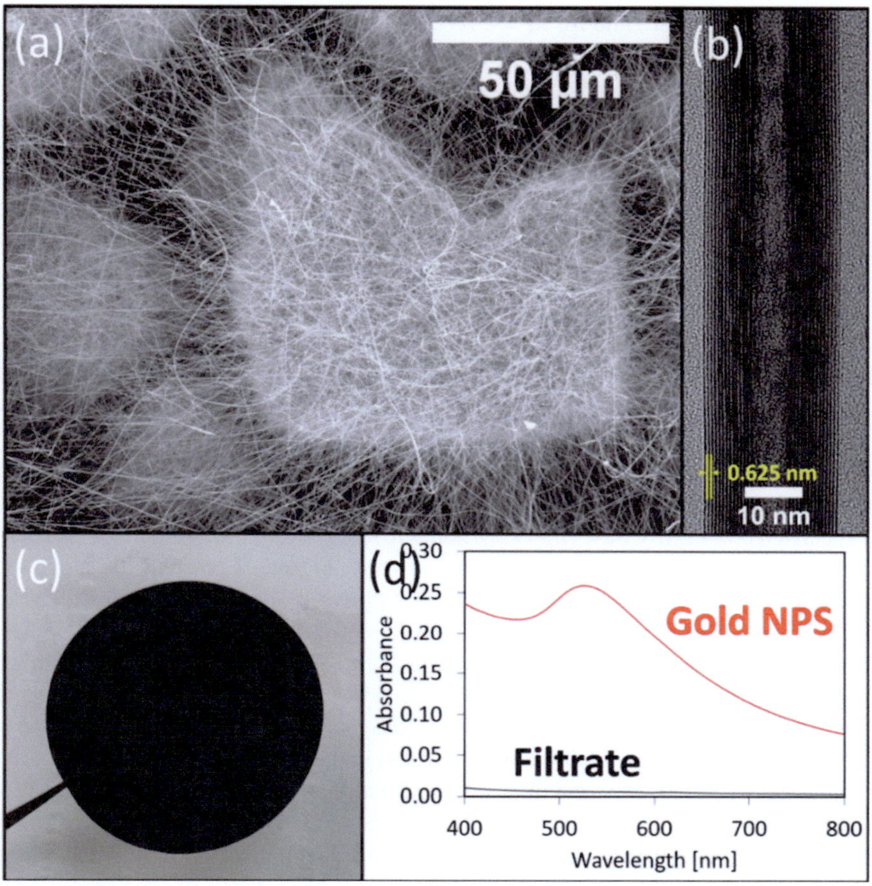

Fig. 2 **a** SEM image of ultralong WS_2 nanotubes produced by sulfidation of W_5O_{14} nanowhiskers. **b** TEM image of ultralong WS_2 nanotube with interlayer distance 0.625 nm. **c** SEM image of “buckypaper” made of WS_2 nanotubes deposited by wet-laying. **d** spectrum of gold colloidal solution before and after filtration through a WS_2-buckypaper membrane.[13]

“Misfit” compounds with layered structure (MLC) have been known in the literature for many years.[14, 15] MLC are incommensurate and non-stoichiometric compounds of the formula $(MX)_{1+y}(TX_2)_m$ or, for brevity (*O*-*T*), where MX (*O*) is a single layer of a compound with (distorted) rocksalt structure like LaS “intercalated” between two TX_2 (*T*) layers of a typical compound with a layered structure, like NbS_2 or TaS_2. MLCs have a common *c*-axis for both subunits, but the (*a*-*b*) plane of the two layers is incommensurate (usually along the *a*-axis). **Figure 1c** displays schematics of the MLC $(SmS)_{1.19}TaS_2$ nanotube. The ratio between the unit cells of the *O* and *T* slabs is irrational, which makes MLCs quasi-periodic compounds. The non-stoichiometric composition is expressed through the parameter $1 + y = 2a_T/a_O$ ($0.08 < y < 0.32$), and m is the repeat unit of the TX_2 layer i.e., for m = 1 the

MLC formula is (*O-T*) and for m = 2 (*O-T-T*). The stability of MLC is attributed in many cases to the charge transfer from the low work function *O* layer to the *T* slab. [16] In analogy to chrysotile and kaolinite discussed above, the strain induced by the incommensurate lattice constants of the *O* and *T* subunits of MLC can be relaxed via the bending of the layers. Such bent structures were mentioned in connection with the MLC mineral cylindrite. [17, 18]

Bernaerts et al.[19] and Gomez-Herrero et al.[20] observed micron-sized cylindrical and conical scrolls of chemically synthesized metal chalcogenide MLCs in $(PbS)_{1+y}(NbS_2)_m$ and $(BiS)_{1+y}(NbS_2)_m$, respectively. However, nanostructures of these compounds (and MLCs in general) were not explored until quite recently. [21, 22] Here, both the Pauling[4] (relaxation of asymmetric layers) and the Kroto-Iijima-Tenne[6] (healing of the dangling bonds of the rim atoms) mechanisms for folding and seaming of the MLC layers into a nanotube are believed to come into play, simultaneously. No wonder in a matter of a decade, many MLC nanotubes were realized. A systematic investigation of the synthetic procedures and structural elucidation of ternary [22] and later quaternary [23] nanotubes from a variety of misfit compounds was carried out. **Figure 3a-e** shows a schematic rendering of the structure of the rock-salt and the hexagonal subunits of a misfit structure. **Figure 3f** presents the concerted mechanism of formation of MLC nanotubes. **Figure 3g** displays a scanning electron microscopy (SEM) image of $(SmS)_{1.19}TaS_2$ (denoted also as $SmS\text{-}TaS_2$) nanotubes. Finally, **Fig. 3h** shows a scanning transmission electron microscopy (STEM) image of one such nanotube. Energy dispersive X-ray spectroscopy (EDS) mapping of the different constituting elements of $SmS\text{-}TaS_2$ nanotubes is also shown, demonstrating the uniform distribution of the elements in the nanotube. The diameter of the nanotubes is 100–200 nm, and they are a few μm long. [24] The magnetic behavior of $SmS\text{-}TaS_2$ nanotubes as a function of temperature was measured, revealing a superconducting transition at 3.9 K. **Figure 4** shows a high-resolution TEM image of a magnified portion of a lamella prepared from a single $SmS\text{-}TaS_2$ nanotube. [24] Structural models of the SmS and TaS_2 slabs are overlayed on the HRTEM image. The perfectness of the periodic structure of the MLC nanotube is revealed. More work is needed to elucidate the physical behavior of such 1D nanoobjects at low temperatures in relation to quantum technologies.

Several intriguing observations regarding the MLC nanotubes were established in recent years. In one such study, the chemical affinity of the rear-earth atoms lanthanum and samarium towards sulfur and that of selenium atoms with respect to tantalum atoms in the MLC nanotubes $LaS\text{-}Ta(Se_xS_{1\text{-}x})_2$ was confirmed.[25] Thus for $x \leq 0.66$ the nanotubes composition was $LaS\text{-}Ta(Se_xS_{1\text{-}x})_2$; for $x = 0.66$ the nanotubes composition was $LaS\text{-}TaSe_2$ and for $x = 0.8$ $LaS\text{-}(TaSe_2)_2$ nanotubes were obtained. Pure selenium-based $LaSe\text{-}(TaSe_2)_m$ nanotubes could not be synthesized. The results of the high-resolution (scanning) transmission electron microscopy (HR(S) TEM) analysis of a lamella prepared from a $LaS\text{-}(TaSe_2)_2$ nanotube were reported.[25] The chemical exclusivity of Ta, La, S, and Se atoms was clearly revealed. This study shows that, notwithstanding the high temperature of the reaction (up to 1100 °C), the entropy term plays a lesser role in this reaction in comparison with the chemical affinity (enthalpy) term.

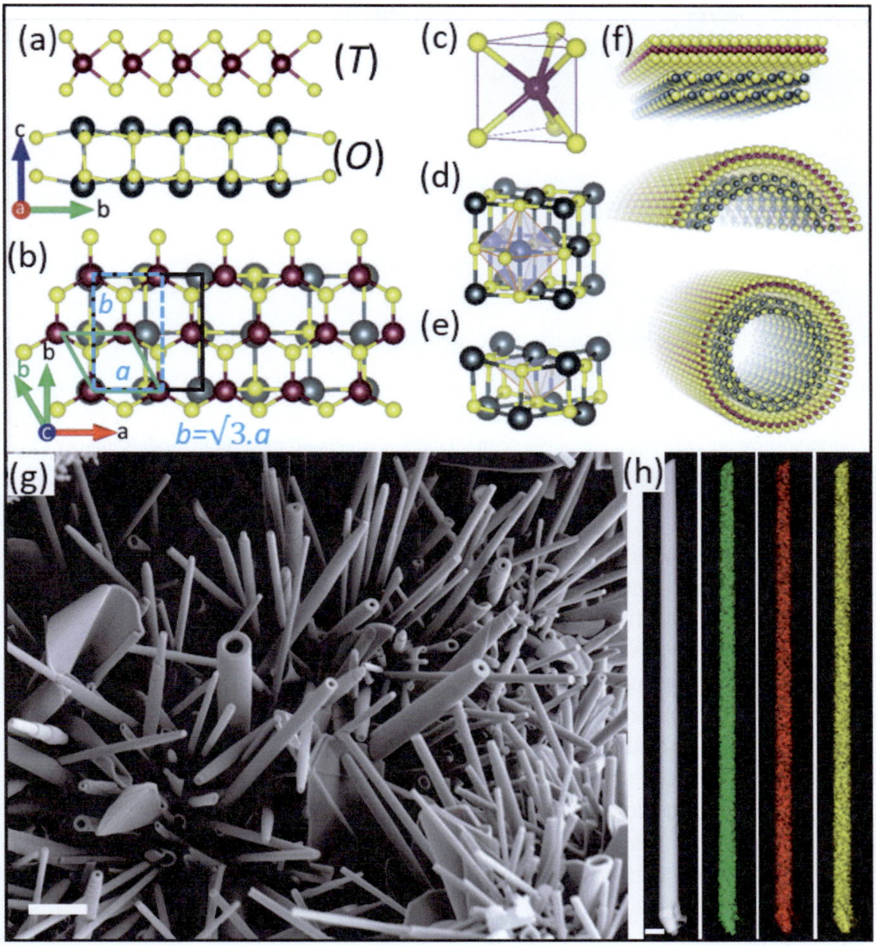

Fig. 3 **a–e** Schematic representation of the structure of the rocksalt and the hexagonal slabs of $(SmS)_{1.19}TaS_2$ misfit structure; **f** presents the concerted action of the Pauling and the Kroto-Iijima-Tenne mechanisms for the formation of MLC nanotubes. **g** SEM image of SmS-TaS_2 nanotubes, scale bar is 2 μm. **h** HAADF-STEM image of one such nanotube. The energy dispersive X-ray (EDS) mapping of the different constituting elements of SmS-TaS_2 nanotubes is also shown, demonstrating the uniform distribution of the elements in the nanotube (Sm, green; Ta, red and S, yellow); scale bar is 200 nm (reproduced from ref. [24]).

Inorganic nanotubes are, in general, metastable structures and are expected to decompose at elevated temperatures. Recently, the stability of SmS-TaS_2 nanotubes at elevated temperatures was examined using different analyses, primarily among them is synchrotron-based X-ray diffraction. [26] Since the different layers of the nanotubes are not commensurate with each other, the (0*kl*) and (*k*0*l*) lines are missing from the diffraction pattern. Using this selection rule, the relative abundance of the

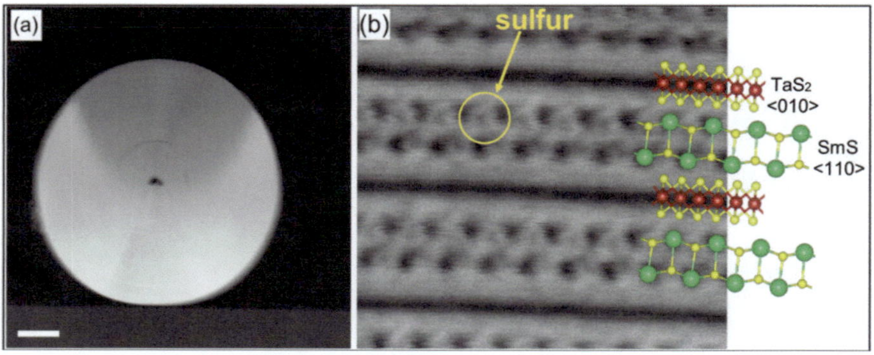

Fig. 4 **a** HRTEM image of a cross-section of SmS-TaS_2 nanotube fabricated by dual-focused beam electron microscope, scale bar 100 nm. **b** Atomic resolution HRSTEM image of the portion of lamella showing the rocksalt and hexagonal layers, corresponding atomic models are overlayed on the HRTEM image (reproduced from ref. [24])

nanotubes with annealing temperature (up to 1200 °C) and annealing time (up to 16 h at 825 °C) were investigated.

Five different decomposition pathways for the nanotubes were found to occur conjointly, each one at a different rate: 1. Coarsening of the nanotubes due to Ostwald ripening. i.e., large diameter tubes grew larger, and nanotubes with small diameter disappeared; 2. transformation to microscopic SmS-TaS_2 flakes; 3. Transformation of the SmS-TaS_2 into the less stable SmS-$(TaS_2)_2$ (and extra binary compound of samarium); 4. Gradual oxidation owing to a side reaction with residual water vapor or the silica walls of the ampule; and 5. Decomposition of the MLC into the binary samarium and tantalum compounds. This study offers a better understanding of the decomposition pathways of misfit nanotubes and metastable nanostructures in general.

In conclusion, nanotubes (and fullerene-like structures) are genuine phases of inorganic compounds with layered structures in the nano-size regime. The chemical synthesis of each kind of nanotubular phase is, in general, unique, but the inference from the synthesis of previously synthesized nanotubes is helpful. In most cases, they come as multiwall nanotubes with a typical diameter distribution of between 15 and 150 nm, and they are a few μm long. Inorganic nanotubes offer various applications in different fields, like reinforced polymer nanocomposites and optoelectronic devices, which were not discussed in depth here.

Acknowledgements RT acknowledges the support of The Estate of Manfred Hecht and the Estate of Diane Recanati. We also acknowledge the Irving and Cherna Moskowitz Center for Nano and Bio-Nano Imaging. The Perlman Family Foundation and the Kimmel Center for Nanoscale Science are greatly acknowledged

References

1. Dickinson, R.G., Pauling, L.: The crystal structure of molybdenite. J. Am. Chem. Soc. 45(12), 1466–1471 (1923).
2. Aminoff, G., Almström, G.K.: Über ein neues mineral von Långban. Zeitschrift für Kristallographie - Crystalline Materials 60, 262 (1924).
3. Bernal, J.D., Bragg, W.L.: The structure of graphite. Proc. Royal Soc. London. Series A 106, 749–773 (1924).
4. Pauling, L.: The structure of the chlorites. Proc. Nat. Acad. Sci. US 16, 578–582 (1930).
5. Turkevich, J., Hillier, J.: Electron microscopy of colloidal systems. J. Anal. Chem. 21, 475–485 (1949).
6. Tenne, R., Margulis, L., Genut, M., Hodes, G.: Polyhedral and cylindrical structures of tungsten disulphide. Nature 360, 444–446 (1992).
7. Margulis, L., Salitra, G., Tenne, R., Talianker, M.: Nested fullerene-like structures. Nature 365, 113–114 (1993).
8. Feldman, Y., Wasserman, E., Srolovitz, D.J., Tenne, R.: High-rate, gas-phase growth of MoS2 nested inorganic fullerenes and nanotubes. Science 267, 222–225 (1995).
9. Feldman, Y., Zak, A., Popovitz-Biro, R., Tenne, R.: New reactor for production of tungsten disulfide onion-like (inorganic fullerene-like) nanoparticles. Solid State Sci. 2, 663–672 (2000).
10. Zak, A., Sallacan-Ecker, L., Margolin, A., Genut, M., Tenne, R.: Insight into the growth mechanism of WS2 nanotubes in the scaled-up fluidized bed reactor. Nano 4, 91–98 (2009).
11. Zak, A., Feldman, Y., Alperovich, V., Rosentsveig, R., Tenne, R.: Growth mechanism of MoS2 fullerene-like nanoparticles by gas phase synthesis. J. Am. Chem. Soc. 122, 11108–11116 (2000).
12. Chithaiah, P., Ghosh, S., Idelevich, A., Rovinsky, L., Livneh, T., Zak, A.: Solving the "MoS2 nanotubes" synthetic enigma and elucidating the route for their catalyst-free and scalable production. ACS Nano 14, 3004–3016 (2020).
13. Kundrat, V., Rosentsveig, R., Bukvisova, K., Citterberg, D., Kolibal, M., Zak, A., Tenne, R.: Submillimeter-long WS2 nanotubes: The pathway to inorganic buckypaper. Nano Lett. 23, 10259–10266 (2023).
14. Wiegers, G.A.: Misfit layered compounds: Structures and physical properties. Prog. Solid State. Chem. 24, 1–139 (1996).
15. Makovicky, E., Hyde, B.G.: Non-commensurate "misfit" layer structures. In Structure and Bonding; Springer Verlag: Heidelberg, Vol. 46, pp. 101–170 (1981).
16. Lorenz, T., Baburin, I.A., Joswig, J.-O., Seifert, G.: Charge Transfer Variability in Misfit Layer Compounds: Comparison of SnS-SnS2 and LaS-TaS2. Isr. J. Chem. 57, 553–559 (2017).
17. Williams, T.B., Hyde, B.G.: Electron microscopy of cylindrite and franckeite. Phys. Chem. Miner. 15, 521–544 (1988).
18. Bernaerts, D., Amelinckx, S., Van Tendeloo, G., Van Landuyt, J.: Microstructure and formation mechanism of cylindrical and conical scrolls of the misfit layer compounds PbNbnS2n+1. J. Cryst. Growth 172, 433–439 (1997).
19. Gomez-Herrero, A., Landa-Canovas, A.R., Otero-Diaz, L.C.: Electron microscopy study of tubular crystals (BiS)1+d(NbS2)n. Micron 31, 587–595 (2000).
20. Hong, S.Y., Popovitz-Biro, R., Prior, Y., Tenne, R.: Synthesis of SnS2/SnS fullerene-like nanoparticles: A superlattice with polyhedral shape. J. Am. Chem. Soc. 125, 10470–10474 (2003).
21. Ohno, Y.: Lamellar and filament-like crystals of misfit-layer compounds containing (Sm, Ta, S) and (Pb, Bi, Nb, S) elements. J. Solid State Chem. 178, 1539–1550 (2005).
22. Panchakarla, L.S., Radovsky, G., Houben, L., Popovitz-Biro, R., Dunin-Borkowski, R.E., Tenne, R.: Nanotubes from misfit layered compounds: A new family of materials with low dimensionality. J. Phys. Chem. Lett. 5, 3724–3736 (2014).
23. Aliev, S. (Y. Elion), Tenne, R.: Quaternary chalcogenide compounds with misfit layered structure. Crystals (MDPI) 10, 468; https://doi.org/10.3390/cryst10060468 (2020).

24. Sreedhara, M.B., Bukvišová, K., Khadiev, A., Citterberg, D., Cohen, H., Balema, V., Pathak, A.K., Novikov, D., Leitus, G., Kaplan-Ashiri, I., Kolíbal, M., Enyashin, A.N., Houben, L., Tenne, R.: Nanotubes from the misfit layered compound (SmS)1.19TaS2: Atomic structure, charge-transfer and electrical properties. Chem. Mater. 34, 1838–1853 (2022).
25. Sreedhara, M.B., Hettler, S., Kaplan-Ashiri, I., Rechav, K., Feldman, Y., Enyashin, A., Houben, L., Arenal, R., Tenne, R.: Asymmetric "misfit" nanotubes: Chemical affinity outwits the entropy at high-temperature solid-state reactions. Proc. Nat. Acad. Sci. US 118, e2109945118 (2021).
26. Sreedhara, M.B., Khadiev, A., Zheng, K., Hettler, S., Castelli, I.E., Arenal, R., Novikov, D., Tenne, R.: Nanotubes from lanthanide-based misfit layered compounds: Understanding the growth, thermodynamic and kinetic stability limits. Chem. Mater., https://doi.org/10.1021/acs.chemmater.4c00481 (2024).

Magneto-Polaron Effects on Resonant Raman Scattering in Transition Metal Dichalcogenides

Vladimir M. Fomin, Darío G. Santiago-Pérez, Dmitry V. Tkachenko, Gilmar E. Marques, and Carlos Trallero-Giner

Abstract First-order resonant Raman spectroscopy in an external magnetic field is a powerful method that allows for exploring configurations of energy levels towards device applications of two-dimensional (2D) transition metal dichalcogenide (TMD) semiconductors. We have unveiled the behavior of the magneto-polaron (MP) resonances as a function of the phonon symmetry inherent in monolayer (ML) TMDs. A large number of avoided crossing points of energy branches involving an optical phonon in the MP spectrum, superposition of the electron and hole states in the excitation branches, and their relationship with optical transitions in different scattering configurations are unique in these 2D structures. To analyze them, the MP resonant scattering in a ML TMD is evaluated as a function of the laser energy and the magnetic field. The MP resonant Raman intensity exhibits three resonant splitting of double avoided-crossing levels. The three excitation branches are present in the MP spectrum due to the coupling of electrons and holes Landau levels through an out-of-plane A_1 optical phonon mode. The energy gaps at the anti-crossing points in the MP Raman spectrum are found as a function of the electron and hole deformation potential constants. An explicit expression for the Raman scattering efficiency is reported, allowing for exploration of the relative electron and hole contributions to polaron formation in the magnetic field. The obtained results are a guideline for control over the MP effects in the magneto-optical properties of TMDs.

V. M. Fomin (✉)
Leibniz IFW, Institute for Emerging Electronic Technologies, Dresden, Germany
e-mail: v.fomin@ifw-dresden.de

Moldova State University, Chişinău, Republic of Moldova

D. G. Santiago-Pérez
Universidad Autónoma del Estado de Morelos, Cuernavaca, México

D. V. Tkachenko
Pridnestrovian State University, Tiraspol, Republic of Moldova

G. E. Marques · C. Trallero-Giner
Universidade Federal de São Carlos, São Paulo, Brazil

A. Vaseashta et al. (eds.), *Proceedings of the Eighth International Symposium on Dielectric Materials and Applications (ISyDMA'8)*, Springer Proceedings in Physics 349, https://doi.org/10.1007/978-3-031-84809-4_2

Keywords Magneto-polarons · Resonant Raman scattering · Transition metal dichalcogenides

1 Introduction

Magneto-polaron effect was predicted in 1961 [1] and experimentally observed in 1965 [2]. This discovery opened a great path of research in optical and transport properties [3]. In Ref. [4], a new series of MP resonances in 2D TMD materials was reported. It is immediately possible to unveil the impact of this result on the magneto-optical properties, particularly, on the resonant magneto-polaron Raman scattering (RMPRS) in these materials. RMPRS measurement is a powerful tool to get accurate information on the electron-hole pair (EHP) MP formation. An explicit expression of the scattering efficiency as a function of a magnetic field (B) in ML TMD is essential to determine the conditions of the MP resonances. In the present paper, a rigorous description of RMPRS in a ML TMD is presented based on Green's function formalism. The peculiarities of the EHP MP spectrum, which includes the coupling of the different Landau states for the electrons and holes with the electron-phonon interaction (EPI), are presented in a systematic way. The RMPRS studies are required for characterizing TMD semiconductors. Our main objective is to describe the effects of a magnetic field on RRS in ML TMD semiconductors.

We investigate RMPRS in 2D TMDs at the K (K')-point of the Brillouin zone (BZ) assuming uncorrelated EHP. At a certain value of the magnetic field, two Landau levels can be coupled by the energy ω_0 of an optical phonon, and the MP resonance takes place. This coupling leads to the appearance of three excitation branches in the renormalized spectrum of the EHP [5]. The superposition of the EHP states in the MP excitation branches exhibit exceptional features in the Raman scattering of ML TMDs. The scattering efficiency is enhanced whenever an interband magneto-optical transition between renormalized Landau levels occurs, and the relative intensities are controlled by tuning the magnetic field. Owing to the possibility to adjust the laser energy and B, RMPRS spectroscopy is a unique tool that allows one to characterize various ML TMD structures.

The Raman cross-section per unit solid angle Ω and unit scattered light frequency ω_s is given by [5]

$$\frac{\partial^2\sigma}{\partial\Omega\partial\omega_s} \propto \left| \sum_{\{\mu_1\},\{\mu_2\}} \left\langle F \middle| H_{E-R}^{(+)} \middle| \Psi_{\mu_2} \right\rangle G_{\{\mu_2\}}(\hbar\omega_l - \hbar\omega_o) \right.$$
$$\left. \langle \Psi_{\mu_2} | H_{E-P} | \Psi_{\mu_1} \rangle G_{\{\mu_1\}}(\hbar\omega_l) \left\langle \Psi_{\mu_1} \middle| H_{E-R}^{(-)} \middle| I \right\rangle \right|^2$$
$$\frac{\Gamma_p/\pi}{(\omega_l - \omega_s - \omega_o)^2 + \Gamma_p^2}, \tag{1}$$

where $\hat{H}_{E-P}$ and $H^{\pm}_{E-R}$ are the electron-phonon interaction (EPI) and the electron-radiation Hamiltonians, $\Psi_{\mu_i}(j = 1, 2)$ are the EHP Landau states, $\mu = (N_e, k_{ey}; N_h, k_{hy})$ represents the set of the EHP states with the Landau quantum numbers N_e, N_h and the wave vectors k_{ey}, k_{hy}; $G_{\{\mu\}}$ is the Green's function at $T = 0K$ considering the MP quasi-particle. The summation in Eq. (1) is carried out over the magneto-polaron states. Figure 1 shows Raman spectra of a ML $MoSe_2$ obtained from Eq. (1) in the backscattering configuration for parallel polarizations of the incident and scattered light at six values of field B. In the figure, the behavior of the peaks as a function of B indicates the characteristic oscillation of the Raman peaks near the two avoided crossing points with the field. This variation immediately follows from the inset of the figure for $Z_l = 2.1$, whereby tracing the spectrum with B, we reach the resonant MP states. The observed maxima in the range of 6.2 T $< B <$ 6.5 T correspond to the incoming resonance near the avoided crossing point between $N = 3$ and the electron-excited state with $p_e = 0$ (blue diamonds).

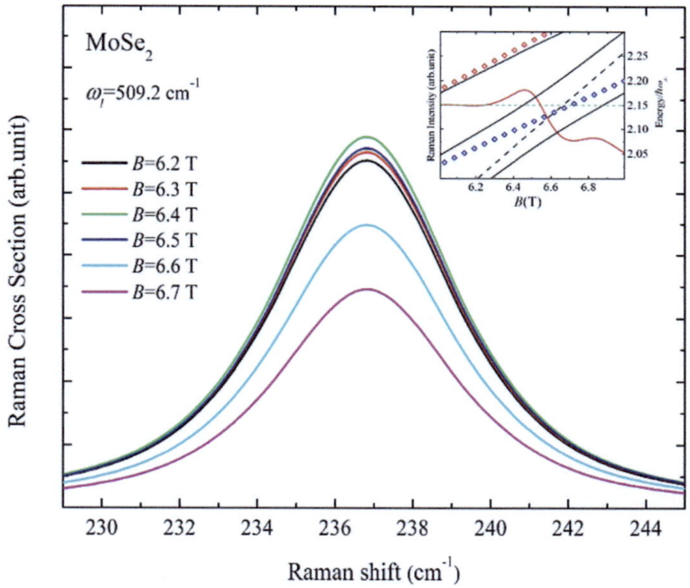

Fig. 1 Calculated Raman spectra for a ML $MoSe_2$ at six values of the magnetic field B and $Z_l = (\hbar\omega_l - E_g)/\hbar\omega_{A_1} = 2.1$ for parallel polarizations of the incident and scattered light. In the inset, shown are the $n = 3$ MP states contributing to the Raman spectra. The maximum occurs at $B = 6.4$ T, where a crossing with the electron renormalized excited state with $p_e = 0$ takes place (see Sect. 2). The calculation is performed via deformation potential (DP) interaction with the Raman-active out-of-plane ZO-mode ($\omega_{A_1} = 237.1$ cm^{-1})

2 Magneto-Polaron Spectra

For the description of the MP resonances, we employed the Green's function method that allows one to consider the valence and conduction bands, the influence of the EPI on the MP quasiparticles and the coupling between the Landau states. For a description of the MP properties, we write Dyson's equation for the full Green's function $G_{\{\mu\}}$

$$G_{n(N)}(E) = G_N^{(0)} + G_N^{(0)} \sum_M S_E(N, M) G_{n(N)}(E),$$

where $G_N^{(0)}$ is the one-particle unperturbed Green's function at $T = 0$ K and $S_E(N, M)$ represents the self-energy due to the EHP optical phonon interaction [5]. Polaron effects are considered when the EHP states are renormalized by their interaction with optical phonons, through $S_E(N, M)$. The pole of the dressed Green's function determines the renormalized resonant polaron energy and the lifetime broadening.

2.1 *Self-Energy*

The MP complex-value energies E_n are the fundamental blocks for the evaluation of the RMPRS. MP effects are considered when the EHP states are renormalized by the EPI through the self-energy $S_E(N, M)$. In 2D TMDs, the out-of-plane A_1- and in-plane LO-modes contribute to the self-energy. For the long-range interaction $S_E(N, M) = 0$. Therefore, the self-energy is reduced to deformation potential contribution $S_E(N, M) = \delta_{N,M} S^{(DP)}(E, N)$. $S_E(N, M)$ describes the electron-electron, electron-hole, and hole-hole coupling via the A_1-optical phonon. Therefore, there occurs a superposition of the valence and conduction bands contributions to the MP spectra. The relative contribution of the electrons or holes to the $S^{(DP)}$ depends on the valence band, D_v, and the conduction band, D_c, deformation potential constants [6].

2.2 *Magneto-Polaron Energies*

The pole of the total Green's function $G_{n(N)}$ provides the complex-valued MP energy $E_{n(N)}$, where the quantum number n labels the new quasi-particle states. By solving the Dyson's equation, we obtain the complex-valued energy

$$E_{n(N)} = \hbar\omega_c \left(N + \frac{1}{2} \right) + \alpha_{DP} \hbar\omega_c \left(1 - \frac{D_v}{D_c} \right)$$

$$\times \sum_{N'=0}^{\infty} \left(\frac{1}{E_{n(N)} - \hbar\omega_{A_1} - \hbar\omega_e(N' + \frac{1}{2}) - \hbar\omega_h(N + \frac{1}{2}) - i\delta} - \frac{1}{E_{n(N)} - \hbar\omega_{A_1} - \hbar\omega_e(N + \frac{1}{2}) - \hbar\omega_h(N' + \frac{1}{2}) - i\delta} \right), \tag{2}$$

where $\alpha_{\mathrm{DP}} = \frac{m_e}{4\pi\rho_m}\left(\frac{D_c}{\hbar\omega_{A_1}}\right)^2$, m_e is the electron effective mass, $\omega_{c_e}\left(\omega_{c_h}\right)$ is the electron (hole) cyclotron frequency, and ρ_m is the 2D-mass density. In ML TMD materials, the electron and hole effective masses are similar to each other. Therefore, in the renormalization process, both particles in a magnetic field B interacting with the A_1-phonon contributes to the MP energy. The above result highlights the fact that the MP quasi-particle results from a combination of two independent electron and hole intravalley transitions described by the first and second terms in Eq. (2). Therefore, there are two independent sets of anti-crossings at the electron and hole cyclotron frequencies $\omega_{c_e} = \omega_{A_1}/(N - p_e)$ and $\omega_{c_h} = \omega_{A_1}/(N - p_h)$, $p_e(p_h) = 0, 1, \ldots, N - 1$. The coupling of the bare excited states with a certain Landau level provokes three MP branches. For a given value of B, the solution of Eq. (2) provides the MP energies $\mathrm{Re}\{E_{n(N)}\}$ and their lifetime broadenings $\mathrm{Im}\{E_{n(N)}\} = \Gamma_n$.

The left panel of Fig. 2 illustrates the MP spectrum calculated according to Eq. (2) for $n = 3$. The dashed black lines represent the unperturbed Landau levels $N = 1$, 2, 3. The MP states are indicated by the color lines, the dressed Landau level by black, the renormalized excited electron energies by red ($p_e = 0$), blue ($p_e = 1$), and magenta ($p_e = 2$), and the hole dressed states are displayed by green ($p_h = 0$), cyan ($p_h = 1$), and olive ($p_h = 2$) lines. We observe anti-crossings at $B = 6.51$, 9.77, and 18.45 T and crossings between the electron and hole MP levels at $B = 8.14$, 11.94, and 23.87 T. At the avoided crossing point, the sequence of three excitations is as follows: the Landau level $N = 3$ plus three associated excited states, $p_e = 0$, 1, and 2 for the conduction band and $p_h = 0$, 1, and 2 for the valence band.

2.3 Lifetime Broadening

The EHP lifetime broadening, $\Gamma_3 = \mathrm{Im}\{E_{n=3}\}$ is related to the imaginary part of the self-energy. In the right panel of Fig. 2, the calculated set of broadening widths for n = 3 MP states are shown as a function of B. The color lines correspond to the MP energies drawn in the left panel of Fig. 2. The anti-crossings in the energy $\mathrm{Re}\{E_{n=3}\} = \varepsilon_{n=3}$ are crossings in Γ_3, while the crossing in the energy occurs where the maxima and minima of the broadening are observed.

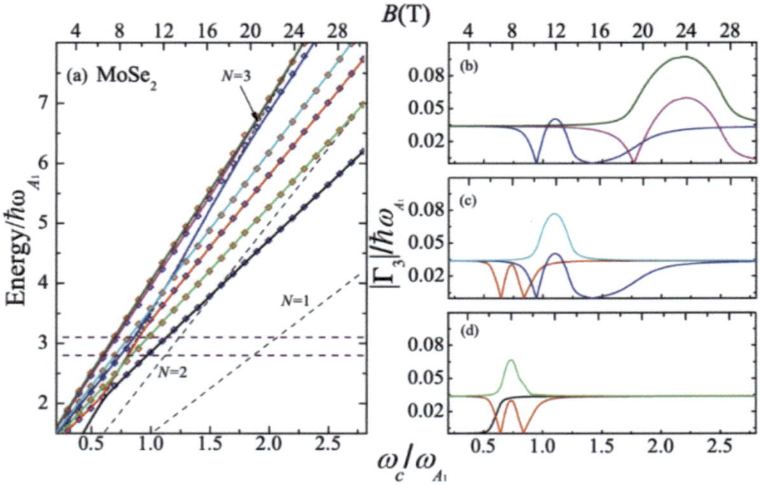

Fig. 2 Magneto-polaron spectrum (left panel) and Γ (right panel) as a function of B and the reduced cyclotron frequency ω_c/ω_{A_1} for the MP states $n = 3$ of a ML $MoSe_2$. Diamonds represent the bare electron and hole excited states

3 Resonant Raman Scattering Efficiency

In the Faraday backscattering configuration from the ML TMD surface, with the parallel polarization $Z(\sigma_s^{\pm}, \sigma_l^{\pm})Z'$, the Raman tensor allows for two optical phonon modes at the Γ-point of the BZ: A_1(ZO) and E(LO) for the longitudinal mode. The out-of-plane ZO-branch corresponds to the short-range DP coupling, while the in-plane LO-mode is mediated by the Pekar-Fröhlich interaction. It is possible to show that the contribution of the Pekar-Fröhlich interaction to the Raman spectrum is zero [6]. Employing the Green's function $G_{\{\mu\}}$, obtained in Sect. 2.2, Eqs. (1) and (2), the Raman scattering efficiency per unit solid angle $d\Omega$ due to the DP interaction is given by

$$\frac{dI^{\mathrm{DP}}}{d\Omega} = \frac{1}{S}\int \frac{\partial^2\sigma}{\partial\Omega\partial\omega_s} d\omega_s \propto$$
$$B^2 \left| \sum_{n(N)} \frac{1}{\left[\hbar\omega_l - E_g - E_{n(N)}(B) - i\delta\right]\left[\hbar\omega_l - E_g - \hbar\omega_{A_1} - E_{n(N)}(B) - i\delta\right]} \right|^2, \tag{3}$$

where $E_{n(N)}$ is the solution of Eq. (2) and the sum runs over all MP states.

Figure 3 shows the scattering efficiency for a ML $MoSe_2$ and its dependence on B for $Z_l = \left(\hbar\omega_l - E_g\right)/\hbar\omega_{A_1} = 3.1$ and 2.8; these two values are indicated in MP spectrum of Fig. 2 by the purple dashed lines. The incoming (green) and outgoing (purple) resonances for the bare Landau levels N are indicated by dashed arrows.

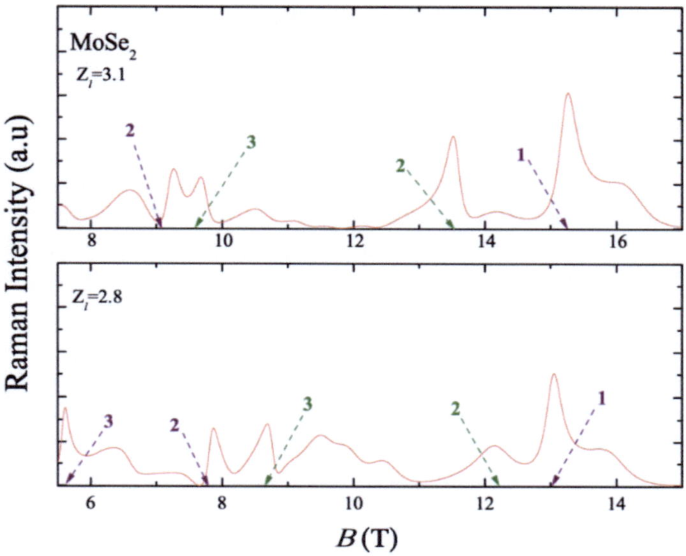

Fig. 3 Profile of the resonant MP Raman scattering intensity for a ML $MoSe_2$

In the calculation, we consider the backscattering configuration with the parallel polarization at the K-valley of the BZ with the effective masses $m_h = 0.59m_0$, $m_e = 0.49m_0$, $E_g = 1.59$ eV, $D_c = 5.2$ eV, $D_v = 4.9$ eV, and $\alpha_{DP} = 0.0577$ [6]. The Raman profile observed in Fig. 3 at $Z_l = 2.8$ in the range 9 T $< B <$ 11 T represents the incoming resonances for three excitation branches of the MP energy spectrum. From the spectrum in Fig. 2, it follows that the incoming resonance at B = 10.5 T corresponds to the renormalized Landau level $N = 3$, asymptotic to the electron undressed excited state with $p_e = 0$ (blue diamonds), while the peaks at 9.8 and 9.3 T account for the polaronic quasi-particles due to renormalized hole (green line) and electron (red line) exited states asymptotic to red- and blue-diamond-curves (undressed excited states) with $p_h = 0$ and $p_e = 1$, respectively. The same analysis might be performed for other values of the magnetic field and laser energies. For the outgoing resonances implied by Eq. (3), there exists the same set of MP energies contributing to the Raman profile shifted by the energy of an optical phonon.

4 Conclusions

We derived analytical expressions for the first-order resonant Raman scattering incorporating MP intermediate states in the K- and K'- valleys of ML TMDs. The model includes the short-range DP electron-phonon interaction due to the out-of-plane A_1-phonon at the Γ-point of the BZ. The A_1-phonon mode breaks the degeneracy between the EHP Landau levels and the bare electron and hole excited states, giving rise to three resonant branches in the MP spectrum, provoking the appearance of two different sets of cyclotron frequencies for the conduction and valence bands. Compact analytical expressions for RMPRS intensity as a function of the magnetic field as well as for MP quasiparticle energy and its lifetime broadening, allow for the analysis of the MP formation in different ranges of the magnetic field and the incoming laser energy. Equations (2) and (3) provide direct experimentally accessible information on the MP states and the band structure at the K- and K'- valleys of ML TMD structures. The main result of the present study lies in the demonstration that the RMPRS uniquely reflects the MP spectrum of a 2D TMD. This finding is fundamental in the search for new optical and photonic devices, along with numerous other potential applications. Understanding the properties of MP states using resonant Raman scattering provides important information for future basic research on a wide range of magneto-optical properties (the MP effect in bilayers, multi-phonon resonant Raman scattering, hot luminescence, and Moiré structures) of TMDs.

References

1. Gurevich, V. L. and Firsov, Y. A.: On the theory of the electrical conductivity of semiconductors in a magnetic field. Sov. Phys. JETP **13**, 137 (1961).
2. Parfen'ev, R., Shalyt, R., S., and Muzhdaba, V.: An experimental confirmation of magnetophonon resonance in n-type InSb. Soviet Physics JETP **20**, 294–298 (1965).
3. Devreese, J. T. (ed.) Polarons in Ionic Crystals and Polar Semiconductors. North Holland, Amsterdam (1972).
4. Trallero-Giner, C., Santiago-Pérez, D. G., and Fomin, V. M.: New magneto-polaron resonances in a monolayer of a transition metal dichalcogenide. Sci. Rep. **13**, 292 (2023).
5. Trallero-Giner, C., Santiago-Pérez, D. G., Tkachenko, D. V., Marques, G. E., and Fomin, V. M.: Raman scattering owing to magneto-polaron states in monolayer transition metal dichalcogenides. Sci. Rep. **14**, 12857 (2024).
6. Trallero-Giner, C., Santiago-Pérez, D. G., Tkachenko, D. V., Marques, G. E., and Fomin, V. M.: Transition metal dichalcogenides: Magneto-polarons and resonant raman scattering. Front. Phys. **12**, 1440069, 1–13 (2024).

Dielectric Barrier Discharge: Generation and Enhancement of Hydrogen Peroxide Action

Sergey Travin, Gheorghe Duca, Ashok Vaseashta, Olga Covaliova, and Lidia Romanciuc

Abstract Barrier dielectric discharge is a type of gas discharge between electrodes, one or both of which are coated with a dielectric layer. Due to the insulating properties of the coating, the avalanche-like development of the discharge is limited in amplitude, because of which cold plasma is formed. Although the phenomenon itself was discovered almost two hundred years ago (1857), the real possibilities for its practical application were discovered only in the last couple of decades. Since then, the number of useful applications has been growing exponentially and this trend will continue for the foreseeable future. Cold plasma technologies cover such areas as surface treatment for electronics and polymer materials, the food industry, biomedical usage, and even wastewater treatment. In the food industry, these methods are used for food decontamination, food quality improvement, toxin degradation, and surface modifications of packaging materials. Cold plasma has proven effective for the inactivation of various pathogens and spoilage organisms without adversely affecting food quality. Cold plasma treatment of water enables to carry out of processes of pesticide decomposition, discoloration of dyes, and general disinfection. Low-temperature plasmas can be generated in usual conditions: normal atmospheric pressure and temperatures below 37 °C, so have opened up a new frontier of applications: biomedical ones. Medical usage of cold plasma includes sterilization, wound healing, disease treatments, and not only. Experimental work conducted at several major universities, research centers, and companies around the world over the recent decade demonstrates that plasma can be used in a variety of medical applications. It is already widely used in surgeries and endoscopic procedures and has been shown

S. Travin
N.N. Semenov Federal Research Center for Chemical Physics, Russian Academy of Sciences, Moscow, Russia

G. Duca · O. Covaliova · L. Romanciuc
Institute of Chemistry, Moldova State University, Chisinau, Republic of Moldova

A. Vaseashta (✉)
International Clean Water Institute, Virginia, VA, USA
e-mail: prof.vaseashta@ieee.org

Institutul de Cercetare, Universităţii din Bucureşti, Bucharest, Romania

A. Vaseashta et al. (eds.), *Proceedings of the Eighth International Symposium on Dielectric Materials and Applications (ISyDMA'8)*, Springer Proceedings in Physics 349, https://doi.org/10.1007/978-3-031-84809-4_3

to control the properties of cellular and tissue matrices, including the biocompatibility of various substrates. Non-thermal plasma has been demonstrated to deactivate dangerous pathogens to stop bleeding without damaging healthy tissue and to treat cancer. Although cold plasma technology has shown promising results, it requires further studies to understand the reactive gas chemistry and toxicological, ecological, and economic impacts of this technology. We realize that mechanisms of plasma interaction with complex natural systems, especially with living tissues must be not very simple, because of the complexity of both: plasma itself and treated objects. So, the discovery of a unified mechanism is unlikely. Anyhow we have a strong belief that the prime cause of miraculous healing is connected with the production of hydrogen peroxide, other forms of active oxygen and its reactive species (radicals and non-radicals), charged particles, photons, and electric fields, which have impactful biological effects.

Keywords Dielectric barrier discharge · Biomedical application · Wastewater purification · Active radicals · Hydrogen peroxide

1 Introduction

Dielectric barrier discharge (DBD) is a discharge in a gas medium filling the gap between electrodes, one or both of which are coated with a dielectric (Fig. 1). Since being discovered in 1857 by Ernst Werner von Siemens these discharges were for the first time used in the production of ozone. Originally called silent (inaudible) discharges, also known as partial discharges, are now used in thousands of applications, from medical aims for sterilizing medical equipment to creating plasma panels and excimer lasers [1, 2].

When an alternating voltage is applied to the electrodes, an electric field occurs in the gas, determined by the voltage applied to the electrodes and charges on the dielectric surface. The discharge occurs if this position exceeds the breakdown field. In numerous experiments, the discharge parameters varied widely: the pressure ranged from units to hundreds of Torr, and the frequency of the applied voltage ranged from units to hundreds of kHz. The distance between the electrodes is usually on the

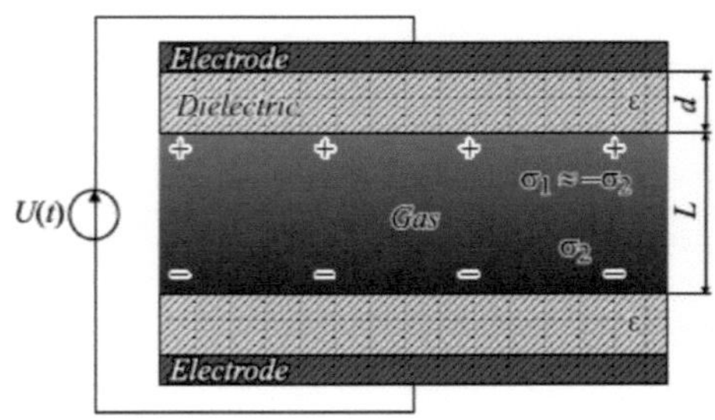

Fig. 1 Schematic diagram of the discharge cell

order of a few millimeters [3]. Such discharge has much in common with discharge between metal surfaces, but there is also a serious difference associated with the current limitation that stops the avalanche-like development of the discharge. Four phases of barrier discharge development can be distinguished after the electric voltage is applied:

1. In the very beginning the number of charged electrons and ions increases exponentially due to the development of an avalanche without distorting the applied electric field—the so-called Townsend phase.
2. The streamer phase is the formation of conductive channels in the gas, inside the gap.
3. The formation of a cathode surface—the current reaches its maximum value.
4. The quenching phase—an electric charge accumulates on dielectric surfaces, which reduces the electric field in the gap below the critical one and thus prevents the formation of new ion-electron pairs in the gas.

A significant difference between a barrier discharge and others is that it cannot develop when a constant electric field is applied, because capacitive coupling through dielectrics requires an alternating field to cause a bias current. From the energetical point of view, DBD has two most important features:

i. low power factor in the range from 0.1 to 0.3 due to capacitive electrical load, and
ii. huge amount of energy stored in an electric field.

Operation with continuous sinusoidal or rectangular waves is mainly used in industrial high-power installations. The impulse operation of DBDs can lead to an increase in discharge efficiency. A significant difference between a barrier discharge and others is that it cannot develop when a constant electric field is applied, because capacitive coupling through dielectrics requires an alternating field to cause a bias current. Barrier discharge can occur in two forms: filamentous and smoldering (uniform). In most cases, it is heterogeneous and consists of a large number of micro-discharges in the gap. The physics of micro-discharge formation is based on the processes of origination and distribution of streamers with subsequent destruction of the plasma channel. A streamer is an ionization wave that moves from the anode to the cathode toward a moving electron avalanche. Streamers develop at a high speed ~108 cm/s and pass the gap in a few nanoseconds. For comparison and awareness: this is 4,000 times the speed of the jetliner; barrier discharge would be able to circumnavigate the world in less than a minute, if not limiting dielectric coats.

Electrons in the conductive channel scatter in about 40 nsecs, while heavy and slow-moving ions remain for a few microseconds. The deposition of electrons on the anode dielectric barrier is completed by creating a negative charge thereon that prevents new avalanches and streamers from forming at the site of charge deposition until the anode and cathode are reversed. After the polarity changes, the deposited negative charge stimulates the development of a new avalanche and streamer at the same place.

As a result, many micro-discharges are formed that are visible as bright threads filling the gap. The terms avalanche, streamer and micro-discharge should be distinguished. The primary electron generated in the gap near the cathode produces secondary electrons by direct ionization, resulting in an electron avalanche. If the avalanche has reached a large enough size, then a streamer appears that moves towards the avalanche. The streamer is a very fast ionization wave that closes the gap in a few nanoseconds and forms a conductive channel with a diameter of ~100 microns, filled with weakly ionized plasma. Significant electric current begins to flow through this channel until the electric field in the gap decreases. The decline of the electric field is caused by the electric charge deposited on the electric barrier and the charge of ions in the gap. The combination of charge-gap processes initiated by an electron avalanche and continuing until the electron current ceases is commonly referred to as a micro-discharge. After the current has stopped, there is no more electron-ion plasma in the discharge channel. However, there are many excited oscillatory electronic states together with the electronic charge deposited on the dielectric surface of the anode and the charge of ions in the gap.

Positive ions slowly move towards the cathode, creating a falling ion current (~10 μs for a gap of 1 mm). The remainder of the micro-discharge stimulates the new micro-discharge at the same location as the polarity of the applied voltage changes. Thus, a filamentous discharge is a group of micro-discharges occurring at the same location as the polarity changes.

Methods for producing cold non-equilibrium plasma: depending on the mechanism for producing low-temperature plasma, the following types of discharges are distinguished: smoldering discharge, discharge with a dielectric barrier, corona discharge, and atmospheric pressure plasma jets (APPJs). The last three types of production of low-temperature plasma have found the greatest distribution in the field of medicine. As a gas mixture, inert gases are most often used—argon, and helium. We need to note that the so-called cold plasma or low-temperature plasma differs from high-temperature plasma in that the proportion of ionized matter in it is relatively small. At the same time, the temperature of the "cold" plasma can reach a million degrees.

1.1 Wastewater and Drinking Water Treatment

One of the important fields of DBD practical application is the purification of various water sources. The situation with anthropogenic environmental pollution in the world currently remains tense and progress in clean and "green" technologies is not expected soon. One of the main causes of environmental degradation medium is the entry of highly toxic chemicals. One of the traditional approaches to polluted water treatment is electrochemical [4]. One of the first reports on using electrical discharge plasmas to directly degrade organic contaminants dates back to the early 1980s. Since then, over a hundred different plasma reactors have been developed to treat a range of contaminants of environmental importance including pharmaceuticals, herbicides, pesticides, warfare agents, bacteria, yeasts and viruses using direct-in-liquid discharges with and without bubbles, and discharges in gas over and contacting the surface of a liquid giving rise to electrohydraulic reactors, bubble reactors, spray reactors, hybrid gas-liquid reactors, falling water film reactors, etc. [5, 6].

Organochlorine compounds belong to the group of persistent organic pollutants, for which the time of natural self-degradation is calculated by tens and hundreds of years. In water bodies, organochlorine compounds can concentrate during transport in trophic chains. Polycyclic aromatic hydrocarbons also belong to the class of compounds widespread in the environment, and several of them have a pronounced carcinogenic effect, therefore their content is both in the atmosphere and in the hydrosphere and should be monitored.

The growing use of atmospheric pressure plasma opened the door for their applications in the treatment of water and air [7]. During the last decade, plasma treatment of water has been investigated for the reduction of pollutants, and inactivation of pathogens in the food cycle, food safety, and water purification [8]. The treatment of air, which traditionally has been focused on the inactivation of airborne pathogens in air flows and on their chemical cleaning, has gained significant renewed interest due to the COVID-19 pandemic.

While there is growing concern about the persistent global water crisis due to these factors, the available potable water is often not completely suitable due to the presence of low quantities of persistent chemicals that are potentially harmful to humans and other living organisms [9, 10]. It is important to ensure that Plasma Water Treatment (PWT) is still a highly energy-intensive process with significantly more complex technology than the existing oxidation processes used. In addition, scientific evidence on the mechanism of plasma effects on natural aquatic environments is scarce and sometimes contradictory.

One of the ways to solve the problem of reducing the negative anthropogenic impact on the environment is to improve existing and develop new highly efficient and energy-saving systems for cleaning pollutant emissions, wastewater, and waste disposal. Recently in environmental protection systems media are widely used methods of high-energy chemistry. In particular, low-temperature plasma, combining both physical (heat, UV radiation, electronic impact), as well as chemical (high concentrations of active particles) exposure, is effectively used to degrade persistent, highly toxic organic compounds [11].

However, the widespread use of high-energy chemistry methods is hindered by the lack of studies on the kinetics and the mechanisms of decomposition of starting compounds and the formation of intermediate and final degradation products, as well as estimates of the impact of process parameters on the degradation efficiency of pollutants. To date, it is possible to more or less clearly define the boundaries of the preferred use of plasma technologies in water treatment and water treatment. Despite the entangled mechanisms and many parameters that affect the performance of plasma reactors, there is growing evidence that existing plasma reactors are superior to other methods in the treatment of surfactant compounds but cannot competitively treat dissolved substances in the water column, which constitute the majority of compounds of environmental importance. It is possible that increasing the treatment efficiency of non-surfactants to levels of surfactants and increasing the energy efficiency and productivity for the treatment of all compounds will long remain unsolved key PWT problems. And if in the case of oxidational and light-induced water purification processes we have an insufficient understanding of free radicals' transformations and detailed mechanisms of ROS action, in the case of plasma technologies one must add the need for a more fundamental understanding of the kinetics and collisional processes in the highly non-equilibrium plasma regime.

Recently, membrane filtration, ozonation, and UV irradiation have also been recognized as effective techniques for achieving microbial inactivation. Anyhow, DBDs such as single and double planar and single and double cylindrical dielectric barrier configurations have been utilized for the efficient degradation of recalcitrant organic pollutants [12]. The overall performance of the different DBD systems varies and depends on several factors. Additionally, the ozone, formed in DBD has attracted scientific attention for many reasons. Firstly, it was used to reduce the utilization of free chlorine in water treatment processes. In the DBD water treatment ozone is formed by the decomposition of diatomic oxygen O_2 according to the following equations:

$$O_2(\text{Electric discharge}) \rightarrow O^{\bullet} + O^{\bullet} \tag{1}$$

$$O^{\bullet} + O_2 \rightarrow O_3 \tag{2}$$

Although it has been established that ozone is an effective oxidant towards hazardous organic pollutants, the whole process has limitations, such as low ozone solubility and short ozone lifetime [13–16]. Therefore, ozonolysis should be combined with other processes using different agents for effective water and wastewater treatment. These agents include hydrogen peroxide and free radicals as products of its decomposition. These combinations are usually referred to as advanced oxidation processes (AOPs).

The literature review indicates that a new generation of these methods is being technologically developed to reduce energy consumption, and operational cost and infinitely enhance the degradation efficiency so that these processes can hopefully meet future regulations [17]. AOPs focus on using hydroxyl radicals as the primary nonselective oxidant of organic pollutants. Such processes often result in a complete mineralization of organic molecules into CO_2, water, and harmless inorganic substances. Literature shows that organic toxins such as pesticides, surfactants, coloring matter, pharmaceuticals, and endocrine-disrupting chemicals have already been treated in wastewater using advanced oxidation techniques (AOTs) [18].

AOPs have been described as water treatment procedures that involve an input of energy into the water matrix to produce strong oxidizers in situ. OH radicals have been identified as the most effective oxidants in decomposing organic compounds into CO_2, water and inorganic salts due to the high standard reduction potential of OH ($E_0 = 2.8$ V), compared to those of atomic oxygen ($E_o = 2.42$ V), ozone ($E_o = 2.07$ V), hydrogen peroxide ($E_o = 1.78$ V) and perhydroxyl radical $HO_2\bullet$ ($E_o = 1.70$ V) [13]. Hydroxyl radical attack occurs by abstraction of the hydrogen atoms (Eq. 3 and 4), followed by electrophilic addition of OH (Eq. 4 and 5), electron transfers (Eq. 5), and radical-radical reactions (Eq. 5 and 6) [19, 20].

$$HR + {}^{\bullet}OH \rightarrow R^{\bullet} + H_2O \tag{3}$$

$$R^{\bullet} + O_2 \rightarrow RO_2 \tag{4}$$

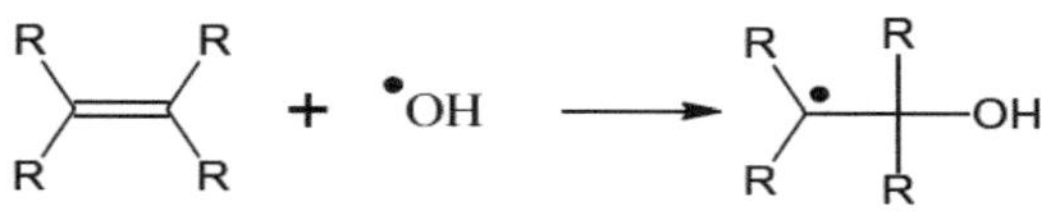

(5)

$$RX + {}^{\bullet}OH \rightarrow XR^{\bullet +} + OH^{-} \tag{6}$$

$$2{}^{\bullet}OH \rightarrow H_2O_2 \quad k = 5.3 \cdot 10^9 Lmol^{-1}s^{-1} \tag{7}$$

$$H_2O_2 + {}^{\bullet}OH \rightarrow HO_2^{\bullet} + H_2O \quad k = 2.7 \cdot 10^7 Lmol^{-1}s^{-1} \tag{8}$$

1.2 Liquid Phase Electrical Discharge: Electrohydraulic Discharge

Even though the application of electrical discharge on the surface of water in water and underwater has been studied for many years, electrical discharge processes in the liquid phase have mostly been explored in recent years for various purposes including water treatment [21], sterilization and decomposition of organic compounds [21–25].

Depending on the electrode configuration, electrohydraulic discharge involves the submersion of two electrodes in water. The high voltage and ground electrodes. Concerning the application of plasma-treated water in the food cycle and food safety fields, plasma parameters affecting the enhancement of plant growth and stability, and healthy management of soil and nutrients, despite the constantly published data on the commercial application of plasma installations, data to optimize the processes used are insufficient. While the influence of plasma-generated nitrate or ammonium as fertilizer is well known, other reactive species of oxygen and especially nitrogen and their potential benefits for plant growth and plant ecosystem have not yet been studied in detail [26, 27]. Concerning food safety, the effects of ROS and RNS in fluids on food quality and nutrients are also not well understood. The use of appropriate controls, such as nitrates, nitrites, and hydrogen peroxide at concentrations consistent with those in PTW, will help identify mechanisms and pave the way for scaling and commercialization efforts.

2 Biomedical Applications of Barrier Discharge

The disinfecting effect of electric current in water has been known for a long time and the first attempt at its usage has been undertaken since the late XIXth century [28, 29]. In 1898 Bredig and then Svedberg in 1909 applied an electric arc for the first time and pulsed electric discharges (PED) to produce water dispersions of metal and organic sol-gels [30]. During the last century, PEDs in water were applied to destroy microbes, spores, protozoa, etc., and to transform organic and inorganic compounds.

There is a certain variation in the arrangement of the electrodes, (Fig. 2), but the general concept of the operation of the DBD remains unchanged. Recently, Fridman et al. [31], developed the concept of floating electrode DBD—floating electrode DBD (FE-DBD). A floating electrode DBD produces plasma between two electrodes, with one of the electrodes coated with dielectric being energized and the second electrode not grounded. This means that the skin or tissues of the animal can play the role of the second electrode.

Commercial LTP (Low Temperature Plasma) decontamination systems, employing low-pressure plasma, are nowadays widely used in healthcare settings

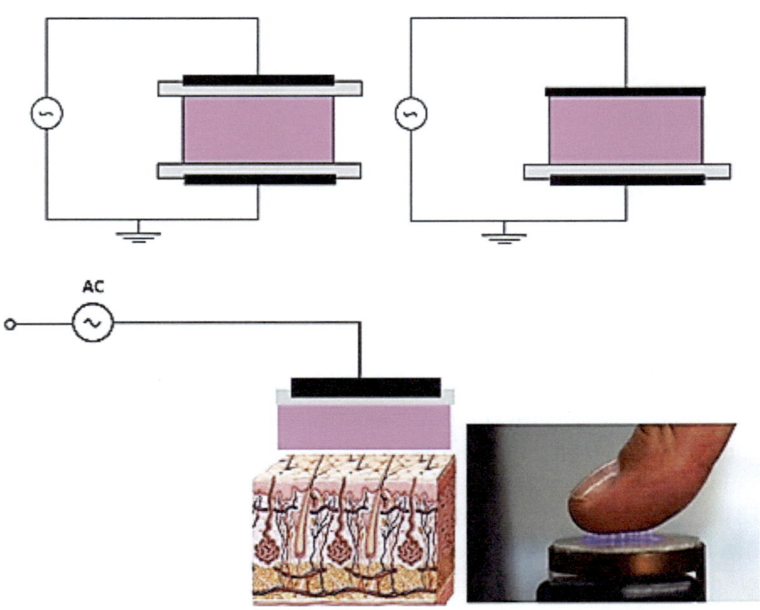

Fig. 2 Adopted from https://vetacademy.ru/obuchenie/stati/plazmennye-tekhnologii/

across the world for the sterilization of temperature-sensitive medical devices [32]. At the end of the twentieth century, interest in the use of plasma for biological decontamination was reignited with a demonstration that LTP generated under ambient pressure conditions could be harnessed for the non-thermal inactivation of microorganisms.

As the area of research continues to expand, it is becoming more diverse and varied. While early works focused primarily on the use of LTP to eliminate planktonic bacteria, recent efforts have focused on more representative contamination such as antibiotic-resistant bacteria, mixed species biofilms, and endospores [33]. Also appeared a formidable challenge to any non-thermal decontamination technology: the global SARS-nCoV-2 pandemic has placed a spotlight firmly on the need to develop fast, efficient, and low-cost viral decontamination techniques that can be easily deployed beyond a healthcare setting.

Nevertheless, the level of understanding of the biomolecular action of the cold plasma is still insufficient. Being able to manipulate and control the fundamental LTP processes to elicit a desired biological response at the target remains the 'holy grail' in the area, yet such control extends beyond our current capabilities. It is known that in natural conditions microorganisms cluster together and form a protective layer for self-preservation, which is called a biofilm. Usually, biofilm is 1000 times more resistant to antimicrobial agents than planktonic cells itself. On the other side, biofilms present ideal conditions for the formation of antimicrobial resistance. And the key question is how and how quickly microorganisms form resistance to plasma—it remains unanswered.

Modern science has a very impressive arsenal including femtosecond techniques for the detection of "thin" molecular movements. Despite this progress, the sheer level of complexity and multiscale nature of plasma-bio interactions dictates that experimental techniques alone will not be sufficient to uncover a full understanding; the introduction of a realistic biological entity into physical models extends beyond our current ability.

To conclude, the development of LTP technology for biological decontamination spans the field of plasma science, engineering, and the life sciences; consequently, a multidisciplinary approach is therefore essential to make progress.

Programmable cell death? It is known to enlightened connoisseurs as apoptosis, but apoptosis is just one of many kinds of programmable cell death. Nature as a divine sculptor creates complex structures and maintains their existence, constantly cutting off the superfluous. The paradox is that for the body to live, its constituent cells must die. And like all vital biological processes, programmable cell death is key to treating many diseases [34].

Initiation of apoptosis, or programmed cell death, is an important issue in cancer treatment as cancer cells frequently have acquired the ability to block apoptosis and thus are more resistant to chemotherapeutic drugs. Targeted and perhaps selective destruction of cancerous tissue is desirable for many reasons, ranging from the enhancement of or aid to current medical methods to problems currently lacking a solution, i.e., lung cancer. Demonstrated in this publication is the inactivation (killing) of human Melanoma skin cancer cell lines, in vitro, by Floating Electrode Dielectric Barrier Discharge (FE-DBD) plasma [34]. Not only are these cells shown to be killed immediately by high doses of plasma treatment, but low doses are shown to promote apoptotic behavior as detected by TUNEL staining and subsequent flow cytometry. It is shown that plasma acts on the cells directly and not by 'poisoning' the solution surrounding the cells, even though a layer of such solution. Potential mechanisms of interaction of plasma with cells are discussed and further steps are proposed to develop an understanding of such systems.

Apoptosis, or programmed cell death, is a complex biochemical process of controlled self-destruction of a cell in a multicellular organism. This process plays an important role in maintaining tissue homeostasis, fetal development, immune cell 'education', development, and aging. Examples of apoptosis that occur during normal body processes include the formation of the outer layer of skin, the inner mucosal lining of the intestine, and the endometrial lining of the uterus, which is sloughed off during menstruation. During apoptosis, cellular macromolecules are digested into smaller fragments in a controlled fashion, and ultimately the cell collapses into smaller intact fragments that can be removed by phagocytosis without damaging the surrounding cells or causing inflammation. In contrast, during necrosis, also termed "accidental cell death", the cell bursts and the cellular contents spill out into the extracellular space, which can cause inflammation. Necrosis is induced by cellular injury, for example, extreme changes in osmotic pressure or heat, that lead to adenosine triphosphate (ATP) depletion of the cell. With cancer cells, however, a problem arises with apoptosis as the tumor cells frequently 'learn' how to turn off apoptosis as one of the processes they employ in evading the immune system and surviving under unfavorable conditions. For this reason, for example, chemotherapy as a means of treatment of breast, colon, and lung cancer met limited success. In general, the employment of systemic chemotherapy drugs to induce apoptosis in cells that try to block it is not an easy task, as these drugs tend to affect all cells in the body. A way to target apoptosis development only in specific areas of the body is needed and a method to do so is offered in this paper where apoptotic behavior is promoted in Melanoma cancer cell lines following low doses of non-thermal plasma treatment insufficient to destroy the cell immediately [34].

In recent years, non-thermal plasma discharges have been gaining popularity in the materials processing industry for their ability to selectively modify a surface with minimal, if any, damage to this surface and practically no change to the bulk material. This way, for example, the surface of an implant may be made biologically compatible to the tissues and cells it will come in contact with, while the bulk material of the implant can be tailored to desirable mechanical properties like high strength, low weight, durability, fracture resistance, etc. Recently, the demand for sterilization and disinfection of various surfaces increased and non-thermal plasmas were found to be an effective solution [35]. Many groups worldwide have successfully demonstrated plasma's ability to treat, disinfect, and sterilize various targets [36–38].

Cell death can be a passive or active process. Passive—death from damage that the cell is unable to repair. An active process takes place when the cell has fulfilled its function and must leave the stage and give way to other cells. A typical example is embryonic development: the formation of an organ in a growing body occurs not only due to the growth and division of cells but also due to the removal of "superfluous".

However, even after damage, an active process of death may be involved: "planned" elimination is preferable to uncontrolled decay.

Any cell population is regulated by three processes equally important: division, differentiation—the transformation of young cells into mature cells (while their number can both increase and decrease), and cell death. The body of an adult consists of tens of trillions of cells, and every day each of us loses tens of billions of them, in terms of weight—about a kilogram. New cells make up for the loss, so we do not lose a kilogram a day. We all know how the superficial layer of the skin, the epidermis, collapses and renews. Among the most actively dying are cells of the intestinal epithelium: throughout a person's life, they are replaced about 4,000 times. If the old cells did not die, then in 70 years our intestines would reach a length of 39 km!

In the individual development of a person or any other creature, it is also impossible to do without programmed cell death. A textbook example of apoptosis is the disappearance of the tail in the tadpole; interestingly, this process, along with other metamorphoses, is regulated by a change in blood levels of the thyroid hormone. And for the animal to form fingers on its paw, the cells located between the beginnings of the fingers must disappear. Programmable cell death is involved in both genital maturation and brain development. The cells of the body that died during apoptosis should be eaten by neighboring cells or macrophages—devouring professionals. Thanks to this, apoptosis is rarely accompanied by inflammation.

The study of programmable cell death has not only fundamental but also applied medical significance. The cause of many chronic diseases was changes in the regulation of the cell death process. Changes can be of genetic or other origin, but one way or another, pathology is characterized by excessive cell death or the survival of defective ones who were supposed to die. The first category includes some neurodegenerative, hematological, immune, infectious, and metabolic diseases. The second category—the appearance of "superfluous," potentially defective cells in the body—is primarily tumors and pre-neoplastic conditions, but also autoimmune, infectious, metabolic, and hematological diseases. To understand the pathophysiology of these numerous diseases, it is fundamentally important to know why and where the failure occurred.

The mechanisms of cell death regulation have proved to be very complex, and despite tremendous progress in this area, much remains incomprehensible. It is believed that there is a basic, core pathway with branches that lead either to specific mechanisms of cell death in individual tissues or to pathologies. Four typical types of cell death are usually listed—apoptosis, necrosis, autophagy, and cornification (keratinization), as well as eight atypical species [39]. At the same time, it cannot be said that typical ones are more important than atypical ones, they are simply better studied.

The most interesting thing is that between the paths of cell death, there are a kind of transfer stations, and this further complicates the picture, making it similar to a subway scheme in a metropolis like New York or Moscow. Why different forms of death work in different cells and tissues, what needs to be done so that, say, a cancer cell, which has one of the death pathways blocked, "made a transplant" and headed along a different path—all these questions are currently being studied.

It is important to emphasize that if you hear about a magic drug that "treats all types of oncology," then you can be sure that this is a bluff. Cancer cannot be treated with a single drug because it has no single cause. This is a systemic disease, and to fight it, it is necessary to fully analyze the system and understand what and where works incorrectly. Only with complex treatment can results be achieved. Even a drug effective for a certain type of cancer is mandatory prescribed not alone but in combination with other substances. It is important to determine which drug combinations are appropriate to use on a case-by-case basis.

Taking into account, that LTP can both accelerate and retard cell death, there are two mainstream directions in LTP medicine—wound healing and cancer treatment. About three decades ago, the application of low-temperature plasma for wound healing was among the first medical samples of successful and promising antimicrobial effects. However, taking the concept into the clinical routine took place only in the last decade.

There is no need to remember that to enter medical practice, plasma devices must be reliable and safe, generate a reproducible plasma, be manufactured based on quality management procedures, receive appropriate accreditation by authorities, and fulfill reasonable criteria for basing business models on existing needs in the health care market.

From the regional perspective, LTP-mediated wound healing in the clinical setting is most advanced in central Europe, especially Germany. All plasma devices that were accredited by official medical authorities to treat chronic wounds within dermatological centers were engineered and put into practice in Germany. Now these devices can be used across the European Union.

From the clinical point of view, several factors are important to explain the benefits of plasma wound healing. Several simultaneously working mechanisms are proposed. Usually, first place is given to the antimicrobial efficacy, but accordingly randomized placebo-controlled, patient-blinded clinical trial comparing plasma vs. antiseptics in a head-to-head study indicated LTP-promoted wound healing independent of antimicrobial efficacy as patients in all groups received antiseptics [40]. In the second place is the effect of LTP-derived electric fields in wound healing, but the scientific basis of this proposal is still completely unclear, especially since there is a lack of studies on intensity-matched exposure of electric fields alone in appropriate in vivo wound models.

And in third place put forward, also without any experimental evidence, the action of reactive oxygen species and nitrogen. As you can see, the preference for this or that explanation of the healing effect of plasma on wound healing is purely subjective. Using our right to our subjective point of view, we speak in favor of the third hypothesis about the miraculous effect of hydrogen peroxide and its decomposition products on all living processes.

Preclinical data on wounds in mice indicate an important role for focal adhesions and matrix remodeling complexes in plasma-treated wounds. Most importantly, a dataset of uninfected wounds of different origins in different species of laboratory animals (e.g., mice, rats, and sheep) and at different locations (e.g., back and ear) showed the efficacy of plasma treatment that promoted wound healing in the absence of infection. This is an indirect argument to support the hypothesis of redox control of wound healing by ROS/RNS derived from plasma.

Especially experienced wound nurses and health care practitioners know that some areas in chronic wounds are more problematic than others, which can be effectively targeted with plasma jets in a loco-regionalized manner and, e.g., with individual treatment times. Regardless of the device type, the plasma therapy complements and does not replace existing wound care measures. Critical steps such as wound debridement and dressing change remain to be vital for therapeutic success. A discussion concerning the plasma methods in dentistry is beyond the scope of this publication, however, it is very important from a technical point of view area. Nevertheless, we will look at some aspects of plasma therapy in the treatment of cancer, although this area is even more complex. But we cannot ignore her, given her paramount vital importance.

The novelty of LTP in cancer treatment lies in its multi-factorial effects that include reactive oxygen and nitrogen species (ROS/RNS) produced in the plasma, physical factors like emitted electromagnetic waves, and the electric fields that are formed when plasma impinges on tissue. Rapid advances in LTP technology have allowed great control of the final composition and properties of plasma for specific biological processes aimed at clinical applications. Specifically in oncology, LTP has enabled scientists to explore the mechanisms of eliminating cancer cells that would be applicable in diverse types of cancers.

Broadly, the experimental approaches encompass two methods of inducing cancer cell death: direct killing of cancer cells by LTP in situ and stimulation of immune responses by inducing controlled oxidative stress in cancer cells. The widely accepted hypothesis of LTP interaction with cells and tissue is based on the notion that chemical species produced by plasma are either stimulating or toxic, depending on their concentration. Various unique combinations of these species might provide great potential for activation of specific signaling pathways in cells where no single component is yet identified as the key effector.

Some studies suggest that non-cancerous cells are less sensitive to the same DBD exposure regimen than tumor cells, although this is not always the case. These results are thought to be related to the chemicals active in plasma. The sensitivity of cancer cells to direct or indirect treatment is hypothesized to be associated with redox imbalance and specific antioxidant systems that are associated with the expression of tumor suppressor genes. Anyhow, plasma-treated solutions (PTS) can also be produced by the gas discharge within the solution and can selectively kill many cancer cell lines in vitro or have anti-tumor effects when directly injected in subcutaneous or intraperitoneal tumor models in vivo [41].

PTS also shows promising application in the treatment of metastatic cancers as shown in a recent study that addresses epithelial-mesenchymal transition (EMT). EMT is a process that endows cancer cells with increased stemness, metastatic potential, and resistance to conventional therapies. In that study, PTS was shown to be significantly more effective against cancer cells that have undergone EMT than their epithelial parent cells. This effect is hypothesized to be due to the increased ROS level in transitioned cells.

The contribution of individual plasma components to its antitumor effects remains largely unknown. While most of the research describes the cause-and-effect relationship between chemical species and tumor toxicity, it is still unclear how much the physical factors of LTP contribute to the cellular responses observed.

When cancer cells or tissues are directly exposed to LTP, they experience both physical and chemical effects simultaneously, and as such, the observed biological response could be attributed to the synergistic effect of the physical and chemical factors. It was shown that in vitro treatment caused a necrotic killing effect on at least six cancer cell lines, including cell lines that are resistant to reactive species.

As in all cases, the main problem is a lack of understanding of the biological effects of LTP on tissues, which is a prerequisite for studying the main mechanism of anticancer effects based on LTP. Besides the well-accepted direct cytotoxic effects of LTP on cancer cells, evidence is now accumulating to show that it triggers unique immune responses and immunogenic cell death in vivo, providing clues to help understand LTP effects at the tissue level.

The goal of plasma oncotherapy is its translation for clinical application, and it remains an important topic for continued exploration and discussion among experts in multiple disciplines that contribute to exciting new developments in plasma medicine. Based on the different ways by which tumor tissue may be exposed to LTP, as described above, it is quite clear that it offers great flexibility in different cancer scenarios. Here we propose our vision for how LTP may be used for cancer therapy.

Note that the UV-irradiation in the range of 100–200 nm, which always accompanies DBD, is absorbed by water forming H_2O_2, and quanta with the wavelength of 200–300 nm are absorbed intensively by the molecules of H_2O_2 [42]. The generation of hydrogen peroxide by the UV irradiation of water is inhibited by anions, such as chlorine, present in water, as shown by the experimental data (Fig. 3).

Contribution subsequent to the decomposition process of H_2O_2 is stipulated by the reaction with impurities, for example, with the iron or copper ions that were formed by the electrode erosion in case, where one of the electrodes is not coated [14]. Especially fast decomposition of hydrogen peroxide takes place in the vicinity of biological tissues due to the presence of the enzyme catalase. Concentration of

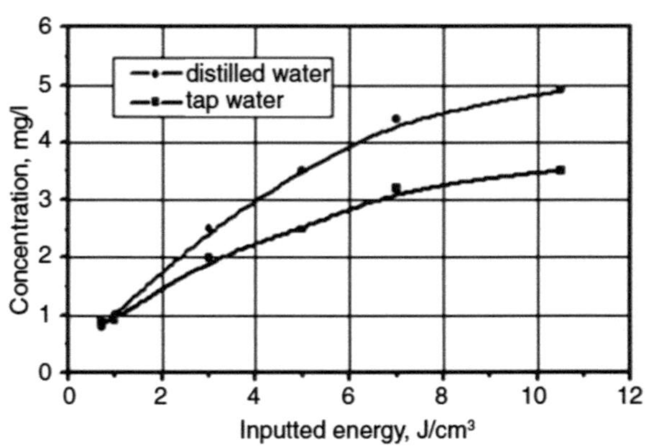

Fig. 3 Concentration of hydrogen peroxide versus input energy

H_2O_2 usually rises in the time range of milliseconds, after which decreases and reaches zero value in a few days maximum even without enzymes and/or metal ions. Otherwise, its lifetime can be as short as some seconds [42, 43].

Another important healing agent is a set of metal nanoparticles generated by a DBD reactor and a spark discharge generator with argon as the working gas. In recent years, there has been a revival of colloid chemistry in a new form: the physical chemistry of nanostructured objects. Due to the demands of nanotechnology, this area is fast developing science. Many thousands of papers are published on this subject annually.

Nanoparticles are colloidal objects having sizes from 1 to 100 nm on at least one of their three dimensions. Recent studies have shown that the physical and chemical properties of nanoparticles differ qualitatively from those of both macro and microscopic objects.

The gas temperature in the discharge region of the DBD reactor remained near room temperature, while that of the spark reactor varied from 470 to 1120 K and generally increased with increasing applied voltage amplitude in the range of 2–10 kV and driving frequency in the range of 1–10 kHz. Compared to spark-generated nanoparticles under the same voltage, frequency, and flow rate, DBD-generated nanoparticles have smaller sizes, better monodispersed, and lower number concentrations.

Unusual properties of nanoparticles relate to the fact that, when the size of the particles becomes comparable with the Bohr radius (or less it), the classic size effects transfer onto the quantum level. Under this, kinetic energy submits to the uncertainty principle. In a metal macro-object, a high density of states at the Fermi level forms a conduction band with continuous energy levels. So, the split of energy into discrete levels takes place in nanoparticles. Under this, nanoclusters behave as quantum dots, i.e., inside them, as well as, in atoms, the free electrons can occupy only certain allowed energy states.

The introduction of low-temperature, atmospheric-pressure plasmas to the biomedical field has ushered in a new revolution in the healthcare arena that promises to introduce plasma-based therapies to combat some thorny and long-standing medical challenges. The multidisciplinary field of biomedical applications of atmospheric pressure low-temperature plasma (AP-LTP) has been over 30 years in progress. Since its introduction in the mid-1990s, the field has expanded tremendously and has reached several critical and very important milestones, which include clinical trials on wound healing and cancer treatment.

A third of a century may seem like a long time for an individual investigator but for an important branch of knowledge, especially concerning living objects, it is quite a shortage to get maturity. Yet, the advances and accumulation of new knowledge about the interaction of plasma with living soft matter have been nothing short of impressive.

LTP produces a rich "cocktail" of reactive species including reactive oxygen species, ROS, and reactive nitrogen species, RNS, which have been shown to play pivotal roles in the observed biomedical outcomes, also it emits photons at various wavelengths/energies, which can also play synergistic biological roles in addition to participation in production ROS and RNS.

The most important goal is to develop control methods that allow for the minimization of the effects of environmental conditions, such as humidity, temperature, air pressure (e.g., altitude), power fluctuations, etc. while at the same time optimizing the treatment as a function of the properties of the target (shape, surface morphology, thickness/depth, conductivity, etc.). These are some of the outstanding issues that are discussed. One should realize that the bio-application of LTP has proved to be of the utmost importance, because of the advent of antibiotic resistance of bacteria and pathogenic proteins (such as prions), and because of viral-driven pandemics, such as the present COVID-19 pandemic, so it becomes the initial seed that was at the genesis of the entire new field of science [44].

Special attention should be paid to the consideration of issues related to LTP-assisted wound healing. This was the first patient-centered medical application of LTP and may be one of the most developed and has undergone extensive animal studies followed by clinical trials at various hospitals, especially in Germany. One of the therapeutic challenges is the healing of chronic wounds, such as diabetic ulcers, which do not heal easily or at all. Thousands of amputations occur every year in the world, because of the inability of conventional methods to satisfactorily deal with this problem. LTP has been shown to safely coagulate blood and enhance the proliferation of cells such as fibroblasts, which made gas plasma an attractive technology for wound healing.

The «Icing on the cake» turned out to be the treatment of cancer and the creation of antitumor methods based on the use of cold plasma, what is referred to as "plasma oncology". This application has seen great progress in the last decade which recently culminated in preliminary clinical trials in a few hospitals in Germany and the USA. In vitro Investigations have shown that LTP can kill cancer cells and in vivo experiments have shown that LTP treatment can shrink tumors' size and/or delay their growth. However, there is still much work to be done to understand the effects of LTP on cancer cells on the sub-cellular and molecular levels and the effects on the tissues surrounding tumors. And several words about the modeling and simulation works that have been conducted so far to further elucidate the plasma-cell and plasma-tissue interactions. The interaction of LTP with entire cells or tissue can up to now only be simulated on a macroscopic scale, obviously providing less detailed information than modeling the interactions with individual cell components, such as the phospholipid bilayer, DNA, or proteins, which is possible on the molecular level. The level of detailed simulation needs accurate input data and experimental validation, and the translation of the obtained insights towards improving the biomedical applications.

Control over plasma composition is mandatory for a targeted medical application. Tremendous progress has been made in controlling electron dynamics e.g., through a multi-frequency approach. Ensuring a reproducible composition by using a tailored gas surrounding the active effluent has opened the door to several promising fundamental biomedical studies. Traditionally, control of LTPs has primarily relied on statistical process control strategies, as widely used in semiconductor manufacturing. However, such control strategies are inherently open loop, in which the lack of online sensing and feedback corrective action can compromise the reliability and repeatability of LTPs due to the intrinsic plasma variabilities and external disturbances. The key challenges in feedback control of LTPs stem from (1) the nonlinear interactions between multiple input variables (i.e., discharge parameters) and multiple output variables of a plasma discharge (i.e., plasma effects), (2) the need to constrain the plasma properties to ensure safe plasma treatment and circumvent adverse plasma-induced surface effects, and (3) the need to realize the proper synergy between the chemical, physical, and electrical effects of LTPs when interacting with complex biological surfaces. The main preliminary stage for medical applications of low-temperature plasma was its use for biological decontamination and disinfection.

2.1 2D-Imaging of Absolute OH and Pulsed Dielectric Barrier Discharge by Photo-Fragmentation Laser-Induced Fluorescence

In the paper of van den Bekerom [45] the pulsed dielectric barrier discharges (DBD) in He-H_2O and He-H_2O-O_2 mixtures were studied in near atmospheric conditions using temporally and spatially resolved quantitative 2-D imaging of the hydroxyl radical (OH) and hydrogen peroxide (H_2O_2). The primary goal was to detect and quantify the production of these strongly oxidative species in water-laden helium discharges in a DBD jet configuration, which is of interest for biomedical applications such as disinfection of surfaces and treatment of biological samples. Hydroxyl profiles are obtained by laser-induced fluorescence (LIF) measurements using 282 nm laser excitation. Hydrogen peroxide profiles are measured by photo-fragmentation LIF (PF-LIF), which involves photo-dissociating H_2O_2 into OH with a 212.8 nm laser sheet and detecting the OH fragments by LIF. Gas temperatures determined from the OH-LIF spectra indicate that gas heating due to the plasma is insignificant. The addition of 5% O_2 in the He admixture decreases the OH densities and increases the H_2O_2 densities (Fig. 4).

The increased coupled energy in the 10-pulse discharge increases OH and H_2O_2 mole fractions, except for the H_2O_2 in the He-H_2O-O_2 mixture which is relatively insensitive to the additional pulses (Fig. 5) [46].

3 Barrier Discharge for Hydrogen Peroxide Production

Expanding upon the transformative research in hydrogen peroxide (H_2O_2) synthesis using non-thermal plasma technology, this innovative approach marks a significant milestone in the pursuit of more sustainable and efficient chemical production methods. The investigation spearheaded by [47] into utilizing a cylindrical barrier discharge for the activation of oxygen/hydrogen mixtures introduces a specialized

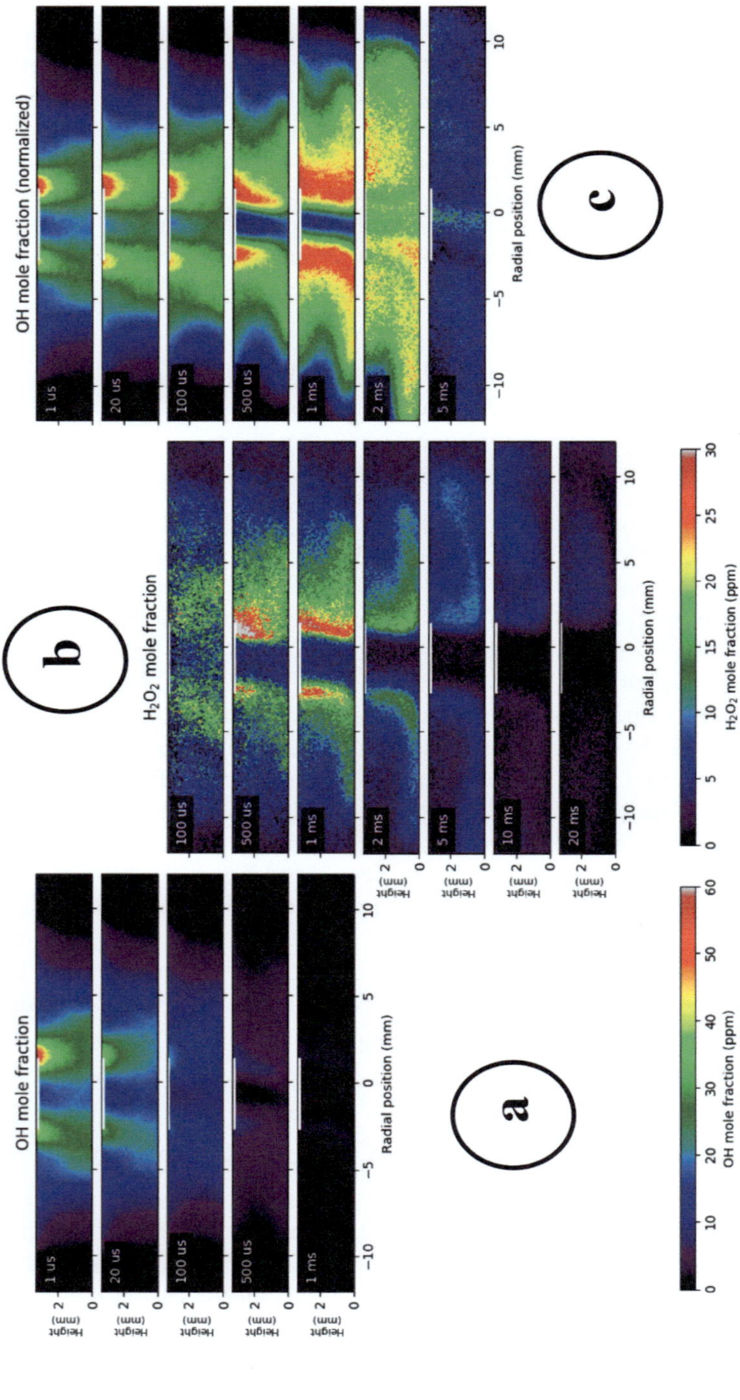

Fig. 4 (**a**) OH and (**b**) H_2O_2 mole fraction profiles for 5-pulse burst in He-H_2O mixture. (**c**) Normalized OH flow profiles

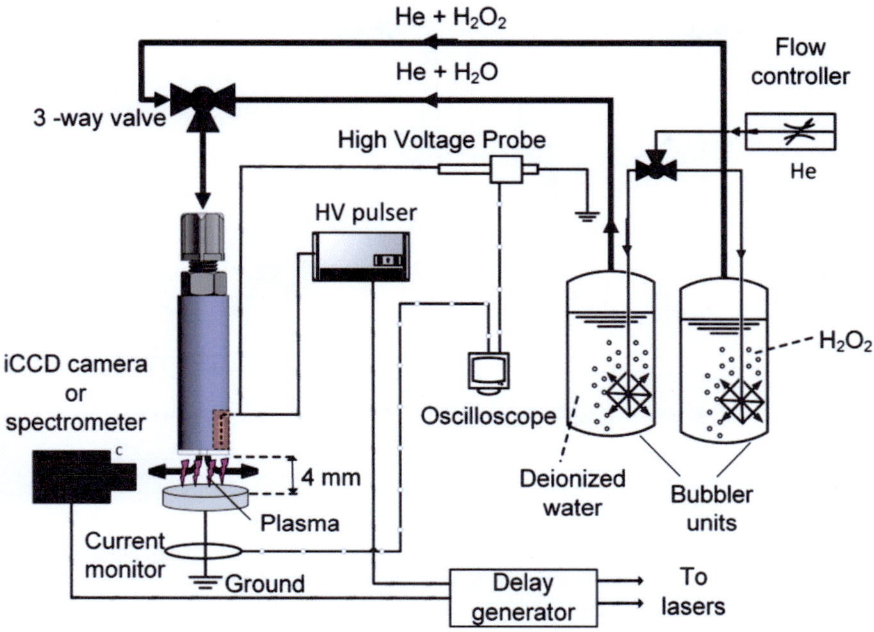

Fig. 5 Schematic image of the plasma discharge setup [46]

experimental setup. This setup is distinguished by its temperature-regulatable liquid ground electrode, which offers a novel means of influencing the reaction environment, a critical factor in the effective synthesis of H_2O_2 as demonstrated in studies by [48–50].

The synthesis process benefits from transitioning from a conventional gas phase discharge, with its inherent limitations in O_2 activation and H_2O_2 selectivity, to a surface discharge approach. This transition is facilitated by the introduction of fibrous materials into the discharge gap, significantly enhancing the efficiency of H_2O_2 production. A pivotal aspect of this enhanced efficiency is the role of temperature in the formation of a water layer on the fiber surface, which has been identified as crucial for the synthesis of H_2O_2 in plasma. Additionally, the incorporation of TiO_2 on the fiber surface further stabilizes hydrogen peroxide, suggesting the formation of a beneficial H_2O_2-TiO_2 complex (Fig. 6). The black curve displays the synthetic spectrum at 300 K; the spectrum at 275 K does not differ appreciably from the 300 K spectrum at this scale. The bottom part of the figure displays the residuals of the fit.

The research meticulously examines the effects of reactant gas composition and the presence of water, concluding that the presence of both H_2 and O_2 is essential for the synthesis reaction, with H_2O_2 primarily forming in the gas phase. This finding underscores the nuanced understanding required to optimize the plasma-based synthesis of hydrogen peroxide, a task that several researchers [51, 52] have

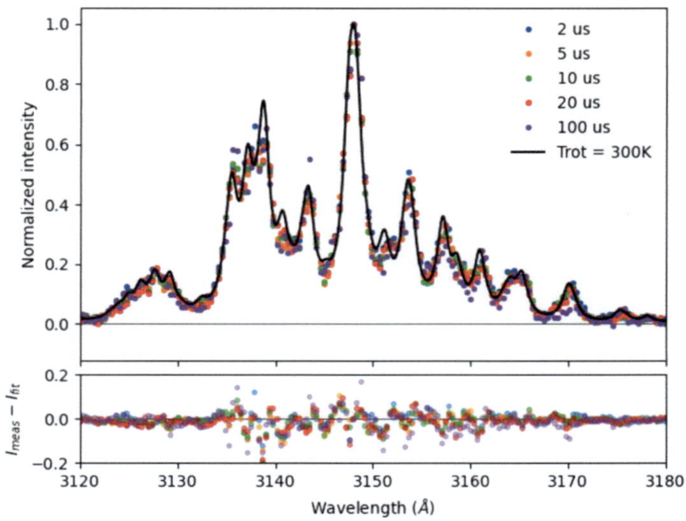

Fig. 6 Measured fluorescence spectra (dots) in the plasma for a range of delay times. Obtained rotational temperatures are within 275–300 K for all recorded time delays

explored, noting the potential of non-thermal plasma as a significant tool for H_2O_2 production, though not yet scalable to industrial processes.

Further analysis by Thevenet et al. [53] reinforces the efficacy of atmospheric pressure non-thermal plasma for H_2O_2 synthesis, particularly highlighting the beneficial role of fibrous materials in improving the synthesis process. Although this method does not currently compete with the anthraquinone self-oxidation process on an industrial scale, its value in generating hydrogen peroxide in situ for biomedical and food preservation applications cannot be overstated. This capability to produce H_2O_2 on-demand offers promising implications for localized treatments and interventions, potentially revolutionizing sectors that rely on hydrogen peroxide's antiseptic and preservative qualities.

The collective insights from these studies not only chart a path for future research into optimizing non-thermal plasma for H_2O_2 synthesis but also illuminate the broader potential of plasma technology in green chemistry. By demonstrating a viable alternative to traditional chemical production methods, this research contributes to the ongoing search for more sustainable, efficient, and environmentally friendly industrial processes [54]. The continual refinement and exploration of non-thermal plasma technology hold the promise of unlocking new applications and efficiencies in chemical synthesis, paving the way for advancements that could have a profound impact on various industries and environmental sustainability efforts.

4 Conclusions

Barrier dielectric discharge and the resultant cold plasma technology have unveiled a revolutionary approach to various applications ranging from surface treatment in manufacturing industries to groundbreaking biomedical applications. Despite being a phenomenon discovered in the mid-19th century, it is only in recent decades that its vast potential for practical applications has been recognized and harnessed. The ability of cold plasma to operate at ambient conditions has broadened its utility across sectors, particularly in food decontamination, and water treatment, and especially in medical fields for sterilization, wound healing, and cancer treatment, among others. The effectiveness of cold plasma in these applications is largely attributed to its generation of bio-active species like hydrogen peroxide and other reactive oxygen species, alongside charged particles and photons, which exhibit significant biological effects. However, the complexity of the interactions between plasma and treated materials, especially living tissues, underscores the necessity for ongoing research to unravel the underlying mechanisms, assess the technology's safety, and evaluate its environmental and economic impacts comprehensively. Despite the challenges and the need for further studies to optimize and understand cold plasma's full capabilities and implications, the optimism surrounding its benefits remains high. Cold plasma technology is poised to continue its trajectory of growth and innovation, offering promising solutions to longstanding problems across various disciplines. The convergence of interdisciplinary research efforts is crucial to unlocking the full spectrum of applications of cold plasma, marking a significant step forward in both scientific exploration and practical utility.

Acknowledgements This article was elaborated as a joint research project of the Moldova State University, Institute of Chemistry; N.N. Semenov Federal Research Center for Chemical Physics, Russian Academy of Sciences; and the International Clean Water Institute, VA, USA.

References

1. Kogelschatz, U. (2003). Dielectric-Barrier Discharges: Their History, Discharge Physics, and Industrial Applications. Plasma Chemistry and Plasma Processing. 23. 1-46. https://doi.org/10.1023/A:1022470901385.
2. Kolawole, E.A.; Lin, T.-C.; Huang, Y.-W.; Locmelis, M.; Han, D. (2024). A Review of Dielectric Barrier Discharge Cold Atmospheric Plasma for Surface Sterilization and Decontamination. IEEE Transactions on Radiation and Plasma Medical Sciences. 99. 1-1. https://doi.org/10.1109/TRPMS.2024.3349571.
3. Fridman, A.; Kennedy, L.A. (2021). Plasma Physics and Engineering. Ed.: New York, London: Taylor & Francis. 724 p. https://doi.org/10.1201/9781315120812.
4. Reimanis, M.; Mezule, L.; Ozolins, J.; Malers, J.; Juhna, T. (2013). Drinking Water Disinfection with Electrolysis. Latvian Journal of Chemistry. 51. 296–304. https://doi.org/10.2478/v10161-012-0016-9.

5. Charlamova,T.A.; Kolesnikov, A.V.; Alaferdov, A.F.; Sarbaeva, M.T.; Gaidukova, A.M. (2013). Perspective Electrochemical Processes for Wastewater Treatment Technology. II. Electrochemical Destruction of Organic Substances; Use of electrolysis in Technology of Water Purification. Electroplating & Surface Treatment. 21(3), 55–62.
6. El-Tayeb, A.; El-Dein, A.Z.; Elnaggar, A.Y.; Hussein, E.E. (2021). Influence of Temperature in Degradation of Organic Pollution Using Corona Discharge Plasma. Sustainability. 13(23), 12971. https://doi.org/10.3390/su132312971.
7. Li S.; Dang, X.; Yu, X.; Abbas, G.; Zhang, Q.; Cao, L. (2020). The Application of Dielectric Barrier Discharge Non-Thermal Plasma in Vocs Abatement: A Review. Chemical Engineering Journal. 388, 124275. https://doi.org/10.1016/j.cej.2020.124275.
8. Zhu, F.; Dong, Z.; Li, X.; Xiong, Q. (2022). Microbial Inactivation Property of Pulsed Corona Discharge Plasma and its Effect on Chilled Pork Preservation. Foodborne Pathogens and Disease. 19(2), 159-167. https://doi.org/10.1089/fpd.2021.0035.
9. Boithias, L.; Acuña, V.; Vergoñós, L.; Ziv, G.; Marcé, R.; Sabater, S. (2014). Assessment of the Water Supply: Demand Ratios in a Mediterranean Basin Under Different Global Change Scenarios and Mitigation Alternatives. Sci Total Environ 470–471, 567–577.
10. Bokhale, N.B.; Bomble, S.D.; Daldhanjan, R.R.; Mahale, D.D.; Hinge, S.P.; Barnerjee, B.S.; Mohod, A.V.; Gogate, P.R. (2014) Sonocatalytic & Sonophotocatalytic Degradation of Rhodamine 6G Containing Wastewaters. Ultrason Sonochem, 21, 1797–1804.
11. Mouele, E.S.M.; Tijani, J.O.; Fatoba, O.O.; Petrik, L.F. (2015). Degradation of Organic Pollutants and Microorganisms from Wastewater Using Different Dielectric Barrier Discharge Configurations - A Critical Review. Environ Sci Pollut Res. 22, 18345-18362. https://doi.org/10.1007/sl1356-015-5386-6.
12. Wenjiao, S.; Wei, L.; Longjie, M.; Danni, J.; Cheng, C.; Qiang, L.; Chen, W.; Cheng, Z.; Min, L. (2021). Research on Different Oxidants Synergy with Dielectric Barrier Discharge Plasma in Degradation of Orange G: Efficiency and Mechanism. Separation and Purification Technology. 277. 119473. https://doi.org/10.1016/j.seppur.2021.119473.
13. Duca, Gh.; Vaseashta, A. (2023a). Environmental and Technological Aspects of Redox Processes. Ed.: U.S.A.IGI Global, 2, 410 p. https://doi.org/10.4018/978-8-3693-0512-6.
14. Duca, Gh.; Vaseashta A. (2023b). Fundamental and Biomedical Aspects of Redox Processes. Ed.: U.S.A. IGI Global. 1, 495 p. https://doi.org/10.4018/978-1-6684-7198-2.
15. Vaseashta, A., Duca, G., & Travin, S. (Eds.). (2022). Handbook of Research on Water Sciences and Society (2 Volumes). IGI Global. DOI.:https://doi.org/10.4018/978-1-7998-7356-3
16. Rip, G.R. (1996) Applications of Ozone for Industrial Wastewater Treatment – A Review. Ozone: Science & Engineering. 18(6), 477515, https://doi.org/10.1080/01919512.1997.10382859.
17. Mouele, E.S.M.; Tijani, J.O.; Badmus, K.O.; Pereao, O.; Babajide, O.; Fatoba, O.O.; Petrik, L.F. (2021). A Critical Review on Ozone and Co-Species, Generation, and Reaction Mechanisms in Plasma Induced by Dielectric Barrier Discharge Technologies for Wastewater Remediation. Journal of Environmental Chemical Engineering. 9(5), 105758. https://doi.org/10.1016/j.jece.2021.105758.
18. Ma, S.; Kim, K.; Chun, S.; Moon, S.Y.; Hong, Y. (2020). Plasma-Assisted Advanced Oxidation Process by a Multi-Hole Dielectric Barrier Discharge in Water and its Application to Wastewater Treatment, Chemosphere. 243, 125377. https://doi.org/10.1016/j.chemosphere.2019.125377.
19. Duca, Gh.; Travin, S.O. (2020) Reactions' Mechanisms and Applications of Hydrogen Peroxide. American Journal of Physical Chemistry. 9(2). 36-44. https://doi.org/10.11648/j.ajpc.20200902.13.
20. Vujevic, D.; Koprivanac, N.; Loncaric Bozis, A.; Locke, B.R. (2004) The Removal of Direct Orange 39 by Pulsed Corona Discharge from Model Wastewater. Environ Technol. 25(7), 791-800. https://doi.org/10.1080/09593330. 2004.9619370.
21. Chang, J.S.; Li, O.L.; Guo, Y. (2010). Pulsed Arc Electrohydraulic Discharge Characteristics, Plasma Parameters, Plasma Parameters, and Optical Emission During Contaminated Pond Water Treatments. Electrical Insulation Magazine, IEEE. 27, 8 - 17. https://doi.org/10.1109/MEI.2011.5739418.

22. Angeloni, D.M.; Dickson, S.E.; Emelko, M.B.; Chang, J.S. (2007). Removal of Methyl-Tertbutyl Ether from Water by a Pulsed Arc Electrohydraulic Discharge System. Journal of Applied. Physics. 45, 8290.
23. Yamatake, A.; Angeloni, D.N.; Dickson, S.; Emelko, M.B.; Yasuoke, K.; Chang, J.S. (2007). UV and Optical Emissions Generated by the Pulsed Arc Electrohydraulic Discharge. J. Journal of App. Phys. 45, 8298.
24. Salomon, E.; Tijani, J.O.; Fatoba, O.; Petrik, L.F. (2015). Degradation of Organic Pollutants and Microorganisms from Wastewater Using Different Dielectric Barrier Discharge Configurations-A Critical Review. Environmental Science and Pollution Research International. 22. https://doi.org/10.1007/s11356-015-5386-6.
25. Cheng, Z. ; Qu, M. ; Chen, D. ; Chen, J. ; Yu, J. ; Zhang, S. ; ... & Wang, J. (2021). Mechanisms of Active Substances in a Dielectric Barrier Discharge Reactor: Species Determination, Interaction Analysis, and Contribution to Chlorobenzene Removal. Environmental Science & Technology. 55(6), 3956-3966. https://doi.org/10.1021/acs.est.0c04914.
26. Kučerová, K.; Henselová, M.; Slováková, Ľ.; Bačovčinová, M.; Hensel, K. (2021). Effect of Plasma Activated Water, Hydrogen Peroxide, and Nitrates on Lettuce Growth and its Physiological Parameters. Applied Sciences, 11(5), 1985. https://doi.org/10.3390/app11051985.
27. Minh, N.P. (2022). Corona Discharge Power of Plasma Treatment Influence on the Physicochemical and Microbial Quality of Enoki Mushroom (Flammulina velutipes). Journal of Pure & Applied Microbiology. 16(1). https://doi.org/10.22207/jpam.16.1.08.
28. Kraft, A. (2008). Electrochemical Water Disinfection: A Short Review. Platinum Metals Review. 52. 177-185. https://doi.org/10.1595/147106708X329273.
29. Ngwenya, N.; Ncube, E.J.; Parsons, J. (2013). Recent Advances in Drinking Water Disinfection: Successes and Challenges. Reviews of Environmental Contamination and Toxicology. 222. 111-70. https://doi.org/10.1007/978-1-4614-4717-7_4.
30. Kolikov, V.; Rutberg, P. (2015). Pulsed Electrical Discharges for Medicine and Biology. Springer, 215 p. https://doi.org/10.1007/978-3-319-18129-5.
31. Fridman, G.; Peddinghaus, M.; Ayan, H.; Fridman, A.; Balasubramanian, M.; Gutsol, A.; Brooks, A.; Friedman, G. (2006). Blood Coagulation and Living Tissue Sterilization by Floating-Electrode Dielectric Barrier Discharge in Air. Plasma Chem Plasma. 3, 425–442. https://doi.org/10.1007/s11090-006-9024-4.
32. Yakushin, R.V.; Kolesnikov, V.A.; Babusenko, E.S.; Brodsky, V.A.; Solov'eva, I.N.; Perfil'eva, A.V.; Golovina, V.A. (2017). Prospects of Water Disinfection by Means of Spark and Barrier Discharges. Chemical Safety Science. 1(1), 92–100. https://doi.org/10.25514/CHS.2017.1.11436.
33. Kelar Tučeková, Z.; Vacek, L.; Krumpolec, R.; Kelar, J.; Zemánek, M.; Černák, M.; Růžička, F. (2021). Multi-Hollow Surface Dielectric Barrier Discharge for Bacterial Biofilm Decontamination. Molecules (Basel, Switzerland). 26. https://doi.org/10.3390/molecules26040910.
34. Fridman, G.; Shereshevsky, A.; Jost, M.; Brooks, A.; Fridman, A.; Gutsol, A.; Vasilets, V.; Friedman, G. (2007). Floating Electrode Dielectric Barrier Discharge Plasma in Air Promoting Apoptotic Behavior in Melanoma Skin Cancer Cell Lines. Plasma Chem Plasma Process. 27. https://doi.org/10.1007/s11090-007-9048-4.
35. Sakudo, A.; & Tsuji, Y. (2021). Hydrogen Peroxide Gas Plasma Sterilizer Combined with Dielectric Barrier Discharge and Corona Discharge Inactivates Prions. Applied Sciences. 11(20), 9777. https://doi.org/10.3390/app11209777.
36. Rybkin, V.V.; Shutov, D.A. (2017). Electric Discharges of Atmospheric Pressure as a Tool for the Chemical Activation of Water Solutions. The Physics of Plasma. 43(11), 929–954. https://doi.org/10.7868/S0367292117110075.
37. Gershman, S.; Harreguy, M.B.; Yatom, S.; Raitses, Y.; Efthimion, P.; Haspel, G. (2021). A Low Power Flexible Dielectric Barrier Discharge Disinfects Surfaces and Improves the Action of Hydrogen Peroxide. Sci. Rep., 11(1), 4626. https://doi.org/10.1038/s41598-021-84086-z.

38. Kun, L.; Duan, Q.; Zheng, Z.; Zhou, Ru.; Zhou, Re.; Tang, W.; Cullen, P.; Ostrikov, K. (2021). Gasphase Peroxynitrite Generation Using Dielectric Barrier Discharge at Atmospheric Pressure: a Prospective Sterilizer. Plasma Processes and Polymers, 18(11), e2100016. https://doi.org/10.1002/ppap.202100016.
39. Nakamura, T. (2010). Comparative Immunohistochemical Analyses on the Modes of Cell Death/Keratinization in Epidermal Cyst, Trichilemmal Cyst, and Pilomatricoma. The American Journal of Dermatopathology. 33. 78-83. https://doi.org/10.1097/DAD.0b013e3181e3aec1.
40. Laroussi, M.; Bekeschus, S.; Bogaerts, A.; Fridman, A.; Lu, X.; Miller, V.; Laux, C.; Walsh, J.; Jiang, C.; Liu, D.; Yusupov, M.; Keidar, M.; Hori, M.; Stapelmann, K.; Mesbah, A.; Thagard, S.; Tanaka, H.; Yan, D.; Ostrikov, K.; Reuter, S. (2021). Low Temperature Plasma for Biology, Hygiene, and Medicine: Perspective and Roadmap. https://doi.org/10.36227/techrxiv.15125520.
41. Tanaka, H.; Bekeschus, S.; Yan, D.; Hori, M.; Keidar, M.; Laroussi, M. (2021). Plasma-Treated Solutions (PTS) in Cancer Therapy. Cancers. https://doi.org/10.3390/cancers13071737.
42. Duca, Gh. (2012). Homogeneous Catalysis with Metal Complexes: Fundamentals and Applications. Ed.: Berlin, Heidelberg: Springer, 102(XII), 478 p. http://www.springer.com/chemistry/book/978-3-642-24628-9.
43. Travin, S.; Duca, Gh.; Gladchi, V. (2019). Self-Purification of Aquatic Media from Hexachlorocyclohexsane in a Radical Process. Chemistry Journal of Moldova, 2019, 14(1), 47–53. https://doi.org/10.19261/cjm.2018.541.
44. Farah, R.I.; Al-Haj Ali, S.N. (2021). Electrolyzed Water Generated On-Site as a Promising Disinfectant in the Dental Office During the COVID-19 Pandemic. Front Public Health. 9, 629142. https://doi.org/10.3389/fpubh.2021.629142
45. Bekerom van den, D.; Tahiyat, M.M.; Huang, E.; Frank, J.H.; Farouk, T.I. (2023). 2D-Imaging of Absolute OH and H_2O_2 Profiles in A He–H_2O Nanosecond Pulsed Dielectric Barrier Discharge by Photo-Fragmentation Laser-Induced Fluorescence. Plasma Sources Science and Technology, 32(1), 015006.
46. Yatom, S.; Dobrynin, D. (2022). Examination of OH and H_2O_2 Production by Uniform and Non-Uniform Modes of Dielectric Barrier Discharge in He/Air Mixture. J. Phys. D: Appl. Phys. 55, 485203. https://doi.org/10.1088/1361-6463/ac95a6.
47. Cameli, F.; Dimitrakellis, P.; Chen, T.-Y.; Vlachos, D. (2022). Modular Plasma Microreactor for Intensified Hydrogen Peroxide Production. ACS Sustainable Chemistry & Engineering. 10.https://doi.org/10.1021/acssuschemeng.1c06973.
48. Dey, G.; Das, T. (2016). Yields of Hydrogen and Hydrogen Peroxide from Argon–Water Vapor in Dielectric Barrier Discharge. Plasma Chemistry and Plasma Processing. 36. https://doi.org/10.1007/s11090-015-9675-0.
49. Chen, Q.; Junshuai, L.; Ostrikov, K. (2022). Recent Advances Towards Aqueous Hydrogen Peroxide Formation in a Direct Current Plasma–Liquid System. High Voltage. 7. 405-419. https://doi.org/10.1049/hve2.12189.
50. Rataj, R.; Werneburg, M.; Below, H.; Kolb, J.F. (2022). Hydrogen Peroxide Production of Underwater Nanosecond-Pulsed Streamer Discharges with respect to Pulse Parameters and Associated Discharge Characteristics. Plasma Sources Science and Technology. 31(10), 105005. https://doi.org/10.1088/1361-6595/ac942a.
51. Zhou, Re.; Zhang, T.; Zhou, Ru.; Wang, S.; Mei, D.; Mai-Prochnow, A.; Weerasinghe, J.; Fang, Z.; Ostrikov, K.; Cullen, P.J. (2021). Sustainable Plasma-Catalytic Bubbles for Hydrogen Peroxide Synthesis. Green Chem., 23, 2977-2985. https://doi.org/10.1039/d1gc00198a.
52. Wapshott-Stehli, H.L.; Myers, B.G.; Quesada, M.J.H.; Grunden, A.; Stapelmann, K. (2022). Plasma-Driven Biocatalysis: In Situ Hydrogen Peroxide production with an Atmospheric Pressure Plasma Jet Increases the Performance of Oletje When Compared to Adding the Same Molar Amount of Hydrogen Peroxide In Bolus. Plasma Processes and Polymers. 19, e2100160, https://doi.org/10.1002/ppap.202100160.

53. Thevenet, F.; Couble, J.; Brandhorst, M.; Dubois, J.-L.; Puzenat, E.; Guillard, C.; Bianchi, D. (2010). Synthesis of Hydrogen Peroxide Using Dielectric Barrier Discharge Associated with Fibrous Materials. Plasma Chemistry and Plasma Processing. 30. 489-502. https://doi.org/10.1007/s11090-010-9234-7.
54. Vaseashta, A. Green Nanotechnologies for Responsible Manufacturing. MRS Online Proceedings Library 1106, 305 (2008). https://doi.org/10.1557/PROC-1106-PP03-05.

Energy Harvesting Efficiency Analysis Using Artificial Intelligence

Souad Touairi and Mustapha Mabrouki

Abstract This paper establishes the emergence of AI technologies tailored to real-world challenges and explores a promising avenue for specific energy harvesting methods, notably the optimization of piezoelectric and triboelectric nanogenerators in terms of recognition, production, and consumption. It proposes an adaptive approach using machine learning and neural network techniques to collect and optimize vibration energy. Recent advancements in computational methodologies, especially within artificial intelligence (AI) and machine learning (ML), have heightened the need for intelligent, self-sustaining devices. Given the global concern over energy consumption, there is an urgent need for solutions that reduce energy usage while maintaining the efficiency of intelligent applications. Energy harvesting technology, which harnesses ambient mechanical vibrations to generate electrical energy, presents a viable solution. This paper aims to demonstrate how cutting-edge AI-driven approaches can enhance energy harvesting efficiency, thereby contributing to the development of sustainable and intelligent energy solutions. Additionally, multiple simulations conducted using MATLAB/Simulink, alongside experimental results obtained with the dSPACE DS1104 board, are discussed to validate the improvements in control and speed estimation.

Keywords Energy harvesting · Sustainable energy · Self-powering systems · Solar power · Piezo transducers

---

S. Touairi (✉)
Mechanical Department, Sciences and Technologies Faculty, Industrial Engineering and Surface Engineering Laboratory, Sultan Moulay Slimane University, Beni Mellal, Morocco
e-mail: touairisouad@gmail.com

M. Mabrouki
Physical Department, Sciences and Technologies Faculty, Industrial Engineering and Surface Engineering Laboratory, Sultan Moulay Slimane University, Beni Mellal, Morocco

A. Vaseashta et al. (eds.), *Proceedings of the Eighth International Symposium on Dielectric Materials and Applications (ISyDMA'8)*, Springer Proceedings in Physics 349, https://doi.org/10.1007/978-3-031-84809-4_4

1 Introduction

In light of recent significant advances in the field of renewable energy, particularly in the area of energy harvesting (HE) using smart materials, the need for intelligent and autonomous devices has become increasingly critical [1]. The quest for sustainable energy solutions has become increasingly urgent in the face of global environmental challenges and the relentless growth in energy demand [2]. An innovative approach gaining considerable attention is the harnessing of energy from highways, a vital component of modern transportation infrastructure [3–8]. This comprehensive review is motivated by the need to integrate environmental sustainability, energy security, and infrastructure optimization [9].

For nearly 150 years, fossil fuels such as coal, oil, and natural gas have supplied approximately 80% of the world's energy needs and powered economies. However, the combustion of these fossil fuels releases carbon dioxide and other greenhouse gases into the atmosphere, causing significant climate alterations [10–18]. The depletion of finite fossil fuel reserves, coupled with the need to mitigate environmental impacts, has spurred the search for renewable energy sources and energy harvesting systems [19].

In this context, energy harvesting systems on highways have emerged as a promising avenue for sustainable energy generation. The imperative to transition towards renewable energy sources is driven by the urgent need to combat climate change and reduce carbon emissions [20]. As countries aim to meet ambitious greenhouse gas reduction targets, exploring novel approaches to energy generation that minimize environmental impact becomes paramount. Highways, spanning both urban and rural landscapes, present a unique opportunity to integrate renewable energy generation technologies [21]. Moreover, the ongoing expansion of urban areas and transportation networks necessitates a re-imagining of highways as multi-functional assets. Beyond their primary role in vehicular transportation, highways can be transformed into energy corridors [22]. This dual-purpose approach aligns with environmental goals and offers a practical solution to meet the rising energy demands of modern societies.

This paper introduces an academic review that comprehensively examines the various energy harvesting methods deployable on highways. The review aims to evaluate the working principles, performance metrics, potential applications, and research gaps associated with each method, providing valuable insights into their practical implementation.

In developing nations, the transportation sector is a major consumer of fossil fuels, accounting for one-third of overall energy consumption and significantly contributing to pollution [23]. In 2020, carbon dioxide emissions were reduced to 7.6 gigatons, similar to 2019 levels due to the COVID-19 pandemic. However, with the resumption of services, emissions grew by 3% in 2022 [24].

A substantial portion of the energy used in transportation is lost as vibrational or thermal energy. According to a study by Zhang, when fossil fuels are utilized on the road, 10 to 16% of the energy is consumed to overcome resistive forces like friction and drag, with the remainder being dissipated as heat, strain, and vibration.

Managing global energy demand concerns without depleting resources remains a formidable challenge. Energy harvesters on highways, such as piezoelectric, electromagnetic induction, solar, thermoelectric, and hybrid systems, offer promising solutions. These technologies aim to convert wasted energy from automotive travel into usable electrical energy, thus reducing dependency on fossil fuels and promoting cleaner energy alternatives [25].

Energy harvesting on highways aims to harness kinetic energy generated by vehicles into sustainable power sources. This energy can be utilized to power roadside infrastructure, and streetlights, charge electric vehicles, or feed into the grid, thereby contributing to a more sustainable energy ecosystem [26].

For instance, simulations by Delorme [27] indicated that class 8 vehicles, carrying loads of 10–100%, traveling at 105 km/h, generate 11–19% of their energy through rolling resistance. In Malaysia alone, with 17,486,589 registered automobiles averaging 15,500 km annually and consuming approximately 1162.5 L of fuel, an estimated 15% of this energy is lost annually through tire inefficiencies and road friction:

$$3.478 * 10^7 * \frac{J}{L} * 1162.5 * 17486589 * 15 = 1.061 * 10^{16} KJ \tag{1}$$

The energy equivalent of one liter of ordinary gasoline is 3.478 × 10^7 J/L. Capturing even a fraction of this energy could significantly benefit by powering low-power systems, reducing reliance on fossil fuels, and promoting cleaner energy alternatives.

Various energy harvesting methods have been proposed and implemented on highways globally, spanning technologies like piezoelectricity, electromagnetic induction, solar power, thermoelectricity, and hybrid systems [12]. Each method has unique characteristics, efficiencies, and challenges, necessitating a comprehensive evaluation for large-scale adoption [13]. This academic review extensively examines different energy harvesting methods applicable to highways, analyzing their principles, performance characteristics, and current applications to provide a comprehensive assessment of their capabilities and limitations.

Despite technological advancements, several research gaps hinder their widespread implementation:

- **Efficiency Optimization**: Significant progress has enhanced the energy conversion efficiency of various methods, but further improvements are needed to convert dissipated energy, including vibration and heat, into useful electrical energy [7]. Studies on thermomechanical, solar, thermoelectric, and wind energy systems are ongoing.
- **Scalability and Integration**: Developing scalable, cost-effective solutions that seamlessly integrate into existing highway infrastructure remains challenging [15]. Future research should focus on modular and adaptable systems suitable for deployment across different highway networks.

- **Reliability and Durability**: Energy harvesting systems must withstand harsh highway conditions, including temperature variations, moisture, vibrations, and mechanical stress, requiring robustness and durability over extended periods [16]. Research efforts should enhance materials, encapsulation techniques, and fault detection mechanisms.
- **Economic Viability**: Comprehensive cost–benefit analyses and life-cycle assessments are crucial to assess the long-term economic feasibility of energy harvesting systems on highways [16]. Despite benefits like clean energy generation, reducing traditional power dependency requires careful economic evaluation.
- **Standardization and Regulations**: The absence of standardized guidelines and regulations poses integration challenges. Developing harmonized standards addressing safety, interoperability, and grid integration is essential for widespread deployment [17].

This review employed a systematic literature review method focusing on the past decade's material using keywords like "road energy harvesting" and "pavement energy harvesting." Selected papers emphasized practical experimentation, ensuring effective measurement based on parameters like cost, effectiveness, and power output. Detailed analysis of research gaps in technology specifics facilitated reliable conclusions on basic principles, energy harvester types, efficiency metrics, and comparative assessments. This paper has been divided into the following subsections:

- Basic principles and types of energy harvesters.
- Efficiency based on cost and energy output.
- Comparison of energy harvesters
- Summary and conclusion

2 The Different Harvesting Systems

Energy harvesting methods and applications have gained significant traction, especially with advancements in microelectronics and MEMS (Micro-Electro-Mechanical Systems). This trend is expected to persist and expand in scope. The power management mechanisms integrated into these energy harvesters enable them to tap into various energy sources, potentially eliminating the need for batteries depending on application requirements. This section of the paper will explore several approaches developed for energy generation through roadways. It will delve into the principles underlying these harvesters and their mechanisms for converting otherwise untapped energies—such as mechanical vibrations, thermal differentials, and solar irradiation—into electrical energy. Each approach will be discussed in terms of its operational principles, efficiency factors, and suitability for different environmental conditions.

3 Electromagnetic Harvesting

Mechanical energy from vehicles on roadways is captured and converted into electricity using electromagnetic energy harvesting technology [18]. Everyday traffic on transportation routes carries various loads from vehicles of different sizes, dissipating energy through road deformation and heat. To harness and utilize this otherwise wasted energy, new electromagnetic technologies are being researched and developed.

These devices operate on the principle of converting mechanical energy generated by highway traffic into a force or movement that acts on a generator, producing electricity. This technology extends to various sources such as mechanical energy from railways [19], vehicle suspensions [20], backpacks [21], and ocean waves [22].

Operating on Faraday's principle, these systems utilize the interaction between an electrical conductor moving within a magnetic field to generate electric current [23]. The amount of electricity produced depends on factors like the conductor's velocity, the strength of the surrounding magnetic field, and the number of coils [24]. As vehicles traverse the road, their mechanical energy interacts with a mechanical system, causing an oscillating conductor within a magnetic field and coil setup to generate electric power [25].

These mechanical systems are categorized into three types:

Rotational Mechanisms: Here, traffic-induced pressure on displacing plates or bumps translates into rotational torque on an electromagnetic generator. Examples include hydraulic systems [26], cam and arm mechanisms [27], rack and pinion setups [11], and chain and sprocket configurations [28]. Hybrid systems, combining mechanisms like hydraulic setups with rack-and-pinion systems, are also explored [29]. The rack-and-pinion configuration, where a top plate supports traffic loads, transfers pressure to an underground rack, generating vertical displacement that rotates a pinion and powers the generator.

Linear Generators: These systems involve the movement between a coil and a magnet due to vertical displacement caused by passing vehicles [31]. The setup typically consists of a stationary stator (magnet or coil) and a translator (magnet or coil) that moves freely in response to traffic-induced surface movements, generating electricity through their interaction [32]. Multiple sets of coils and magnets are often employed to maximize power output.

Rotational Input Systems: In this category, moving traffic imparts rotational input to the harvester, turning a generator through a chain-and-sprocket mechanism [33].

The authors in [34] proposed an electromagnetic system for high-speed railways featuring a double rocker mechanism, a flywheel, and a generator. The rocker mechanism converts pavement vibrations into linear motion of a generator shaft, while the flywheel smoothens the generator's output power. Despite generating an average output power of 1.034 W under field conditions, long-term durability, cost implications, and environmental impact remain critical considerations unaddressed by the study.

In a similar electromagnetic experiment, the authors [35] developed a Simple Power Road (SPR) capable of harvesting energy from moving objects. The SPR comprises a polymer plate and an electrode array, generating significant electric charge when objects move across it. The SPR's potential was demonstrated with measurements of open and short-circuit voltage and current, albeit showing lower efficiency compared to established technologies.

Researchers in [36] introduced optimized electromagnetic vibration energy harvesters based on new damping models, predicting and maximizing power production through genetic algorithms. While effective, these models are computationally expensive and limited by input parameter accuracy. Additionally, The authors in [37] implemented a hydraulic mechanical harvester within speed bumps, utilizing vehicle-induced fluid movement to rotate a flywheel connected to electromagnetic generators, thereby producing electricity.

4 Piezoelectric Harvesting

Piezoelectric materials react to applied stress and strain by generating electrical energy, a phenomenon known as the piezoelectric effect [38]. These materials can also function as electrically actuated stress or strain sensors. When connected to an external load, the electrical energy generated can be utilized [39]. The piezoelectric effect is described by Eqs. (2) and (3), which illustrate the relationship between electric displacement (D), strain (S), electric field (E), and piezoelectric coefficients:

$$\begin{cases} D = \epsilon^{\sigma} + d\sigma \\ S = S^{E}\sigma + dE \end{cases} \tag{2}$$

D–Electric displacement S – Strain

E–Electric field

D–Piezoelectric coefficient

σ–Stress

ϵ^{σ}–Dielectric permittivity

S^E–Material's mechanical compliance matrix

Piezoelectric materials exhibit two modes of coupling, referred to as 31 and 33, characterized by their respective piezoelectric coefficients d31 and d33 [40]. The difference lies in the orientation of tension direction and polarity: d31 involves perpendicular directions, whereas d33 involves parallel directions. Research indicates that piezoelectric materials with d33 coupling are more efficient for energy-harvesting applications [41]. Researchers in [40] compared energy harvesters utilizing 31 and 33 modes and found that those employing 33-mode piezoelectric elements demonstrated superior performance when subjected to pressure from vehicles on roads.

Figures 1, 2 illustrate a piezoelectric energy harvesting system on a road. As trucks move over the road surface, vibrations activate embedded piezoelectric cells, generating electricity to illuminate streetlamps. Additionally, the stress and strain forces produced by the tires on the road surface also activate these piezoelectric cells.

Piezoelectric materials occur naturally and can also be manufactured. Quartz is a common natural piezoelectric element, while synthetic alternatives include polymers, ceramics, and composites [42]. The most frequently used composites are lead

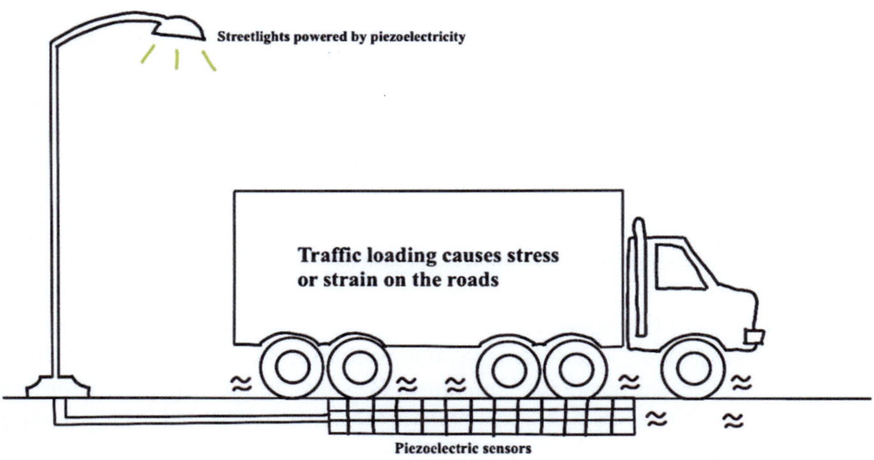

Fig. 1 The concept of Piezo energy harvesting on roads

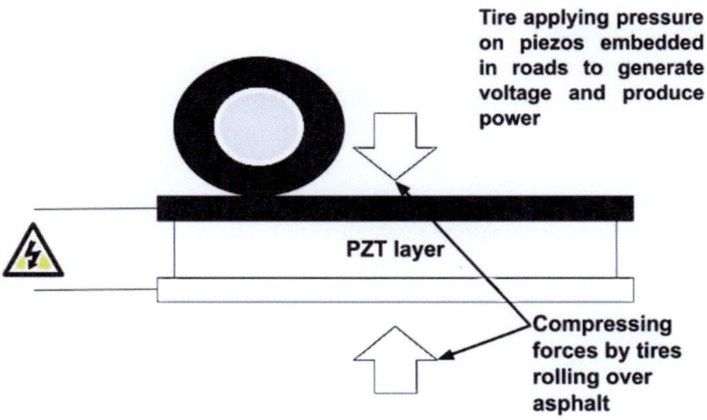

Fig. 2 The concept of Piezo modules

zirconate titanate (PZT), a piezo-ceramic, and polyvinylidene fluoride (PVDF). PZT-based piezoelectric are particularly prevalent in industrial applications. For energy harvesting, these materials are fabricated into various transducers such as cymbals, bimorphs, multilayers, moonies [43], thin-layer uniform ferroelectric drivers and sensors (THUNDER), macro-fiber composites (MFC) [44], and reduced and internally biased oxide wafers (RAINBOW) [45]. Each piezoelectric material has distinct characteristics that impact its performance and efficiency. Lead zirconate titanate (PZT), an industrial ceramic, is known for its brittleness and high piezoelectric coefficient. In contrast, polyvinylidene fluoride (PVDF) is more robust and flexible, offering higher chemical and thermal resistance, but it has a significantly lower piezoelectric coefficient compared to PZT [45]. An energy harvesting system using piezoelectric elements is developed in two vital parts: the mechanical platform and the electrical circuitry. The mechanical platform generates electricity, while the electrical circuitry stores and conditions the generated electricity [46]. The efficiency of the device is significantly influenced by the integration and interaction of these components. Additionally, various factors such as piezoelectric constants, electrical permittivity, ceramic tolerance, geometry, and ceramic delamination affect the performance of piezoelectric devices in energy harvesting [47].

When implementing piezoelectric devices on roads, stringent requirements for rigidity, durability, and strength must be considered. Piezoelectric components need to be enclosed in well-designed structures made from materials like steel, plastic, or concrete to protect them from automotive stresses and harsh environmental conditions [48]. Several studies have focused on the design of piezoelectric element enclosures, mainly investigating cantilever beam arrangements. These studies concluded that cantilever structures are generally too fragile to withstand vehicle loads [49]. Nonetheless, research has shown that energy harvesting through piezoelectric cantilever devices is feasible for road applications. Experimental results indicated that while trapezoidal bridge structures can handle heavier loads, arch bridge structures offer the highest energy efficiency in electricity generation. Two research groups evaluated various configurations of piezoelectric applications on roads based on efficiency, stiffness, and brittleness, and found that neither design was optimal for road energy harvesting. Bauer claimed that a fabricated piezoelectric ceramic disc cymbal-shaped transducer, enclosed by two metal end caps, performed effectively in pavement applications [50]. Wang proposed an experimental investigation based on traffic flow for piezoelectric energy harvesting. They built a harvester comprising a piezo stack and a load cell, which was mounted on a road surface and subjected to vibrations from passing vehicles [51]. The results demonstrated that at a traffic flow of 20 km/h, the average power output was 100 mW. The study concluded that the power output is proportional to traffic speed and vibration amplitudes, highlighting the potential for developing and harvesting energy with this technology [52]. The prototype showed a significant power generation capability at low traffic speeds, but its effectiveness is limited by traffic speed [53]. Notably, the system could withstand large vibrations and was durable. Another energy harnessing method was introduced by [54], who proposed a system to sense axle loads using cylindrical piezoelectric components to generate power. This system, known as a weight-in-motion (WIM) system, consists of a series of piezoelectric elements embedded in the road surface [55]. When a vehicle passes over these elements, they are compressed, generating an electrical signal, which is then amplified and processed to determine axle loads and vehicle speed. The piezoelectric elements, 1 mm thick and 5 mm in diameter, offer a low power consumption of 10 mW on average and do not require an external power source, making the system more sustainable and cost-effective than traditional WIM systems. Laboratory tests showed the system accurately measured axle loads and vehicle speed with an accuracy of 0.5% and 0.1 km/h, respectively, and could operate under various environmental conditions, including different temperatures and road surfaces. Batra investigated the impact of the size of piezoelectric energy harvesters on their electrical properties [56]. The study found that the diameter size had the greatest effect on output power, while the number of stacked layers and thickness also played a role. The optimal size for a piezoelectric energy harvester for road applications was determined to be a diameter of 15–20 mm and a thickness of 30 mm, with 20–40 stacked layers.

Under these conditions, the harvester achieved a maximum output power of 39.2 mW and an energy output efficiency of 21.4%. Additionally, the output power was found to increase proportionally with load and frequency. The output power of piezoelectric energy harvesters is influenced by vibrational amplitude, which is related to load and frequency. This has significant implications for designing and optimizing piezoelectric energy harvesters for road applications. The optimal size of the energy harvester can be determined based on vibration load and frequency, and adjusting the number of stacked layers can further enhance output power. However, while this study focused on the module size, it did not consider how material properties and module geometry could significantly impact system performance.

Another study investigated the structural optimization of piezoelectric energy harvesters for road applications. The authors found that the energy harvester's structure has a significant impact on output power. They determined that a cantilevered structure with a rigidly constrained end provides a high-stress concentration, leading to higher output power. Additionally, dividing piezoelectric layers into blocks further improves power output by reducing the charge cancellation effect, resulting in a higher output voltage.

Beyond energy harvesting, piezoelectric transducers can provide valuable data on transport and road usage, converting standard roads into smart roads. For instance, in 2019, a self-powered piezo sensing system tested on-site utilized 80 piezoelectric modules to measure vertical displacement caused by different vehicles. This system could measure strain, temperature, and power indicators. More recently, in 2022, a piezoelectric energy harvester was used to predict slippery regions on roads due to ice buildup and identify danger zones.

Wang tested the relationship between the size of piezoelectric elements and their electrical power output. In their experiment, various sizes of piezoelectric elements were placed under different traffic conditions, and the resulting voltage and power were analyzed. They developed an energy generation calculation method and provided output efficiency metrics. Their experiment concluded that a 15 mm diameter and 30 mm height piezoelectric element had the highest influence on power generation under random vehicle loads. The maximum voltage and power under an SUV were 86 V and 39.2 mW, respectively. Within 2 s, the generated energy and output electric energy were 69.2 millijoules and 14.7 millijoules, respectively. The system's output efficiency reached 21.4%.

5 Thermoelectric Harvesting

Thermoelectric Generators (TEGs) harness the Seebeck effect to generate voltage across two junctions made of different semiconductors: an N-type and a P-type material. The interplay between these two types of semiconductors is crucial for the effective functioning of TEGs. The thermoelectromotive forces generated in this system are influenced by the temperature differential [41]. Researchers have explored combining these semiconductors and enclosing them with thermally sensitive yet electrically insulating ceramic plates to regulate power output [40]. Kim [42] reported that up to 250 mW of electrical power can be generated from a 5 °C temperature difference. TEGs have been applied in various settings, includissng generating electricity from automotive exhaust pipes [43], building walls, and soils [44], demonstrating the potential of converting heat into electrical power. Some studies have investigated their effectiveness in pavements. Effective thermoelectric materials require high electrical conductivity and thermal insulation to efficiently convert temperature differentials into electrical energy [61]. Lead telluride (PbTe), bismuth telluride (Bi2Te3), and silicon germanium (SiGe) are three semiconductors known for possessing these properties [61, 62]. TEGs typically include heat sinks and thermally conductive plates as part of their thermal energy harvesting systems [62]. Thermal cycle tolerance is critical for the efficiency of TEGs, given their exposure to varying temperature conditions. Metals are often used for heat-conducting plates [63], which absorb and transfer heat from the surroundings to where TEGs are positioned.

To minimize heat loss during heat transfer, areas around these heat sinks should not be insulated. Heat sinks can take various shapes, including L-shapes and Z-shapes. Insulation and waterproofing of the entire system are crucial for maximizing efficiency [61]. Figures 3, 4 illustrate the concept of a thermoelectric energy production system, where heat from car tires heats up the road surface, creating a temperature gradient that allows embedded thermoelectric cells to generate electricity from the produced heat energy. Weng [64] investigated temperature variations between ambient air and pavement surfaces to control a TEG system deployed on different types of pavements, including asphalt and Portland concrete. It was found that the Portland concrete pavement generated more energy.

Yuan [65] conducted an experimental investigation on road thermoelectric generator systems focused on temperature regulation. The system comprised a heat sink, thermostat controller, and a TEG, tested across various ambient temperatures and levels of solar irradiation. Results indicated the system could maintain an operating

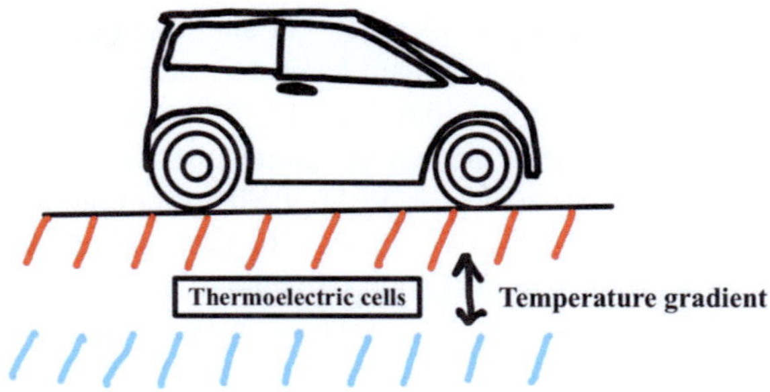

Fig. 3 The concept of thermoelectric energy harvesting on roads

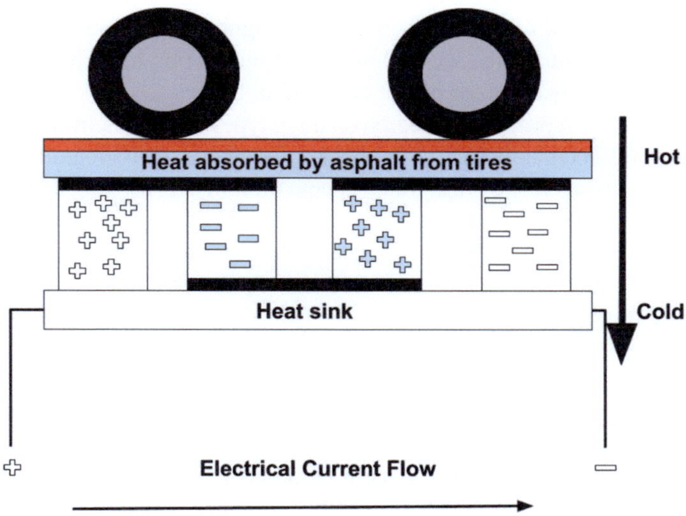

Fig. 4 A more detailed concept of TEG modules

temperature of 50 °C when the ambient temperature was 40 °C and solar irradiance was 1000 W/m2. However, given the laboratory setting, real-world conditions may yield different outcomes, particularly in terms of longevity and durability against traffic vibrations. Zha [66] introduced a novel thermoelectric module designed to harvest energy from pavements while also reducing surface temperatures. This module consisted of composite blocks made from a blend of cement, metal oxides, and expanded graphite. Expanded graphite provided high thermal conductivity, while metal oxides enhanced the Seebeck coefficient. Installed on pavements, the module generated electric current from temperature differentials induced by passing vehicles. Experimental tests demonstrated a peak power generation of 1.5 W/m2 and a reduction in pavement temperature by 1 to 3 °C.

Zou developed a thermoelectric power generation device suitable for integration into roads, evaluating multiple thermoelectric generator units (TEGUs) under various conditions [67]. Four prototypes, each comprising 10 TEGUs connected in series, were encased in ceramics with an aluminum plate acting as both an upper cover and heat carrier to safeguard against vehicle damage. Tested under field conditions during a sunny winter day (9 to 17 °C), these prototypes, positioned 2 cm beneath the road surface at median and shoulder zones, demonstrated optimal performance with a load resistance of 50 Ω, yielding an output voltage of 5.582 V and power output of 0.623 W. Researchers [6] proposed an advanced thermoelectric energy harvesting system integrated into a multilayer structural board, featuring aluminum fins for efficient heat dissipation. This design allowed for the integration of multiple TEGs without increasing road area usage, significantly enhancing power generation efficiency. Finite element heat transfer analysis optimized the structural size, demonstrating that such a system could efficiently collect 16.94 kWh of power over 6 h from a 1 km road with a 10 m width. The study highlighted that a multilayer structural board with 10 aluminum fins, each 10 mm wide and 50 mm long, provided optimal heat dissipation and maximized power generation.

These authors [68] proposed research on three types of road thermoelectric generator systems (RTEGs) based on the Seebeck effect: pavement-subgrade, pavement-ambient, and pavement-flowing water. They thoroughly investigated the output voltage and temperature characteristics of these RTEGs under both indoor and outdoor conditions. Their findings indicated that the output power followed the order: of pavement-flowing water > pavement-ambient > pavement-subgrade. Specifically, their developed pavement-flowing water RTEGs demonstrated a maximum power density of 4.28 mW/cm3.

The pavement-subgrade RTEGS were noted to melt snow in winter approximately 60 min faster than conventional pavement. On the other hand, the pavement-ambient RTEGS were effective in reducing summer road-surface temperatures by up to 10 °C. Meanwhile, the pavement-flowing water RTEGS could reduce summer road-surface temperatures by up to 5.46 °C. These results highlight the diverse applications and benefits of thermoelectric generator systems in various road environments.

6 Pyroelectric Harvesting

Pyroelectricity involves generating electrical energy from thermal energy without relying on a temperature differential, instead utilizing the material's response to heating or cooling rates [45]. Unlike thermoelectricity, which depends on temperature differences, pyroelectric materials produce electrical energy or a pyroelectric current based on the rate of temperature change within the material.

Several piezoelectric materials also exhibit pyroelectric properties, such as PZT [69], PVDF [70], and MFC [71]. Petrini [72] examined nanofiber composite cement and lithium tantalate, finding them suitable for energy harvesting on roads. Other authors studied a cement composite embedded with steel/carbon nanofibers, which enhanced its electrical properties [73]. Guan, and Liao explored a hybrid material of PZT and PVDF-TrFE, noting increased pyroelectric and dielectric properties compared to PVDF-TrFE alone [74]. Another study demonstrated that a composite material consisting of polyvinylidene fluoride-hexafluoropropylene (PZT-PVDF-HFP) and Lead Zirconate Titanate powder increased its pyroelectric constant at high temperatures [75]. The authors [76] developed a sensor that utilizes the pyroelectric effect to power itself, while Hwang [77] highlighted PVDF, barium strontium titanate, lithium tantalate, modified PZT, and triglycine sulfate as materials suitable for generating electrical power through the pyroelectric effect due to their electromechanical properties. To effectively harness pyroelectric energy, setting up a complete bridge diode is recommended because it can accommodate the changing polarity of the generated electricity in response to fluctuations in heat flow direction [72]. This method ensures efficient conversion and utilization of pyroelectric energy for powering various devices.

7 Photovoltaic Harvesting

An American Electrical Engineer pioneered the installation of photovoltaic (PV) cells on pavements in 2009 [46]. This innovative approach involves using double-layered toughened glass hexagonal solar panels embedded with solar cells as replacements for standard pavements. PV panel technology has since been widely adopted in various modern applications for collecting solar energy. Its versatility allows for use across industries ranging from aerospace to transportation.

Solar energy is a renewable and environmentally friendly energy source known for its longevity and cleanliness. It serves as a significant energy resource that has played a crucial role in reducing energy shortages, lowering emissions, combating climate change, and improving overall energy efficiency. Light, abundant in the atmosphere, can be harnessed by technologies like photovoltaic (PV) cells to convert it into usable electrical energy. At an environmental condition of 1000 W/m^2 sun exposure, monocrystalline silicon planar PV cells exhibit an efficiency of 20%, which can increase up to 50% in multilateral planar wafers, providing a wide range of conversion efficiencies [47]. This efficiency is due to the photovoltaic phenomenon in doped silicon, which allows these cells to recover a significant amount of power compared to other types of harvesters available.

PV cells consist of N-type and P-type semiconductors, where electricity generation occurs through the movement of free electrons between these types of semiconductors. Under sunlight exposure, P-type semiconductors acquire positively charged electrons, while N-type semiconductors receive negatively charged electrons [78]. Figure 5 illustrates the concept of photovoltaic energy harvesting systems on roads, where solar cells integrated into the road surface absorb sunlight, convert it into electrical energy, and illuminate LED boards or streetlights along the road.

This technology utilizes the abundant solar energy in the atmosphere, converting it directly into usable electrical energy through the photoelectric effect in PV cells embedded in the road infrastructure.

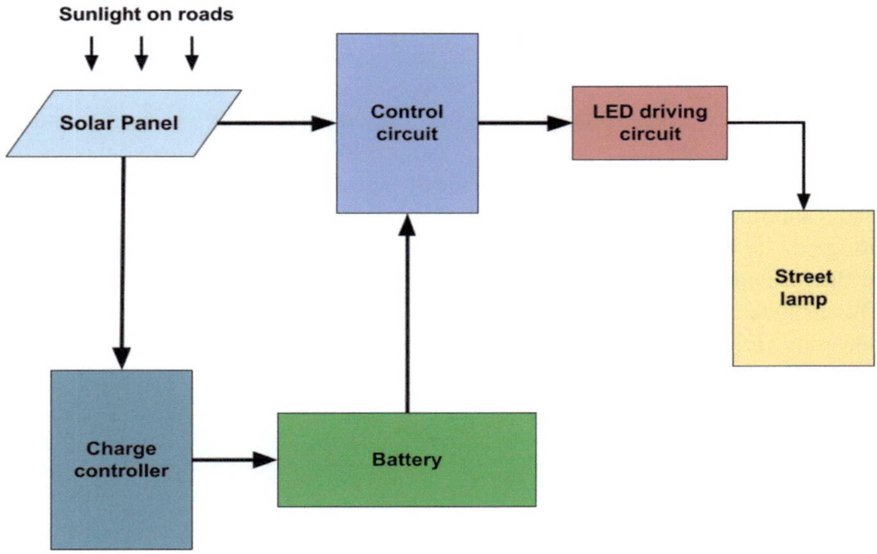

Fig. 5 A photovoltaic energy harvesting system on roads

There are several common materials used in PV cell fabrication, including monocrystalline silicon cells, dye-sensitized cells, polycrystalline silicon cells, thin-film solar cells, and organic thin-film solar cells [79]. Among these, monocrystalline silicon cells are known for producing the highest energy output.

Touairi designing PV systems, a critical factor to consider is avoiding shading, as it significantly impacts PV performance [80]. Factors such as using transparent surfaces and designing robust support systems are crucial when evaluating the structural properties of these panels. These considerations ensure that PZT cells receive maximum exposure to sunlight throughout the day, optimizing their energy production capabilities. Another critical design consideration is ensuring that PZT materials provide adequate grip suitable for vehicle braking [81]. This requirement suggests a need for surface roughness, which can sometimes conflict with the desire for surface transparency [78]. Transparent surface layer options such as acrylic, tempered glass, and polycarbonate can be textured while retaining their transparency. Despite their higher compression strength compared to concrete, questions remain about their tensile strength durability under heavy traffic and various climatic stresses. The tensile modulus of the panel support network is also crucial due to traffic-induced tensile stresses [81]. Waterproofing and flexibility are essential, especially considering varying foundation support conditions, to prevent hydroplaning. Incorporating sloping edges on PV cell surfaces facilitates water drainage, enhancing safety [81].

Touairi [82] conducted a simulation and experimental study on the energy performance of prefabricated photovoltaic pavements (PVPH). This system included a PZT module, thermal insulation layer, and concrete base. A thermal-electrical mathematical model was used for simulation, and experimental data was validated using a PVP testbed. Results indicated an estimated annual electricity generation of 1.33 kWh/m2 and a maximum surface temperature reduction of 8.6%. The system demonstrated potential applications such as snow melting and traffic signal power supply. However, the study's reliance on lab conditions raises questions about real-world applicability, especially concerning cost and lifespan considerations.

Another study [83] conducted experiments in both lab and field settings, focusing on a solar energy harvester using PV modules embedded on pavement surfaces. This module featured a thin-film-coated PV solar panel covered by a wooden-framed transparent cover. Lab results indicated 699 Wh of power generated over a seven-hour sunlight exposure, whereas field site conditions yielded 500 Wh under similar conditions, achieving an efficiency of 15%. However, the study primarily addressed power generation without discussing cost-effectiveness or long-term durability, prompting feasibility concerns. This letter conducted a performance comparison of three pavement energy harvesters using solar energy: pavement-integrated photovoltaic (PIPV) modules, pavement-integrated solar thermal (PIST) modules, and pavement-integrated photovoltaic terminals (PIPVT) modules. Transient models were employed to simulate power and heat production across four days in different seasons in Shanghai. The study also investigated the modules' impact on the urban heat island (UHI) effect.

8 Efficiency

This section of the paper describes the efficiency of each harvesting technology conducted with a practical approach by other paper publications. These efficiencies will be depicted by cost consideration and the amount of power generated.

9 Electromagnetic Harvesting

Several researchers have investigated various electromagnetic energy harvesters and reported varied output powers. For instance, the author [11] developed a speed bump-style rack-and-pinion and actuator as a mechanical energy harvesting prototype, achieving a maximum power output of 200 W. They observed peaks of 647 W per passing axle and tested a novel cam-arm mechanism for capturing energy from high-speed vehicles, achieving a maximum of 24 W. In China, a pinion and bevel gear system generated 66.025 W with a 2 Hz loading cycle in both laboratory and field testing [11]. Author [25] experimented with a simple rack-and-pinion device under different load conditions, recording a maximum output power of 3.21 mW.

They concluded that both methods are viable for harvesting energy from highways. Several studies have also examined the economic feasibility of these mechanical energy harvesting technologies, including applications in railroads. For example, a rack-and-pinion harvesting system for railroad applications costs approximately

$2135, covering materials and labor, indicating financial viability similar to systems installed on roadway pavements. Despite being in the early stages, harvesting mechanical energy from roads using electromagnetic techniques appears promising due to their ability to generate sufficient electrical energy to offset costs.

10 Piezoelectric Harvesting

Previous research on the use of Piezoelectric Energy Harvesting (PEH) in pavements can be categorized into numerical simulations [26], practical findings [29], lab results [26, 31], and mathematical analyses. Numerical simulations estimate the electrical output of PZT using finite element methods, optimizing parameters such as PZT shape and size. Laboratory and practical studies involve testing various device and system designs, including different piezoelectric configurations, rectifier types, cantilever setups, ceramic disk-shaped PZT elements, and manufactured energy harvesting systems. Mathematical models are employed to analyze PEH variables such as vehicle speed, PEH placement, pavement conditions, and loads. Consequently, the electrical energy generated varies significantly among different harvesting methods.

JR East, for example, has deployed commercial piezoelectric harvesters. At a Tokyo railway station, JR East installed a piezoelectric array energy collection system capable of powering a 100 W lamp for 80 min. However, energy production declined after three weeks due to a system failure.

Comparing these costs with traditional electric grid prices in the United States (approximately $0.15 to $0.30 per kWh), piezoelectric technology appears relatively expensive. However, it's crucial to note that this technology is still in development and has not yet benefited from industrial economies of scale. Moreover, it may prove more cost-effective in remote areas lacking power infrastructure. Improvements in system efficiency and increased vehicle traffic could potentially reduce the cost of power generation from piezoelectric materials.

Laboratory tests with four types of asphalt mixes and varying loading conditions showed significant differences in harvested power. Dense-graded asphalt mix under high-frequency loading yielded the best results. The study concluded that while there is promising potential for road-based piezoelectric energy harvesting, further research is necessary to optimize harvester design and loading conditions.

11 Thermoelectric Harvesting

Wu and Yu investigated the energy harvesting capabilities of Thermoelectric Generators (TEGs) embedded in pavement structures. They placed the TEG module on the pavement surface and used an aluminum rod to thermally connect its bottom side to lower pavement layers. This setup generated 300 mV and 0.05 mW of electricity with a temperature difference of 20° C between the top and bottom pavement layers, sufficient to power a flashing LED light. A 4 cm by 4 cm by 0.5 cm TEG unit under a 6.44 °C thermal gradient achieved a peak output power of 0.02 W, generating 1000 J per day, suitable for applications like traffic monitoring. The electronic system efficiency was reported at 41.3%, resulting in an overall TEG performance of 1.6%. Jiang explored thermal gradients between asphalt and the surrounding air, conducting tests indoors and outdoors. Voltage outputs ranged from 0.4 V for a 15 °C gradient in winter to 0.6–0.7 V for a 25–30 °C variation in summer. They estimated that a one-kilometer-long, one-meter-wide road could generate 160 kWh over eight hours. In another study, Jiang developed a novel Road Thermoelectric Generator System (RTEGS) yielding 0.564 V in field tests. They projected that covering an area of 10,000 m2 could produce 33 kWh daily. Rahman [63] created a prototype producing an average of 29 mW per day with a maximum output of 34 mW under extreme temperature gradients (up to 34 °C in hot Texas summers). Xie [46] evaluated a TEG mounted on paved surfaces, reporting a maximum output voltage of 0.123 V.

12 Photovoltaic Harvesting

PV solar panel installations typically involve configuring multiple panels in series or parallel to achieve specific voltage or current requirements. Touairi [81] investigated two additional prototypes: a solar panel and a solar pavement. Solar panels consisted of solar cells embedded between rubber and plexiglass layers, while solar pavement had solar cells integrated between two porous rubber layers. Measurements of voltage and current outputs allowed for the calculation of power conversion efficiency (PCE). The solar panel achieved a maximum output current of 39.1 mA and voltage of 3.16 V, whereas the solar pavement produced a maximum current of 24.5 mA and voltage of 3.04 V. The PCE of the solar pavement was 5.336% higher than that of the solar panel. SolaRoad represents a commercially available PV panel designed for use on highway surfaces. In 2014, a 70-m bike path was constructed as a test site by SolaRoad, which reported that the initial design functioned well after six months, despite minor degradation in the slip-resistant surface layer.

Amorphous-Si solar cells consistently produced similar power output levels across all three light sources, indicating their suitability for energy harvesting applications. Mathews compared GaAs and GaInP solar cells under LED lights and compact fluorescent lamps (CFL) at intensities of 200 and 1000 lx. GaAs cells demonstrated higher power density at 1000 lx, producing 92.2 W/cm2 compared to GaInP cells at 87.2 W/cm2.

Under CFL lighting at 200 lx, cells treated with plasma-enhanced atomic layer deposition achieved a maximum power density of 5.4 W/cm2 and a maximum efficiency of 24%, demonstrating robust performance under varying conditions. While the literature provides insights into the performance of PV technologies, there is limited information on the cost-effectiveness of integrating photovoltaic cells into road surfaces and the economic value of the electrical energy generated.

For instance, the construction of the 70-m SolaRoad testing site cost $3.7 million, highlighting the significant upfront investment required due to specific structural requirements. However, these costs are not necessarily indicative of large-scale highway surface implementations. This latter compared three technologies: PIPT had the highest overall energy efficiency at 37.31%, followed by PIPV at 34.73%, and PIST with 29.52%, underscoring the variability in efficiency among different pavement-integrated photovoltaic modules.

13 Comparison of These Technologies

This section of the study examines various energy harvesting systems, categorizing them into two parts. The first part discusses the advantages and disadvantages of each technique, while the second part analyzes different types of energy harvesting systems, including their peak power output and power density. Figure 6 illustrates the contrast in manufacturing costs among these energy harvesting systems.

A comparison of various energy harvesting systems involves examining numerous aspects, primarily energy output as shown in Table 1, and cost considerations. Table 2 displays the energy output when these technologies are applied in road applications. The energy production of a technology defines its potential, while cost determines its economic feasibility. However, both factors are largely influenced by technological maturity. The Technology Readiness Level (TRL) is a scale that assesses the maturity and readiness of a technology for widespread adoption, ranging from 1 (early stage) to 9 (fully operational). Hasani [11] reported TRLs of 3 to 4 for piezoelectric and thermoelectric harvesting devices in roadways, indicating moderate development stages. Xie [46] mentioned a TRL of 5 to 6 for their thermoelectric generator system. In terms of energy cost comparison, the Levelized Cost of Energy (LCOE) is widely used.

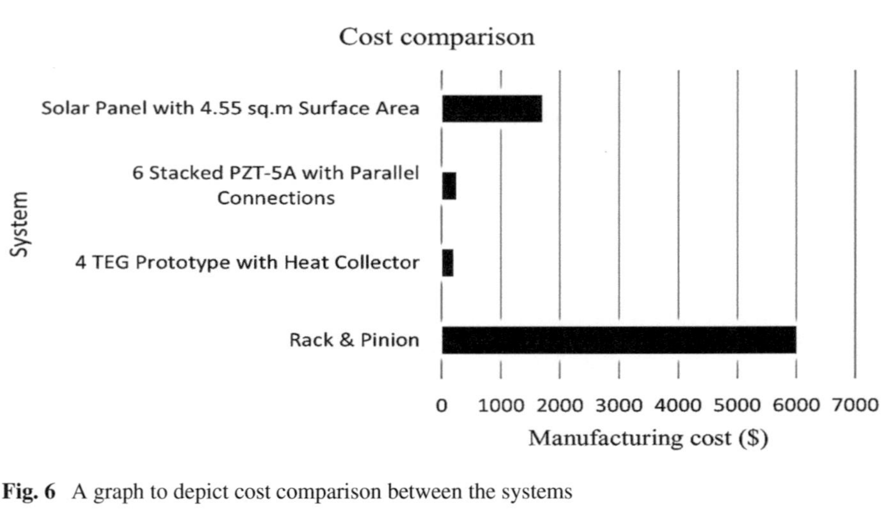

Fig. 6 A graph to depict cost comparison between the systems

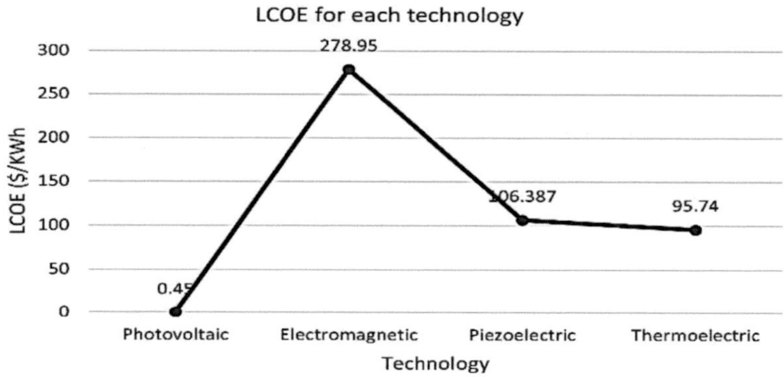

Fig. 7 LCOE of the systems

The LCOE provides a standardized method for evaluating and comparing energy systems, considering costs over different life spans and ranges of energy output. It encompasses the total cost of energy production, including initial construction, deployment, maintenance costs over the system's life, and decommissioning costs.

Calculating the Levelized Cost of Energy (LCOE) requires considering the total energy production of the system over its entire life cycle. Figures 7, 8 illustrate the relationship between the Technology Readiness Level (TRL) and LCOE for the various harvesting systems reviewed in this paper. To better understand the scale of highway energy harvesting systems, estimating the Levelized Cost of Energy (LCOE) per lane mile of roadway is crucial. The cost of electricity varies significantly depending on the Technology Readiness Level (TRL). Mature technologies with higher TRLs generally have lower energy costs. For instance, photovoltaic technology, with a TRL of 9, has a relatively low LCOE (0.45$/kWh). In contrast, thermoelectric technology, at TRL 3, has a higher LCOE (95.74$/kWh). Technologies like photovoltaics have achieved high TRLs through extensive research and optimization for cost-effectiveness before widespread adoption. In contrast, piezoelectric and thermoelectric harvesting technologies are still in the early stages with lower TRLs, impacting their cost-effectiveness. Roadway energy collection systems tend to be more expensive than conventional power generation systems. The calculation of LCOE can also vary based on the specific study inputs and assumptions.

14 Summary

The majority of the energy harvesting systems examined can gather energy from highways. A few are more adapted than others, and others are more developed. A number of these technologies, such as piezo-electric and pyroelectric materials, can complement one another. The most potential technologies are listed below in brief, viz. Photovoltaic (PV) Systems: Mature technology with high efficiency in

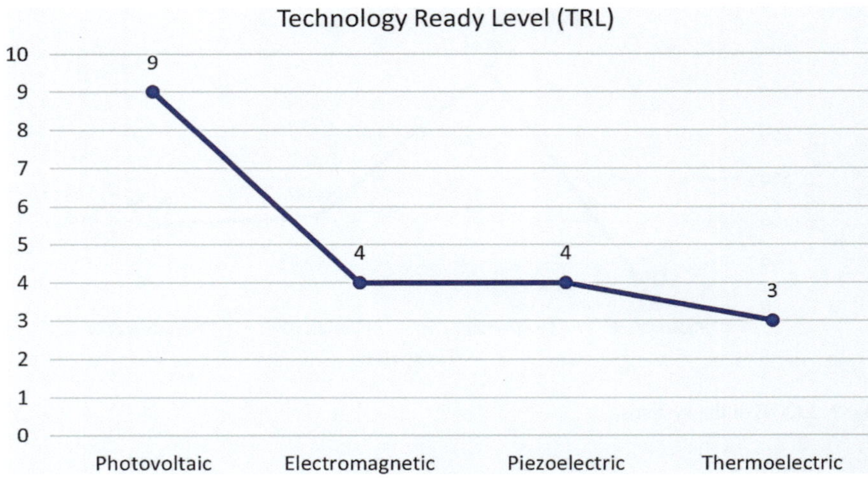

Fig. 8 Technology readiness level of the systems

converting sunlight into electricity, suitable for large-scale deployment on highway surfaces like the SolaRoad project. Piezoelectric Systems: Capable of converting mechanical vibrations from passing vehicles into electrical energy, though currently at early stages (low TRL). Thermoelectric Generators (TEG): Convert temperature differentials between pavement and ambient air into electricity, suitable for applications where temperature variations are significant. Pyroelectric Systems: Utilize rapid temperature changes to generate electrical energy, often considered in hybrid systems with piezoelectric materials. Hybrid Systems: Combining multiple energy harvesting technologies (e.g., piezoelectric and pyroelectric) to improve overall efficiency and reliability. These technologies show varying levels of maturity, with PV systems being the most mature and widely implemented on highways, while others like piezoelectric and thermoelectric systems are still in early development stages. Hybrid systems are increasingly explored to maximize energy capture from diverse environmental conditions and mechanical inputs.

15 Conclusion

All these energy harvesting technologies produce sustainable green energy that can be utilized in roadside systems to power LED lights, sensors, and microprocessing units for various data-gathering applications. Their benefits are evident: they produce no greenhouse gas emissions associated with fossil fuel consumption and eliminate the need for electrical transmission and distribution systems to supply grid power to highways. As the demand for smart highway infrastructure, including connectivity and signage, continues to grow, energy-harvesting technology is poised to become a necessity. Thermoelectric and piezoelectric harvesters are among the technologies closest to commercialization. Despite their lower power generation capacity compared to advanced PV technology, they can be discreetly integrated into pavement layers. The economic viability of these customized solutions will need reassessment as they are scaled up to ensure that increased deployment reduces unit costs.

References

1. Aswathaman, V., Priyadharshini, M.: Every speed breaker is now a source of power. In: Proc. of 2010 International Conference on Biology, Environment and Chemistry IPCBEE, vol. 1 (2011)
2. Obeid, H., Alhameedi, A.k., Hassan, N.: Design and motion modeling of an electromagnetic hydraulic power hump harvester. Advances in Mechanical Engineering 2014, 1–11 (2014). https://doi.org/10.1155/2014/150293
3. Xiong, H., Wang, L.: Piezoelectric energy harvester for public roadway: On-site installation and evaluation. Applied Energy 174, 101–107 (2016). https://doi.org/10.1016/j.apenergy.2016.04.031
4. Ullah, K.M., Ahsan-uz-Zaman, K.M., Hosen, S., Khan, R.H., Parvin, S.: Electrical power generation through speed breaker. In: 2016 9th International Conference on Electrical and Computer Engineering (ICECE), pp. 30–33 (2016). https://doi.org/10.1109/ICECE.2016.7853848
5. Ting, C.-C., Tsai, D.-Y., Hsiao, C.-C.: Developing a mechanical road-way system for waste energy capture of vehicles and electric generation. Applied Energy 92, 1–8 (2012). https://doi.org/10.1016/j.apenergy. 2011.10.006
6. Sarma, B., Jyothi, V., Sudhir, D.: Design of power generation unit using roller mechanism. IOSR Journal of Electrical and Electronics Engineering 9, 55–60 (2014). https://doi.org/10.9790/1676-09315560
7. Kim, H.S., Kim, J.-H., Kim, J.: A review of piezoelectric energy harvesting based on vibration. International Journal of Precision Engineering and Manufacturing 12 (2011). https://doi.org/10.1007/s12541-011-0151-3
8. Sodano, H., Inman, D.: A review of power harvesting from vibration using piezoelectric materials. The Shock and Vibration Digest 36, 197– 205 (2004). https://doi.org/10.1177/0583102404043275
9. Cao, H., Kong, L., Tang, M., Zhang, Z., Wu, X., Lu, L., Li, D.: An electromagnetic energy harvester for applications in a high-speed rail pavement system. International Journal of Mechanical Sciences 243, 108018 (2023)
10. Cao, X.: A power road by harvesting electromagnetic energy from moving object. Nano Energy 112, 108486 (2023)
11. Hasani, M., Rahaghi, M.I.: The optimization of an electromagnetic vibration energy harvester based on developed electromagnetic damping models. Energy Conversion and Management 254, 115271 (2022)
12. Obeid, H.H., Jaleel, A.K., Hassan, N.A.: Design and motion modeling of an electromagnetic hydraulic power hump harvester. Advances in Mechanical Engineering 6, 150293 (2014)
13. Zhang, Z., Xiang, H., Shi, Z.: Modeling on piezoelectric energy harvesting from pavements under traffic loads. Journal of Intelligent Material Systems and Structures 27 (2015). https://doi.org/10.1177/1045389X15575081
14. Zhao, H., Yu, J., Ling, J.: Finite element analysis of cymbal piezoelectric transducers for harvesting energy from asphalt pavement. Journal of the Ceramic Society of Japan 118, 909–915 (2010). https://doi.org/10.2109/jcersj2.118.909

15. Dineva, P., Gross, D., Mu¨ller, R., Rangelov, T.: Dynamic Fracture of Piezoelectric Materials: Solution of Time-Harmonic Problems Via BIEM. Solid Mechanics and Its Applications. Springer, ??? (2014). https://books.google.com.my/books?id=r4OzngEACAAJ
16. Dogan, A.: Flextensional "moonie and Cymbal" Actuators. PhD thesis, Pennsylvania State University (January 1994)
17. Roundy, S., Wright, P.K., Rabaey, J.M.: Energy Scavenging for Wireless Sensor Networks: With Special Focus on Vibrations. Springer, ??? (2004). https://books.google.me/books?id=PmmvRsfsUKcC
18. Starner, T., Paradiso, J.: Human-generated power for mobile electronics. Low Power Electron. Des. 1990 (2003). https://doi.org/10.1201/9781420039559.ch45
19. Jiang, X.Z., Li, Y.C., Wang, J., Li, J.C.: Electromechanical modeling and experimental analysis of a compression-based piezoelectric vibration energy harvester. International Journal of Smart and Nano Materials 5(3), 152–168 (2014) https://arxiv.org/abs/https://doi.org/10.1080/19475411.2014.919971
20. 19475411.2014.919971. https://doi.org/10.1080/19475411.2014.919971
21. Inman, D.: A distributed parameter electromechanical model for cantilevered piezoelectric energy harvesters. Journal of Vibration and Acoustics 130 (2008). https://doi.org/10.1115/1.2890402
22. Ali, S.F., Friswell, M.I., Adhikari, S.: Piezoelectric energy harvesting with parametric uncertainty. Smart Materials and Structures 19(10), 105010 (2010). https://doi.org/10.1088/0964-1726/19/10/105010
23. Moure, A., Rodr´ıguez, M.A., Rueda, S., Gonzalo Mart´ın, A., Rubio- Marcos, F., Urquiza, D., P´erez-Lepe, A., Fern´andez, J.: Feasible integration in the asphalt of piezoelectric cymbals for vibration energy harvesting.
24. Energy Conversion and Management 112, 246–253 (2016). https://doi.org/10.1016/j.enconman.2016.01.030
25. Gu, L.: Low-frequency piezoelectric energy harvesting prototype suitable for the MEMS implementation. Microelectronics Journal 42(2), 277–282 (2011). https://doi.org/10.1016/j.mejo.2010.10.007
26. Mitcheson, P.D., Yeatman, E.M., Rao, G.K., Holmes, A.S., Green, T.C.: Energy harvesting from human and machine motion for wireless electronic devices. Proceedings of the IEEE 96(9), 1457–1486 (2008). https: doi.org/https://doi.org/10.1109/JPROC.2008.927494
27. Ye, G., Yan, J., Wong, Z.J., Soga, K., Seshia, A.: Optimization of a piezoelectric system for energy harvesting from traffic vibrations. In: 2009 IEEE International Ultrasonics Symposium, pp. 759–762 (2009). https://doi.org/10.1109/ULTSYM.2009.5441942
28. Zhao, H., QIN, L., LING, J.: Test and analysis of bridge transducers for harvesting energy from asphalt pavement. International Journal of Transportation Science and Technology 4, 17–28 (2015). https://doi.org/ https://doi.org/10.1260/2046-0430.4.1.17
29. Zhao, H., Ling, J., Yu, J.: A comparative analysis of piezoelectric transducers for harvesting energy from asphalt pavement. Journal of the Ceramic Society of Japan 120, 317–323 (2012). https://doi.org/10.2109/jcersj2.120.317
30. Datta, U., Dessouky, S., Papagiannakis, A.: Harvesting of thermoelectric energy from asphalt pavements. Transportation Research Record Journal of the Transportation Research Board 2628 (2017). https://doi.org/10. 3141/2628–02
31. Wang, J., Qin, X., Liu, Z., Shi, K., Ding, G., Li, X., Cai, G.: Experimental field study on a full-scale road piezoelectric energy harvester. Smart Materials and Structures 31(5), 055003 (2022)
32. Khalili, M., Vishwakarma, G., Ahmed, S., Papagiannakis, A.T.: Development of a low-power weigh-in-motion system using cylindrical piezo-electric elements. International Journal of Transportation Science and Technology 11(3), 496–508 (2022)
33. Wang, S., Wang, C., Yuan, H., Ji, X., Yu, G., Jia, X.: Size effect of piezoelectric energy harvester for road with high-efficiency electrical properties. Applied Energy 330, 120379 (2023)
34. Wang, C., Zhou, R., Wang, S., Yuan, H., Cao, H.: Structure optimization and performance of piezoelectric energy harvester for improving road power generation effect. Energy 270, 126896 (2023)

35. Cho, J.Y., Kim, K.-B., Hwang, W.S., Yang, C.H., Ahn, J.H., Do Hong,
36. S., Jeon, D.H., Song, G.J., Ryu, C.H., Woo, S.B., et al.: A multifunctional road-compatible piezoelectric energy harvester for autonomous driver-assist led indicators with a self-monitoring system. Applied energy 242, 294–301 (2019)
37. Do Hong, S., Ahn, J.H., Kim, K.-B., Kim, J.H., Cho, J.Y., Woo, M.S.,
38. Song, Y., Hwang, W., Jeon, D.H., Kim, J., et al.: Uniform stress distribution road piezoelectric generator with free-fixed-end type central strike mechanism. Energy 239, 121812 (2022)
39. Wang, S., Wang, C., Yuan, H., Ji, X., Yu, G., Jia, X.: Size effect of piezoelectric energy harvester for road with high efficiency electrical properties. Applied Energy 330, 120379 (2023).https://doi.org/10.1016/j.apenergy.2022.120379
40. Tahami, S.A., Gholikhani, M., Nasouri, R., Dessouky, S., Papagiannakis, A.: Developing a new thermoelectric approach for energy harvesting from asphalt pavements. Applied Energy 238, 786–795 (2019). https://doi.org/10.1016/j.apenergy.2019.01.152
41. Tahami, Seyed Amid, Gholikhani, Mohammadreza, Nasouri, Reza, Dessouky, Samer: Evaluation of a novel road thermoelectric generator system. MATEC Web Conf. 271, 08002 (2019). https://doi.org/10.1051/matecconf/201927108002
42. Kim, J., Lee, S.-T., Yang, S., Lee, J.-J.: Implementation of thermal- energy-harvesting technology on pavement. Journal of Testing and Evaluation 45, 582–590 (2017). https://doi.org/10.1520/JTE20140396
43. Cuadras, A., Gasulla, M., Ferrari, V.: Thermal energy harvesting through pyroelectricity sens. Sensors and Actuators A: Physical 158, 132–139 (2010). https://doi.org/10.1016/j.sna.2009.12.018
44. Yuan, D., Jiang, W., Xiao, J., Ling, X., Zhang, Y., Lu, R.: Experimental study on the temperature-regulating function of road thermoelectric generator system. Journal of Cleaner Production 384, 135586 (2023)
45. Wei, J., Zhou, Y., Wang, Y., Miao, Z., Guo, Y., Zhang, H., Li, X., Wang, Z., Shi, Z.: A large-sized thermoelectric module composed of cement-based composite blocks for pavement energy harvesting and surface temperature reducing. Energy 265, 126398 (2023)
46. Xie, Z., Shi, K., Song, L., Hou, X.: Experimental and field study of a pavement thermoelectric energy harvesting system based on the Seebeck effect. Journal of Electronic Materials 52(1), 209–218 (2023)
47. Wei, J., Hui, J., Wang, T., Wang, Y., Guo, Y., Zhang, S., Zhang, Y., Qiao, X.: A thermoelectric energy harvesting system for pavements with a fin cooling structure. Sustainable Energy & Fuels 7(1), 248–262 (2023)
48. Yuan, D., Jiang, W., Sha, A., Xiao, J., Wu, W., Wang, T.: Technology method and functional characteristics of road thermoelectric generator system based on Seebeck effect. Applied Energy 331, 120459 (2023). https://doi.org/10.1016/j.apenergy.2022.120459
49. Batra, A., Sampson, J., Davis, A. et al. Electrospun nanofibers doped with PVDF and PLZT nanoparticles for potential biomedical and energy harvesting applications. J Mater Sci: Mater Electron 34, 1654 (2023). https://doi.org/10.1007/s10854-023-11066-6
50. Bauer, S.: Piezo-, pyro- and ferroelectrics: soft transducer materials for electromechanical energy conversion. IEEE Transactions on Dielectrics and Electrical Insulation 13(5), 953–962 (2006). https://doi.org/10.1109/TDEI.2006.247819
51. Bhattacharjee, S., Batra, A.K., Cain, J.: Energy Harvesting from Pavements Using Pyroelectric Single Crystal and Nano-Composite Based Smart Materials, pp. 741–750. https://doi.org/10.1061/41167(398)71
52. Wen, S., Chung, D.: Pyroelectric behavior of cement-based materials. Cement and Concrete Research - CEM CONCR RES 33 (2003). https://doi.org/10.1016/S0008-8846(03)00147-9
53. Dietze, M., Es-Souni, M.: Structural and functional properties of screen-printed PZT–PVDF-TrFe composites. Sensors and Actuators A: Physical 143(2), 329–334 (2008). https://doi.org/10.1016/j.sna.2007.11.016
54. Malmonge, L.F., Malmonge, J.A., Sakamoto, W.K.: Study of pyroelectric activity of PZT/PVDF-HfP composite. Materials Research 6, Mat. Res., 2003 6(4), (2003). https://doi.org/10.1590/S1516-14392003000400007

55. Yang, Y., Zhou, Y., Wu, J.M., Wang, Z.L.: Single micro/nanowire pyro-electric nanogenerators as self-powered temperature sensors. ACS Nano 6(9), 8456–8461 (2012) https://doi.org/10.1021/nn303414u. PMID: 22900676
56. Batra, A.: Simulation of energy harvesting from roads via pyroelectricity. Journal of Photonics for Energy 1, 014001 (2011). https://doi.org/10.1117/1.3656395.
57. Partain, L.D., Fraas, L.M.: Solar Cells and Their Applications. John Wiley & Sons, (2010)
58. Northmore, A., Tighe, S.: Innovative pavement design: Are solar roads feasible? (2012)
59. Aranguren, P., Araiz, M., Astrain, D., Mart´ınez, A.: Thermoelectric generators for waste heat harvesting: A computational and experimental approach. Energy Conversion and Management 148, 680–691 (2017). https://doi.org/10.1016/j.enconman.2017.06.040
60. Dezfooli, A., Moghadas Nejad, F., Zakeri, H., Kazemifard, S.: Solar pavement: A new emerging technology. Solar Energy 149, 272–284 (2017). https://doi.org/10.1016/j.solener.2017.04.016
61. Ahmad, M.M., Khan, F.U.: Review of vibration-based electromagnetic–piezoelectric hybrid energy harvesters. International Journal of Energy Research 45(4), 5058–5097 (2021) https://arxiv.org/abs/https://onlinelibrary.wiley.com/doi/pdf/https://doi.org/10.1002/er.6253. https://doi.org/10. 1002/er.6253
62. Zhang, Y., Ma, T., Yang, H., Li, Z., Wang, Y.: Simulation and experimental study on the energy performance of a prefabricated photovoltaic pavement. Applied Energy 342, 121122 (2023)
63. Rahman, M., Mabrouk, G., Dessouky, S.: Development of a photovoltaic-based module for harvesting solar energy from pavement: A lab and field assessment. Energies 16(8), 3338 (2023)
64. Ma, T., Li, S., Gu, W., Weng, S., Peng, J., Xiao, G.: Solar energy harvesting pavements on the road: comparative study and performance assessment. Sustainable Cities and Society 81, 103868 (2022)
65. Hu, H., Zha, X., Li, Z., Lv, R.: Preparation and performance study of solar pavement panel based on transparent resin concrete. Sustainable Energy Technologies and Assessments 52, 102169 (2022)
66. Azzouz, S., Chatterjee, A., Rorabaugh, R., Venegas, C., Duke, K., Weller, J., Smith, C.: Active road rumble energy harvesting panels. (2013). https: //doi.org/https://doi.org/10.1115/IMECE2013-64483
67. Zou, D., Liu, G., Rao, Z., Tan, T., Zhang, W., Liao, W.-H.: A device capable of customizing nonlinear forces for vibration energy harvesting, vibration isolation, and nonlinear energy sink. Mechanical Systems and Signal Processing 147, 107101 (2020). https://doi.org/10.1016/j.ymssp.2020.107101
68. Iswarya, M., Lakshmi, G.R.P.: Generation of electricity by using speed breakers. In: 2017 IEEE International Conference on Power, Control, Signals and Instrumentation Engineering (ICPCSI), pp. 896–899 (2017). https://doi.org/10.1109/ICPCSI.2017.8391841
69. Das, C.K., Hossain, S.M., Hossan, M.S.: Introducing speed breaker as a power generation unit for minor needs. In: 2013 International Conference on Informatics, Electronics and Vision (ICIEV), pp. 1–6 (2013). https://doi.org/10.1109/ICIEV.2013.6572532
70. Gkoumas, K., Petrini, F., Bontempi, F.: Piezoelectric vibration energy harvesting from airflow in HVAC (heating ventilation and air conditioning) systems. Procedia Engineering 199, 3444–3449 (2017). https://doi.org/ https://doi.org/10.1016/j.proeng.2017.09.496
71. Petrini, F., Gkoumas, K.: Piezoelectric energy harvesting from vortex shedding and galloping induced vibrations inside HVAC ducts. Energy and Buildings 158 (2017). https://doi.org/10.1016/j.enbuild.2017.09.099
72. 73. Demir, D.; Bolgen, N.; Vaseashta, A. Electrospun Nanofibers for Biomedical, Sensing, and Energy Harvesting Functions. Polymers 2023, 15, 4253. https://doi.org/10.3390/polym15214253
73. 74. Guan, M., Liao, W.-H.: Design and analysis of a piezoelectric energy harvester for rotational motion system. Energy Conversion and Management 111, 239–244 (2016). https://doi.org/10.1016/j.enconman.2015.12.061
74. Cascetta, F., Lo Schiavo, A., Minardo, A., Musto, M., Rotondo, G., Calcagni, A.: Analysis of the energy extracted by a harvester based on a piezoelectric tile. Current Applied Physics 18 (2018). https://doi.org/ https://doi.org/10.1016/j.cap.2018.04.015

75. Sharpes, N., Vuˇckovi´c, D., Priya, S.: Floor tile energy harvester for self-powered wireless occupancy sensing. Energy Harvesting and Systems 3(1), 43–60 (2016). https://doi.org/10.1515/ehs-2014-0009
76. Hwang, S., Jung, H., Kim, J.-H., Ahn, J., Song, D., Song, Y., Lee, H., Moon, S., Park, H., Sung, T.: Designing and manufacturing a piezoelectric tile for harvesting energy from footsteps. Current Applied Physics 15 (2015). https://doi.org/10.1016/j.cap.2015.02.009
77. 78. Pourashraf, T., Bonello, P., Truong, J.: Analytical and experimental investigation of a curved piezoelectric energy harvester. Sensors 22, 2207 (2022). https://doi.org/10.3390/s22062207
78. 79. Datta, R., Ranganathan, V.T.: Variable-speed wind power generation using doubly fed wound rotor induction machine-a comparison with alternative schemes. IEEE Transactions on Energy Conversion 17(3), 414–421 (2002). https://doi.org/10.1109/TEC.2002.801993
79. Touairi, S., & Mabrouki, M. (2022). Chaotic dynamics applied to piezoelectric harvester energy prediction with time delay. International Journal of Dynamics and Control, 1–22.
80. 81. Touairi, S., & Mabrouki, M. (2021). Vibration harvesting integrated into vehicle suspension and bodywork. Indonesian Journal of Electrical Engineering and Computer Science, 23(1), 188-196.
81. Touairi, S., & Mabrouki, M. (2021). Control and modeling evaluation of a piezoelectric harvester system. International Journal of Dynamics and Control, 1–17.
82. Touairi, S., Zekraoui, M., & Mabrouki, M. (2024). Compensation of Harvestable Electrical Power Produced by Piezoelectric Systems Under Variable Excitation: Design and Experimental Validation. Proceedings of the Sixth International Symposium on Dielectric Materials and Applications (ISyDMA'6), 2022.
83. Touairi, S., & Mabrouki, M. (2022). Maximizing Electric Power Recovery through Advanced Compensation with MPPT Algorithms. Modeling and Simulation in Engineering, 2024

Spintronics and Spinquant 2D Structures Prepared Using Laser Plasma Method

Paata J. Kervalishvili

Abstract The transformation of digital information devices from large installations to portable systems has been enabled by advanced processing technologies that allow ultrahigh integration of devices with atomic-scale dimensions that increasingly possess properties that are dominated by quantum physics. The quantum particles - Leptons are good candidates to be quantum information carriers (qubits). After all, they also have an associated anti-particle, which has the same mass but opposite charge. Along with the quantum properties and potential of 2D materials as solid-state platforms for quantum-dot qubits, single-photon emitters, superconducting qubits, and topological quantum computing elements it is necessary to select the best method of their preparation. The paper aims to elaborate on the concept of usage of leptons-based spintronic and spinqubit systems obtained by Laser Plasma technology of 2D materials preparation. We are looking for further development of LP processes aimed at preparing the next (higher) level of spintronic structures based on diluted semiconductors. Our work shows that the LP method and technology are very useful for the preparation of new highly effective multiqubit elements.

Keywords Spinqubit · Spinquant · 2D Structures

1 Introduction

If we consider the universal introduction of quantum information science as one of the last remarkable achievements of the end of the twentieth century, we should recognize the development of quantum information technologies as the most important achievement of the first decades of the twenty-first century. Now, in the quantum era, the most interesting prospect is the creation and use of quantum information devices and their networks. As the Moore law gradually loses its effect, conventional charge-based electronics will come to an end soon. Developing alternative high-speed

P. J. Kervalishvili (✉)
Georgian Technical University, Tbilisi, Georgia
e-mail: kervalpt@yahoo.com

A. Vaseashta et al. (eds.), *Proceedings of the Eighth International Symposium on Dielectric Materials and Applications (ISyDMA'8)*, Springer Proceedings in Physics 349, https://doi.org/10.1007/978-3-031-84809-4_5

and low-energy-consuming information technology is urgently needed. Up to now, many new methodologies have been proposed, such as molecular electronics, nano-electronics, spintronics, quantum information techniques, etc. In parallel, quantum information science has emerged as an alternative to conventional transistor technology, promising new paradigms in computation, communication, and sensing. The convergence between quantum materials properties and prototype quantum devices is especially apparent in the field of 2D materials, which offer a broad range of material properties, high flexibility in fabrication pathways, and the ability to form artificial states of quantum matter.

Based on the difference from conventional electronics which uses the electron's charge degree of freedom for information processing, spintronics is devoted to incorporating the electron's spin degree of freedom. In an ideal situation, there will be purely spin current and no charge current in the spintronic circuit, thus practically no heat will be created and wasted [1]. Meanwhile, information will be transmitted at a high speed owing to the spin coherence effect.

Although a photon is a boson with spin "one", which is a typical boson spin, it is also a good candidate to be a quantum information particle (qubit). If we consider qubit itself as a symmetrical quantum system it could be represented by two optical modes, mainly spatial or polarization ones. Using the single photon with two optical modes of internal polarization degree of freedom we are coming to the term "polarization qubit" which together with so-called "dual-rail" qubits is the main engine for the organization of nonlinear quantum computing [2–5].

In linear optical quantum computing the basic building blocks are beam splitters, half- and quarter-wave plates, phase shifters, etc. The main characteristic of the photonic computing process is the effect of the nonlinearity of computation. The design of photonic computing devices which was inspired by the theoretical Ising model is based on lasers, mirrors, and other optical components commonly found on an optical table. Basic features for the possible realization of the quantum optical computing device using quantum photonic treatment are connected with the corpuscular nature of light. Photon refraction phenome-non when travel through the interface between two different transparent solids is simulated, and photons, as signal carriers are tracked by controlling Short Range Interatomic Forces.

Leptons are fermions—particles with ½ spin and they are an important part of the Standard Model. Following the spin-statistics (spin-communication theorem), the theorem states that one fermion can exist in a given quantum state and no two leptons of the same species can be in the same state at the same time. Following this, a lepton can have only two possible spin states–up or down. The charged lepton is the electron; the next lepton to be observed was the muon, which was classified as a meson at the time. After investigation, it was realized that the muon has not the expected properties of a meson, but rather behaved like an electron, but with higher mass. Another lepton, the first neutrino, the electron neutrino, was proposed to explain certain characteristics of beta decay. The muon and the tau neutrinos were discovered later up to the end of the twentieth century. [6–8].

An important property of Leptons is chirality, which in turn is closely related to a more easily visualized property called helicity. The helicity of a particle is the direction of its spin depending on its momentum; particles with spin in the same direction as their momentum are called right-handed and particles with different direction spins are called left-handed [9]. In field theory (quantum electrodynamics and quantum chromodynamics) left- and right-handed fermions are identical. However, the Standard Model's weak interaction treats left-handed and right-handed fermions as asymmetrical: Only left-handed fermions (and right-handed anti-fermions) participate in the weak interaction. Following these right-handed neutrinos and left-handed anti-neutrino have no possible interaction with other particles.

Today usage of electron spins as quantum bits for quantum information processing in so-called quantum computers, where a qubit exists in more than one state simultaneously (until its state is), is clear. Qubits in this state display a degree of correlations impossible in classical physics. This phenomenon is called entanglement and is a crucial property of quantum computing [10, 11]. It is also possible to create qubits as the up/down and spin-states of electrons on quantum dots. In this case, it is impossible to control the state of a single electron well enough to perform calculations using them.

According to the known properties of neutrinos: each neutrino flavor state is a linear combination of the three discrete mass eigenstates, the flavor eigenstates (creation and annihilation combinations) are not the same as the neutrino mass eigenstates and they are in associated specific quantum superposition. Following this it is possible to consider this family as a good candidate to be the qubit for quantum information systems.

Perfection and ultra-purity are not the only parameters that characterize material usefulness for quantum devices. Modification of properties by different structural non-perfections (structural defects: impurities, isotopes, etc.) is the smart instrument for the regulation of their characteristics. (Fig. 1).

The potential of the laser plasma process for 2D materials preparation, particularly its usefulness for the organization of nanostructures applicable in spintronic and quantum computing devices nowadays is actively developing. Laser plasma formed under the ionizing effect of powerful laser radiation on the thing. The usage of resonance light heat creates the opportunity to energize the selected atoms as well as their groups (assemble) and to produce plasma with the necessary properties relevant to structures that must be prepared.

Finally, when we are choosing the particles for quantum computing we should consider that the candidate for a qubit generally needs to have the quantum properties of superposition and entanglement. There are also the main technical requirements of quantum computation which are: scalable physical systems with well-characterized qubits (Zeeman Splitting); long decoherence time higher than gate operation one; existence of qubits at the ground state; set of quantum gates; measurement capabilities, etc. Leptons—fermions (electrons, protons, neutrons, muons, tauon, and even neutrinos) as we know have those kinds of properties. Concerning the photons—bosons particles with frequency-dependent energy collecting into the same energy state (Bose–Einstein condensation), they also could be acting as a qubit because of the polarization effects they characterized. The usefulness of the other boson particles

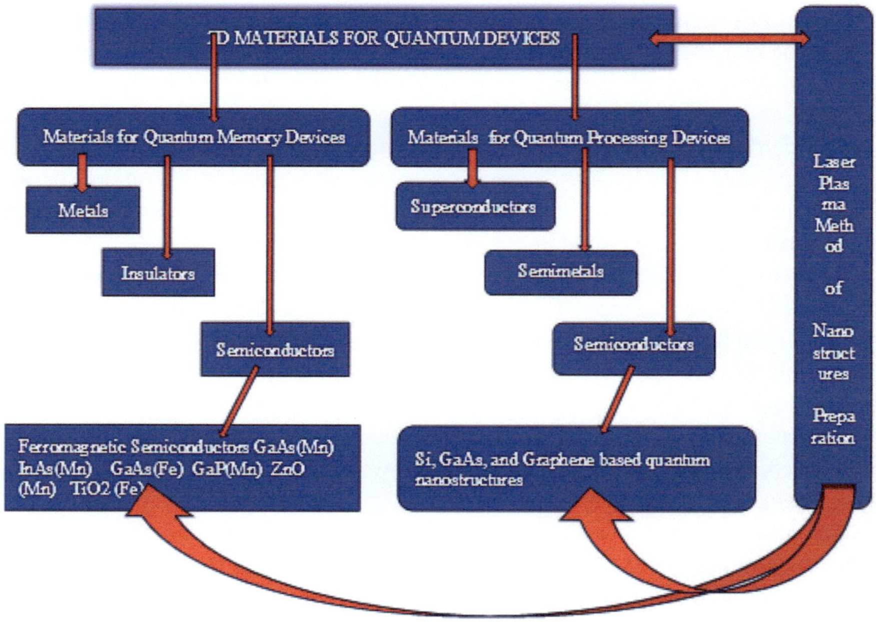

Fig. 1 Roadmap of works related to spin electronic nanostructures for quantum devices

as quantum information carriers is an exciting task for current and future research works.

2 Discussion

Pure spin generation and injection mainly depend on the degree of spin polarization in the used semiconductors or metals. Thanks to the discovery of carbon-based nanomaterials such as graphene and carbon nanotubes, the challenge of long-distance spin transport is likely to be solved in the near future. Because of their very weak spin–orbit coupling (SOC), carbon-based nanomaterials can have a long spin coherence length up to a few micrometers and, thus are very good spin transportation materials. As with spin manipulation and detection, there are some methods based on the coupling effects between spin and light, magnetic field, and electric field. According to their electronic and magnetic properties, spintronics materials can be classified as magnetic metals, topological insulators, and magnetic semiconductors In a spintronic device, magnetic metals and topological insulators, serve as spin sources and drains, while magnetic semiconductors constitute the central region of the device.

At the same time usage of electron spins as quantum bits for quantum information processing in so-called quantum computers, where a qubit exists in more than one state simultaneously, is clear. Qubits in this state display a degree of correlations impossible in classical physics. The main requirements of quantum computation are Scalable physical systems with well-characterized qubits (Zeeman Splitting); Long decoherence time: Existence of qubits at the ground state; Set of quantum gates; measurement capabilities, etc. Candidate for a qubit needs longer decoherence time than gate operation time [12].

The transformation of digital computers from bulky machines to portable systems has been enabled by new materials and advanced processing technologies that allow ultrahigh integration of solid-state electronic switching devices. As this conventional scaling pathway has approached atomic-scale dimensions, the constituent nanomaterials increasingly possess properties that are dominated by quantum physics.

Along with the quantum properties and potential of 2D materials as solid-state platforms for quantum-dot qubits, single-photon emitters, superconducting qubits, and topological quantum computing elements it is necessary to select the best method of their preparation.

The potential of laser plasma process for 2D materials preparation, particularly its usefulness for the organization of nanostructures applicable in spintronic and quantum computing devices nowadays is actively developing. Laser plasma formed under the ionizing effect of powerful laser radiation on the thing. For example, LP arises during optical breakdown in gaseous media, laser radiation on top solid body, in laser thermonuclear targets.

LP can exist in a wide range of temperatures—from 1 to 104 eV (104–108 K) and arise as a result of ionization of the electron impact with the subsequent image electronic avalanche, or as a result of many photons ionization. In LP. experimentally, self-focusing of the laser beam is observed. The impact of a light wave on LP leads to the formation of plasma waves (a colony of electronic and ionic densities), which interact with the primary and scattered light. As a result, electric magnetic waves are formed with a frequency that is a multiple of the frequency of the incident light of this wave (the so-called harmonics). The probability of generating high harmonics increases with an increase in the intensity of laser radiation.

For the preparation of 2D materials (semimetals, semiconductors) for novel quantum devices based on our previous investigations we will develop and use the Laser-Plasma method which enables the preparation of nanostructured layers with fine and perfect structures and high purity. The properties of nanomaterials prepared by laser plasma technique are unique, and they are not reproducible by any other method including chemical ones. The usage of resonance light heat creates an opportunity to energize the selected atoms as well as their groups (assemble) and to produce plasma with the necessary properties relevant to structures that must be prepared This technique was successfully used by the authors of the paper to study the conditions for obtaining diamond-like films, as well as thin layers of boron carbide [13]. See Fig. 2.

In the last two decades, the Laser plasma method was used to form both homogeneously doped GaAs:Mn layers and two-dimensional structures, including a δ-doped GaAs:Mn layer and a $InxGa_{1-x}As$ quantum well separated by a GaAs spacer with a thickness of $d = 3–6$ nm. It is obvious that only Mn ions, which are part of

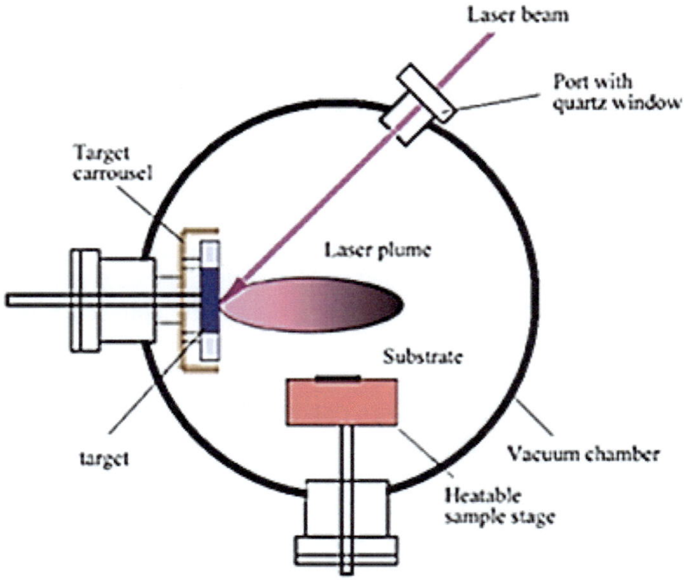

Fig. 2 The scheme of experimental laser-plasma technology setup

the GaMnAs solid solution and are distributed almost uniformly in it, can noticeably exchange with quantum well carriers, leading to their spin polarization and, consequently, to the anomalous Hall effect [14, 15].

The organization of spin arrangements manifests diverse spin textures, encompassing ferromagnetism, antiferromagnetism, and chiral vortices. However, mapping these spin textures in ultrathin magnetic multi-layers has remained elusive. A reconfigurable chiral spintronic emission method is introduced through investigations on a model system of synthetic antiferromagnet and the spin information of individual ferromagnet (FM) layers. Upon femtosecond photoexcitation of the synthetic antiferromagnet (FM1/Ru/FM2), the ferromagnets generate a pair of linearly polarized ultrafast spin currents, which, after relaxation at the FM/Ru interface, emit corresponding THz pulses [16]. The Ruderman–Kittel–Kasuya–Yoshida interactions between the two FMs give rise to magnetic-field-controlled spin textures causing a spin relaxation imbalance at the interfaces and inducing a phase shift between the orthogonal components of the emitted terahertz fields, consequently reconfiguring the polarization of the emitted terahertz field from linear to circular. In the case of Graphene's multilayers, they are mainly forming stack–clusters with sublattices, which is the most common arrangement of nearest neighbor layers observed in nature. [17–22].

3 Conclusions

Selection of laser sources and their parameters gives the possibility to vary the energy of ionized atoms in plasma plum, activate them to the necessary level, and deposit the hot atoms and their clusters on substrates of different origin (semiconductors: Silicon, GaAs, etc.; Metals: Fe, Ni, etc.; Insulators: Al_2O_3, etc.). For the organization of these processes, it is also possible to use the resonance wavelength of the light sources to have a direct and strong interaction with the electron's bonding energies.

Concerning the graphene-silicon nanosystems, the long coherence times of single spins in silicon quantum dots make these systems highly attractive for quantum computation, but the scalability of these kinds of structures steel needs to be developed. As a first step, it is necessary to organize the strong coupling of a single electron spin and a single microwave photon. In these conditions, the electron spin must be trapped in a silicon double quantum dot and the microwave photon must be stored.

For the particles–candidates of qubits they generally need to have the quantum properties of superposition and entanglement as well as a long decoherence time higher than gate operation one. Leptons—fermions (electrons, protons, neutrons, muons, tauon and even neutrinos) have that kind of properties. Photons—bosons also could act as qubits because of the polarization effects they characterized. The usefulness of the other boson particles as quantum information carriers is a very interesting task for current and future research works.

References

1. Fert, A., The origin, development, and future of spintronics, Nobel Lecture, December 8, Stockholm, Sweden, (2007).
2. Kervalishvili, P., Nuclear spin-based model of quantum information system. Book of abstracts of International Conference of Nanosensory Systems and Nanomaterials, June 6–9, 2013, EU-ISTC-GTU, Tbilisi, Georgia. 28–31. (2013).
3. Pieter Kok, W. J. Munro, Kae Nemoto, T. C. Ralph, Jonathan P. Dowling, and G. J. Milburn, Rev. Mod. Phys. 79, 135 – Published 24 January 2007.
4. Kervalishvili, P., Photons, Information Transfer and Speed of Light (Invited Lecture), International Conference eRA -7 The Synergy Forum, TEIPIR, Athens Greece, (2012).
5. Kervalishvili, P., Quantum information technology: Theory and applications, 2015 IEEE Seventh International Conference on Intelligent Computing and Information Systems (ICICIS), IEEE Xplore, https://doi.org/10.1109/IntelCIS.2015.7397187, IEEE. 15p. (2016).
6. Dekel, A., Ostriker, J.P. (eds). Formation of structure in the universe, Cambridge University Press, (1999).
7. Kervalishvili, P., Is information a matter and does it have a mass? Abstracts of Scientific Conference (NATO ARW "Philosophy and synergy of information: Sustainability and Security"). Tbilisi – Spring 2011, Georgian Technical University, Tbilisi, Georgia. (2011).
8. Kervalishvili, P., Leptons Based Quantum Computing. Acta Scientific, Computer Sciences. Published: June 23, Volume 5(7), 12–14. (2023).
9. Dartora C.A. and Cabrera, G.G., Can we explain why leptonic electroweak interactions are chiral? Euro physics Letters, Vol. 101(5), Published 13 March (2013).
10. Noë, A., The Entanglement: How Art and Philosophy Make Us What We Are, Princeton University Press, Published (US): Jun 27, 2023, (UK): Aug 22, (2023).

11. Hill, S; Wootters, W. K. "Entanglement of a Pair of Quantum Bits". Phys. Rev. Lett. 78 (26): 5022–5025. (1997).
12. Shlimak, I., Safarov, V., Vagner, I., Isotopically engineered silicon/silicon-germanium nanostructures as basic elements for a nuclear spin quantum computer Journal of Physics: Condensed Matter 13 (26), 6059. (2001).
13. Kervalishvili, P., Mikelashvili, V., Shalamberidze, S., Properties of Boron Carbide Films prepared by Laser Plasma Deposition Method. Georgian Eng. News, 3, (2004).
14. P. Kervalishvili and A. Lagutin, Nanostructures, magnetic semiconductors and spintronics, Microelectronics Journal, 39. 1060–1065. (2008).
15. Kervalishvili, P., Quantum processes in semiconducting materials and spin electronics. Rev. Adv. Mater. Sci. (RAMS), Vol 14(1), (2007).
16. Agarwal, P., Mishra, S.S., Medwal, R., Mohan, J.R., Asada, H., Fukuma, Y., Singh, R., Reconfigurable Chiral Spintronic THz Emitters. Adv. Optical Mater. 2024, 12, 2303128. https://doi.org/10.1002/adom.202303128.
17. Novoselov, K.S., Geim, A.K, Morozov, S.V., et al. Electric field effect in atomically thin carbon films. Science. 306: 666–9. (2004).
18. Kervalishvili, P., Semiconductive nanostructures - Materials for spin-electronics: New Data Bank Requirement. Data Science Journal, Volume 6, (2007).
19. Kervalishvili, P.J., Laser technologies of nanosystems preparation. Conference "Nanotechnology", Georgian Technical University, 4 – 7 October (2021), Tbilisi, Georgia, p. 62.
20. Kervalishvili, P.J., Novel nanostructured semiconductive materials for spin electronics applications, Materials of ISTC Conference "Spinelectronics 07" Tbilisi State University, October 22–24 (2007). Tbilisi, Georgia.
21. Bakhtiari, H., Berberashvili, T., Kervalishvili, P., Preparation of Graphene Structures by Continuous Wavelength Laser Deposition Method, American Journal of Condensed Matter Physics, 12(1) 2-11. (2022).
22. Kervalishvili, P., Spintronic and spinqubit nanostructures for quantum devices. Sustainability through Science and Technology, SIPS Conference, Panama. 27 November-1 Dec. (2023).

Invited Articles

New Functions of a Semiconductor Photodetector with a High-Resistivity Layer

Surik Khudaverdyan, Ashok Vaseashta, Mane Khachatryan, Lenrik Matevosyan, Elya Makarryan, and Gagik Ayvazyan

Abstract In semiconductor structures with high-resistivity layers between oppositely directed potential barriers, under conditions of longitudinal illumination and barrier closure, a new functionality is revealed: the mutual compensation of photocurrents from the potential barriers and, consequently, the short-wave and long-wave spectral maxima. The long-wavelength maximum coincides with the intrinsic absorption band of the semiconductor. Certain technological solutions result in abnormally high photosensitivity at spectral peaks. In some cases, this is expressed as a short-wavelength maximum, while in others, it appears as a long-wavelength maximum, which is the result of internal photocurrent amplification. Notably, the mutual compensation of oppositely directed photocurrents results in very small dark currents, on the order of 10 nA. This, in turn, provides a low signal-to-noise ratio and enables the detection of weak light signals.

Keywords Semiconductor structures · High photosensitivity · Contact point · Signal-to-noise

1 Introduction

One of the effective ways to combat man-made environmental issues is to create modern technical monitoring tools that can be used in various fields of industry, agriculture, and transport to control the pollution of fresh water with oil products,

S. Khudaverdyan (✉) · M. Khachatryan · E. Makarryan · G. Ayvazyan
Armenian National Polytechnic University, Yerevan, Armenia
e-mail: khsuren@seua.am

A. Vaseashta
International Clean Water Institute Manassas, Manassas, VA, USA

Institutul de Cercetare, Universității din București, Bucharest, Romania

L. Matevosyan
Institute of Radiophysics and Electronics, Ashtarak, Armenia

A. Vaseashta et al. (eds.), *Proceedings of the Eighth International Symposium on Dielectric Materials and Applications (ISyDMA'8)*, Springer Proceedings in Physics 349, https://doi.org/10.1007/978-3-031-84809-4_6

phenols, nitrogen, pesticides, and heavy metal compounds. Semiconductor photosensors play a key role in monitoring systems [1–7]. The main requirements for these sensors are high photosensitivity, low noise, stability, speed, and repeatability.

In terms of ensuring high photosensitivity and low noise, the lifetime of the minority charge carriers of the original semiconductor and the width of the band gap is decisive. In that regard, wide-band-gap semiconductors are preferable, since the lifetime of minority charge carriers reaches 10^{-3} s, which leads to high inertia (about 10 kHz) [8–11].

However, the existing highly sensitive photosensors based on wide-band-gap semiconductors have limited applicability because of the stability, speed, and repeatability issues. In addition, the wide-band-gap semiconductors are very expensive.

Silicon photosensors do not have these problems [12–16]. In this regard, the Hamamatsu photosensors have excellent characteristics, with the maximum photosensitivity reaching 0.7 A/W [17]. They ensure the minimum dark current of 10 pA with the pulse rise time of the order of 1 μs at room temperature1. However, the innovative structural, technological, and schematic solutions applied in these sensors as well as in the other popular sensors complicate the manufacturing process leading to an increase in price.

In our previous works, we proposed a novel silicon photodetector structure with a high-resistivity layer based on a double-barrier design that provides counter-currents [18–21]. These photodetectors are fabricated using simple, traditional silicon technology. In this study, we investigate the possibility of achieving high spectral photoresponse sensitivity of these photodetectors in both the short-wave and long-wave regions.

2 Structure and Experiment

The object of research was the n^+-p-n^+ structure. Figure 1 shows the distribution of the potential energy of holes in the valence band and the direction of the photocurrents in the base of the silicon sample.

High-resistance p-layers are located between low-resistance n^+ regions. This provides the structure with several functionalities. The structure has three layers and two outputs: one in the near-surface region and the other in the rear region. The base thickness d and the concentration of impurities in certain regions are designed to ensure the closing of the potential barriers at the point x_m.

The samples are made on an n-type (100) orientation silicon substrate with an impurity concentration of 1×10^{18} and 1×10^{15} cm^{-3} and a thickness of 460 nm. A base approximately 5.8 and 4.5 μm thick with a cathode depth of 0.3 μm is epitaxially grown. The effective film thickness is ~5.5 and 4.2 μm. The impurity concentration in the base is 1.2×10^{14} cm^{-3} and 8×10^{14} cm^{-3}, and the impurity concentration in the cathode is 5×10^{18} cm^{-3}. Thus, it is possible to cover the base with depleted areas of oppositely directed potential barriers. The position of x_m is 3.9 and 2.9 μm,

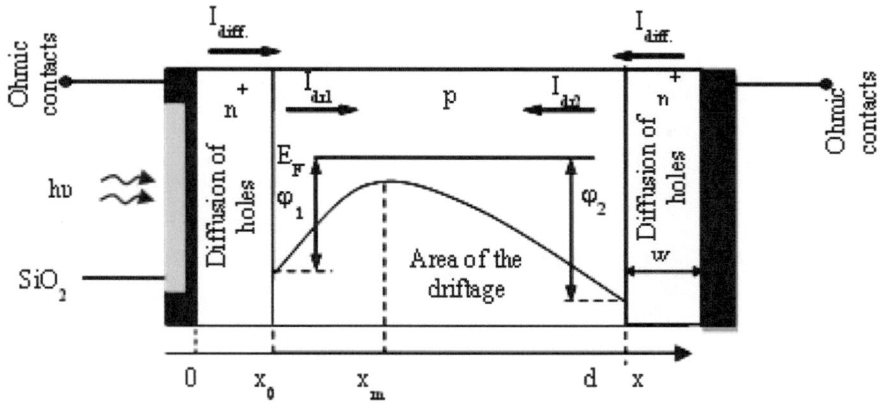

Fig. 1 The direction of diffusion and drift currents in the silicon n^+-p-n^+ structure

respectively. An antireflection 50 nm thick SiO_2 film ensures effective radiation transmission.

The photodetector samples are irradiated through a monochromator. Total radiation with a wavelength of $\lambda = 300$–1100 nm is used. To obtain the current–voltage characteristics and spectral characteristics, a Keithley 6340 Sub-Femtoamp Remote Source Meter Instrument (Cleveland, Ohio, USA) is used to step the voltage supply to the photodetector in steps of 1 mV. The device can measure the current with a sensitivity of 1 pA.

3 Results and Discussions

The spectral dependence of the current photosensitivity (Si) of the samples with $x_m = 2.9$ μm at various bias voltages is presented in Fig. 2.

A significant feature is the unusually high photosensitivity in the spectral regions of the short-wave (~6 A/W at 570 nm) and long-wave (~11.6 A/W at 870 nm) maxima. Considering that the best Hamamatsu photodiodes have photosensitivity of 0.72 A/W in the region of the intrinsic absorption of silicon, this feature could be utilized for recording weak but important optical signals. When the active base thickness is $d = 4.5$ μm, the depth $x_m = 2.9$ μm is the photocurrent generation region of the first barrier. The photocurrent generation region of the second barrier is much wider and is equal to $d\text{-}x_m + L_p = 2.6 + L_p$, where $L_p \sim 40$ μm is the diffusion length of the minority carriers in the rear diffusion region (Fig. 1). Under such conditions, the short waves are predominantly absorbed in the region of the near-surface barrier, creating a short-wave maximum, while the long waves are predominantly absorbed in the region of the rear barrier (including the rear diffusion region), creating a long-wave maximum.

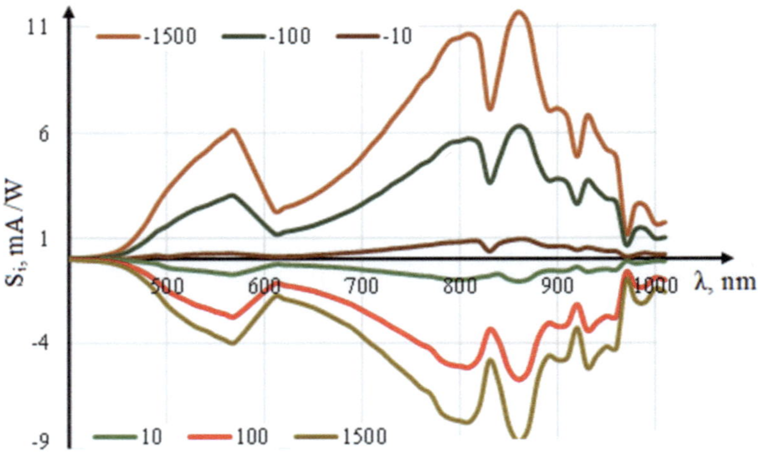

Fig. 2 Spectral dependence of the photosensitivity of the samples with $x_m = 2.9$ μm at different bias voltages (in mV)

The drop in the value of the spectral photocurrent in the intermediate spectral region (around $\lambda = 620$ nm, Fig. 2) results from the mutual compensation of the photocurrents from the oppositely directed potential barriers. Although the absorption of electromagnetic waves occurs according to the exponential law, it is estimated that the section of the rear barrier responsible for photocarrier generation, with a length of 42.6 μm in the long-wave absorption region, can generate more photocurrent than the section of the near-surface barrier responsible for photocarrier generation, with a length of 2.9 μm. This results in a greater value of long-wave photosensitivity compared to short-wave photosensitivity.

Figure 3 shows the spectral dependence of the current photosensitivity of the samples with $x_m = 3.9$ μm at various bias voltages. The advantage is that in samples manufactured using the same technology, with an active base thickness of $d = 5.5$ μm, the depth $x_m = 3.9$ μm corresponds to the photocurrent generation region of the first barrier, which is much wider than in the first case. This results in a significant increase in short-wave photosensitivity (~4.2 A/W) compared to long-wave photosensitivity (~1.4 A/W).

When the near-surface barrier is reverse biased (the curves at negative voltages in Figs. 2 and 3), the spectral sensitivity is significantly higher than when the rear barrier is reverse biased. This is due to an increase in the compensation capacity of the photocurrent of the rear barrier and a decrease in the number of quanta absorbed with depth.

The maximum of the mutual compensation of the photocurrents (the minimum value of the photosensitivity) S_{min} with a base thickness of $d = 5.5$ μm occurs in the wavelength region of $\lambda = 600$ nm (Fig. 4, red line). As the base thickness decreases to $d = 4.5$ μm, this minimum, along with other peaks, shifts towards

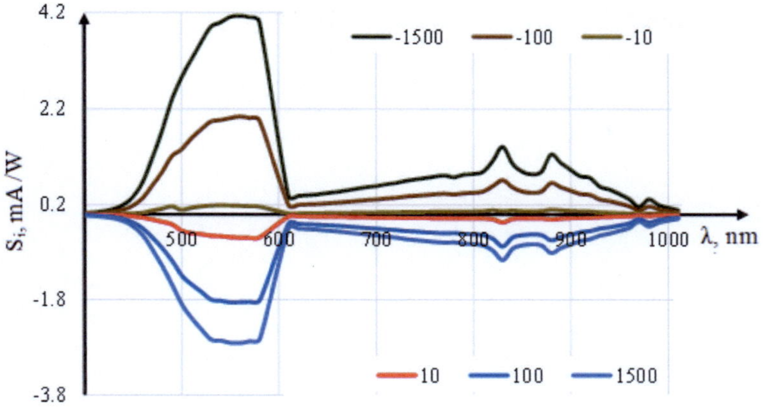

Fig. 3 Spectral dependence of photosensitivity of the samples with $x_m = 3.9$ μm at various bias voltages (in mV)

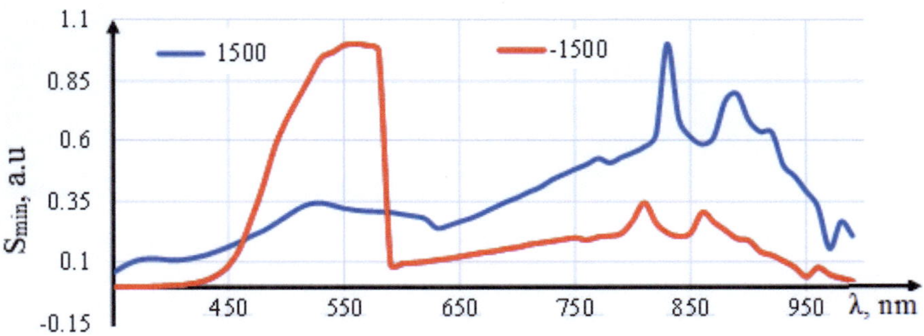

Fig. 4 Minimum value of the photosensitivity (in a.u.) at bias voltages 1500 and -1500 mV

the long wavelengths (Fig. 4, blue line). This is likely related to the proportion of short-wave and long-wave photocurrents in the total photocurrent.

4 Mechanism of the Internal Amplification of the Photocurrent

The external bias voltage is applied to the depleted regions of the n-p junctions. Electrons from the first diffusion region of the forward-biased n-p junction are injected as minority carriers into the base, and then from there through the reverse-biased junction into the second diffusion region (the rear barrier). Simultaneously, due to photocarrier generation, the heights of the surface and rear potential barriers are reduced, the bias voltages are redistributed onto the reverse-biased n-p junction, and

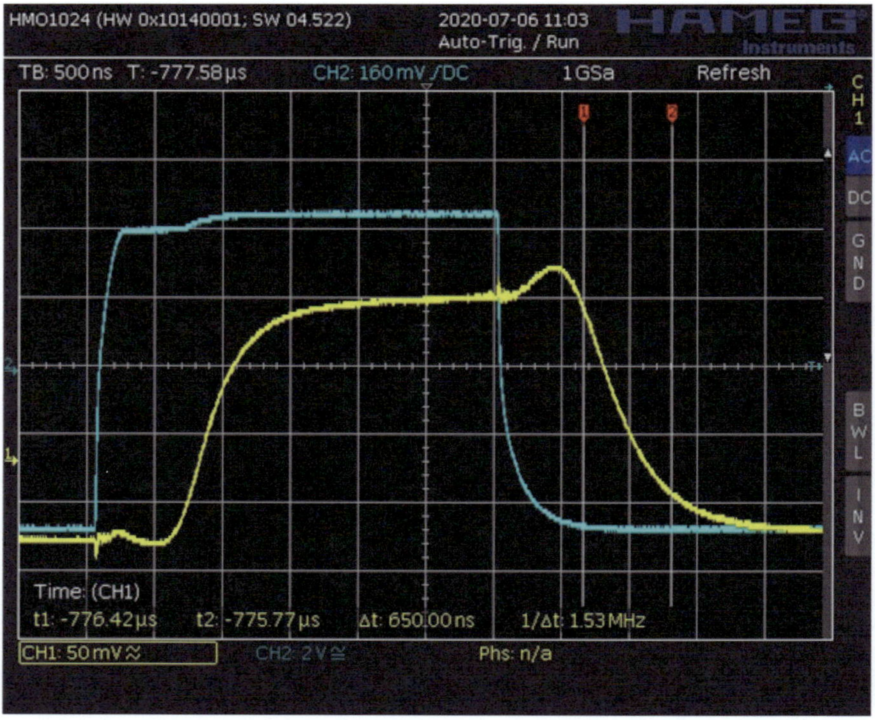

Fig. 5 The rise and droop times of the investigated samples with a base thickness of 4.5 μm

the field intensity that pulls carriers into the rear region increases. This positive feedback creates an unusually high photosensitivity. The mechanism works both with the reverse bias of the near-surface barrier and with the reverse bias of the rear barrier (Figs. 2 and 3). This process creates additional inertia in the photo signal passage.

The response to the radiation pulse determines the rise and droop times of the pulse, which are 0.6 and 1.2 μs, respectively (Fig. 5), corresponding to a response speed of 0.83 MHz. This is inferior to the response speed of the best Hamamatsu photodiodes, which provide a response speed of 25 MHz [17].

5 Noise Equivalent Power of the Investigated Structures

In the high mutual compensation mode of the photocurrents of the oppositely directed barriers, the low dark currents (~10 nA) [21] and noise is ensured. Figure 6 shows the spectral dependence of Noise Equivalent Power (NEP) in the short wavelength region. The initial, comparatively high value of NEP is the result of the absorption of the short waves in the region of the near-surface barrier, the photocurrent of which is large and is weakly compensated by the weak photocurrent of the rear barrier.

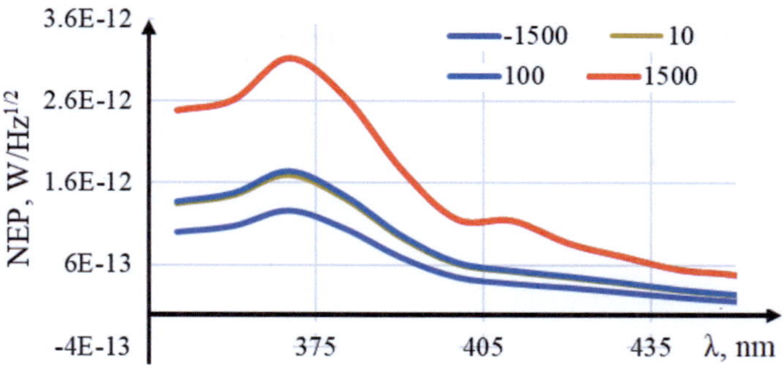

Fig. 6 The spectral dependence of NEP in the short wavelength region at various bias voltages (in mV)

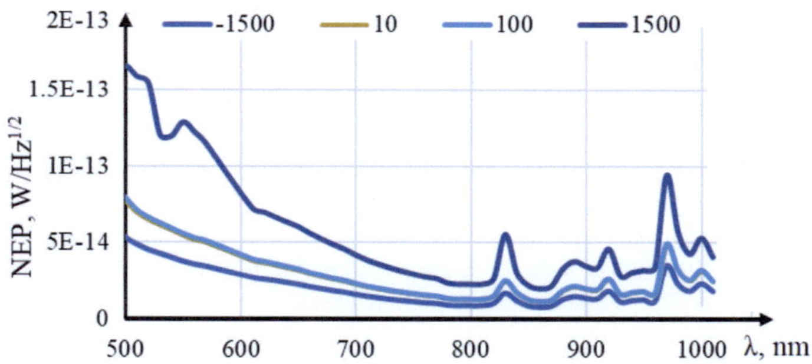

Fig. 7 The spectral dependence of NEP in the long wavelength region at various bias voltages (in mV)

In the case of the long waves (Fig. 7), with the increase in the wavelength, the fraction of the rear photocurrent and the mutual compensation increase and pass through the maximum, where NEP has a minimum value. The rear photocurrent prevails and leads to an increase in NEP. The value of the latter within the wavelength range of $\lambda = 700$–950 nm is lower than 5.10^{-14} W/Hz$^{1/2}$ which is close to the values of Hamamatsu photodiodes [17].

6 Conclusion

A silicon double-barrier structure with a high-resistivity layer has been developed, providing unusually high spectral photosensitivity in both short-wave and long-wave regions, with mutually consistent structural parameters. High photosensitivity is achieved by introducing carriers through a forward-biased barrier into the base and then passing them through a second, reverse-biased barrier, which is reduced by light. The greater the mutual compensation of the photocurrents from both barriers, the lower the noise of the photodetector. It is expected that such devices will find applications in high-efficiency sensors, detectors, and spectroscopy.

Acknowledgment The research was supported by the Higher Education and Science Committee of MESCS RA (Research project № 25RG-2B010).

References

1. Chen J, Kuang H, Wang Y, et al (2024) Design and optimization of a multidirectional photodetector in optoelectronic integration. Opt Lett 49:997. https://doi.org/10.1364/OL.514161
2. Tian J (2023) An overview of pyroelectric photodetector: Photoresponse mechanisms and applications. AIP Advances 13:050701. https://doi.org/10.1063/5.0146279
3. Gundepudi K, Neelamraju PM, Sangaraju S, et al (2023) A review on the role of nanotechnology in the development of near-infrared photodetectors: materials, performance metrics, and potential applications. J Mater Sci 58:13889–13924. https://doi.org/10.1007/s10853-023-08876-8
4. Seo SG, Kim SY, Jeong J, Jin SH (2021) Progress in light-to-frequency conversion circuits based on low dimensional semiconductors. Nano Res 14:2938–2964. https://doi.org/10.1007/s12274-021-3586-6
5. Choi GI, Choi HW (2022) A Study to Improve the Performance of Mixed Cation–Halide Perovskite-Based UVC Photodetectors. Nanomaterials 12:1132. https://doi.org/10.3390/nano12071132
6. Vaseashta A (2006) Carbon Nanotubes Based Devices and Sensors. Materials and Manufacturing Processes 21:710–716. https://doi.org/10.1080/10426910600613595
7. Vaseashta A, Khudaverdyan S (2013) Advanced Sensors for Safety and Security. Springer Netherlands, Dordrecht
8. An B-G, Kim H-R, Chang YW, et al (2021) Photosensors-based on cadmium sulfide (CdS) nanostructures: a review. J Korean Ceram Soc 58:631–644. https://doi.org/10.1007/s43207-021-00141-5
9. Asriyan HV, Shatveryan AA, Aroutiounian VM, et al (2005) Semiconductor-metal interface influence on the bulk low-frequency noise behavior and role of the phonons refraction points. In: Bergou JA, Smulko JM, Dykman MI, Wang L (eds). Austin, TX, p 192
10. Zhao L, Hu L, Fang X (2012) Growth and Device Application of CdSe Nanostructures. Adv Funct Materials 22:1551–1566. https://doi.org/10.1002/adfm.201103088
11. Wassel AR, Mansour SA, Mohamed FM, El-Mahalawy AM (2024) Exploration of doping-dependent structural, optical, and electrical properties of thermally evaporated CdSe thin films for improved photodetection performance. Journal of Alloys and Compounds 976:173052. https://doi.org/10.1016/j.jallcom.2023.173052
12. Ayvazyan G, Ayvazyan K, Hakhoyan L, Semchenko A (2023) NO_2 Gas Sensor Based on Pristine Black Silicon Formed by Reactive Ion Etching. Physica Rapid Research Ltrs 17:2300058. https://doi.org/10.1002/pssr.202300058
13. Zhao Z, Zhang Z, Jing J, et al (2023) Black silicon for near-infrared and ultraviolet photodetection: A review. APL Materials 11:021107. https://doi.org/10.1063/5.0133770
14. Ghandiparsi S, Mayet AS, Bartolo-Perez C, Islam MS (2023) Efficient surface nano-textured CMOS-compatible photodiodes for Optical Interconnects. In: Photodetectors. Elsevier, pp 437–466
15. Park J, Kang I-S, Sim G, et al (2021) Broadband responsivity enhancement of Si photodiodes by a plasmonic antireflection bilayer. Opt Express 29:26634. https://doi.org/10.1364/OE.432689

16. Ozel K (2024) Ag nanoparticles decorated AZO/p-Si heterojunction UV photodetector. Optics & Laser Technology 174:110621. https://doi.org/10.1016/j.optlastec.2024.110621
17. Hamamatsu (2022) Compact spectrometers with built-in Hamamatsu image sensor, optical element, etc. Mini spectrometer
18. Khudaverdyan S, Vaseashta A, Ayvazyan G, et al (2022) On the Selective Spectral Sensitivity of Oppositely Placed Double-Barrier Structures. Photonics 9:558. https://doi.org/10.3390/photonics9080558
19. Vaseashta A, Khudaverdyan S, Ayvazyan G, et al (2023) On the unusually high photosensitivity of two barrier structures. Appl Phys B 129:101. https://doi.org/10.1007/s00340-023-08048-1
20. Khudaverdyan S, Vaseashta A, Ayvazyan G, et al (2024) Functional Capabilities of Two-Barrier Semiconductor Structures. In: Sontea V, Tiginyanu I, Railean S (eds) 6th International Conference on Nanotechnologies and Biomedical Engineering. Springer Nature Switzerland, Cham, pp 12–21. https://doi.org/10.1007/978-3-031-42775-6_2
21. Khudaverdyan S, Vaseashta A, Ayvazyan G, et al (2021) On the Semiconductor Spectroscopy for Identification of Emergent Contaminants in Transparent Mediums. In: Vaseashta A, Maftei C (eds) Water Safety, Security and Sustainability. Springer International Publishing, Cham, pp 663–689

Superhydrophobic Coating Based on Functionalized SiO_2 Nanoparticles for Self-Cleaning and Anti-Icing

Brahim Nomeir, Sofia Boukheir, Sara Lakhouil, Mustapha Aitali, and Sanae Naamane

Abstract Developing transparent and superhydrophobic coatings is challenging due to their conflicting properties, which often require expensive materials and limit their application to a laboratory scale. In this study, we attempted to create a low-cost transparent superhydrophobic coating using a combination of silica nanoparticles and PMMA resin. SiO_2 nanoparticles were produced using the sol–gel technique and functionalized using MTES to lower their surface energy. The hydrophobic nanoparticles were mixed with a PMMA solution and deposited onto a glass substrate using a centrifuge and immersion method. The resulting coating had a transmittance of over 79% in the visible spectrum, a contact angle of over 150°, a slip angle of 2°, and advanced self-cleaning capabilities. The coating also demonstrated strong thermal stability and UV resistance. PMMA's low cost makes it a potential candidate for producing transparent superhydrophobic coatings with other polymers in perspective.

Keywords Superhydrophobicity · Self-cleaning · Transparency · Spin-coating

1 Introduction

In recent years, the importance of superhydrophobic surfaces has greatly increased due to their interesting properties such as anti-icing [1, 2], water/oil separation [3, 4], anti-biofouling [5], corrosion resistance [6, 7], reduction of resistance to drag[8], dirt-repellent textiles [9, 10], and above all, self-cleaning [11, 12]. In general, superhydrophobic surfaces are defined as anti-wetting surfaces that exhibit a contact angle >150° with a slip angle of 10° to provide the surface with self-cleaning properties.

B. Nomeir · S. Boukheir (✉) · S. Lakhouil · S. Naamane
Moroccan Foundation for Advanced Science Innovation and Research, Mohammed VI Polytechnic University (UM6P), Benguerir, Morocco
e-mail: s.boukheir@mascir.ma

B. Nomeir · M. Aitali
Faculty of Sciences Semlalia, Molecular Chemistry Laboratory, Unit Coordination and Catalysis Chemistry, Cadi Ayyad University, Marrakech, Morocco

A. Vaseashta et al. (eds.), *Proceedings of the Eighth International Symposium on Dielectric Materials and Applications (ISyDMA'8)*, Springer Proceedings in Physics 349, https://doi.org/10.1007/978-3-031-84809-4_7

To obtain a surface with these two characteristics, there are two key conditions to achieve:

- Hierarchical morphology with a rough enough nanostructure capable of trapping the air,
- Low surface energy that prevents the formation of bonds between the drop of water and the surface to allow the drop of water to adopt a spherical shape, which facilitates its sliding on the surface.

In some applications, such as the field of solar energy, the glazing of buildings, or the manufacture of optical devices, the application of a self-cleaning coating requires transparency as an essential parameter with superhydrophobicity to be accepted in the industrial scale. Superhydrophobicity and transparency are two competing properties, but the increase in roughness that generates superhydrophobicity leads to a decrease in transparency and vice versa [14], which requires an in-depth study to find a compromise equilibrium point that ensures the two properties at the same time. In general, a surface's roughness must be less than 100 nm to make it extremely translucent to visible light. In this respect, a variety of techniques, including spin coating [15, 16], dip coating [17], chemical vapor deposition [18], spray coating [19], photolithography [20], the sol-gel method [21, 22], and electrospray [23], have been used to create transparent superhydrophobic surfaces. For example, Bui et al. [24] prepared a flexible polyurethane acrylate (PUA) film by the photolithography method; the surface showed a contact angle of 146° with a 3–5% decrease in transmittance compared to uncoated glass. Yoon et al. [25] prepared a transparent anti-fog superhydrophobic coating even. The design of this coating was carried out using the lithography technique. The coating showed a higher transmittance of 90% and a contact angle of 162°. Chen et al. [26], prepared a superhydrophobic coating based on hydrophobic silicon dioxide (SiO_2) via the sol–gel method, the coating showed a transparency of 97.8%. Liu et al. [27] used the chemical vapor deposition technique to prepare a superhydrophobic coating with a contact angle of 170° and a transmittance of 90%. Nevertheless, despite the variety of methods available for creating superhydrophobic surfaces, the majority of these techniques require time-consuming processes, complicated synthesis protocols, and expensive materials, such as Polydimethylsiloxane (PDMS), polyurethane (PU), and polystyrene (PS), to ensure improved transparency, which makes the application of these types of coatings unprofitable and limited to a laboratory scale. Poly(methyl methacrylate) (PMMA) is a highly transparent, hydrophilic polymer having excellent UV protection, fascinating mechanical, and thermal stability, being reasonably priced, and functioning with different nanobatteries. PMMA has only recently been recognized by scholars as a possible substitute for more expensive polymers because of these features. Sutar et al. [28] prepared a superhydrophobic coating based on PMMA and SiO_2 using the dip-coating technique. The surface presented a contact angle of 167° with acceptable mechanical stability. Sanjay et al. [29] also prepared a superhydrophobic surface based on PMMA and SiO_2 with the same deposition technique, the obtained surface showed a contact angle of 159° with a low slip angle. On the other, Yilgör et al. [30] also combined PMMA and silica nanoparticles for preparing a superhydrophobic covering using the spin-coating method. They applied the covering in up to five layers to achieve a high contact angle that can reach 163°. Meena et al. [31] also prepared a superhydrophobic coating based on PMMA and SiO_2 using spin-coating; the coating presented a contact angle greater than 160° with very interesting thermal and mechanical stability. Pan et al. [32] also reported a simple method using the spray-coating technique to prepare a superhydrophobic coating using SiO_2 in combination with PMMA. Wong et al. [33] also prepared a superhydrophobic coating based on PMMA and SiO_2 using spin-coating with a contact angle greater than 160° and very interesting thermal and mechanical stability. However, despite the successful preparation of superhydrophobic coatings using PMMA as a hydrophilic resin, they failed to retain transparency. To address the issue of transparency, we present a straightforward and economically viable industrial process for producing a superhydrophobic nanocomposite coating made of PMMA and SiO_2, with high-light transmission. This study compares and describes the two deposition methods used to prepare the PMMA-based coating: spin coating and dip coating. The research focuses on the coating's wettability, surface morphology, transmittance, and stability under challenging conditions. Furthermore, the coating underwent self-cleaning experiments and anti-icing tests to evaluate its effectiveness.

2 Experimental

2.1 Materials

Glass substrate of 3 cm thickness purchased from a local store, Ammonia (NH4OH), Methyltriethoxysilane (MTES), absolute Ethanol (EtOH), and Tetraethylorthosilicate (TEOS), Poly(methyl methacrylate) (PMMA), Toluene, Propanol, Tetrahydrfurane (THF) and chloroform. All the chemicals were provided by Sigma Aldrich and were used as received without further purification.

2.2 Preparation and Hydrophobic Modification of SiO_2 Nanoparticles

In a flask, 5 ml of TEOS was dissolved in 38.53 ml of ethanol mixed with 0.4 ml of water, and then 3.15 ml of NH_4OH was subsequently added to the reaction medium. The mixture was stirred for 1 h at 60 °C. The solution was degassed for 5 min to replace oxygen with nitrogen, and then MTES was added dropwise under nitrogen to the mixed solution and left to stir overnight at 25 °C. The product was then centrifuged, and washed with ethanol at least three times to eliminate the residuals and the excess of silane molecules. Then the supernatant was eliminated, and the resulting white powder was hydrophobic nano-SiO_2 was dried at 80 °C (Fig. 1).

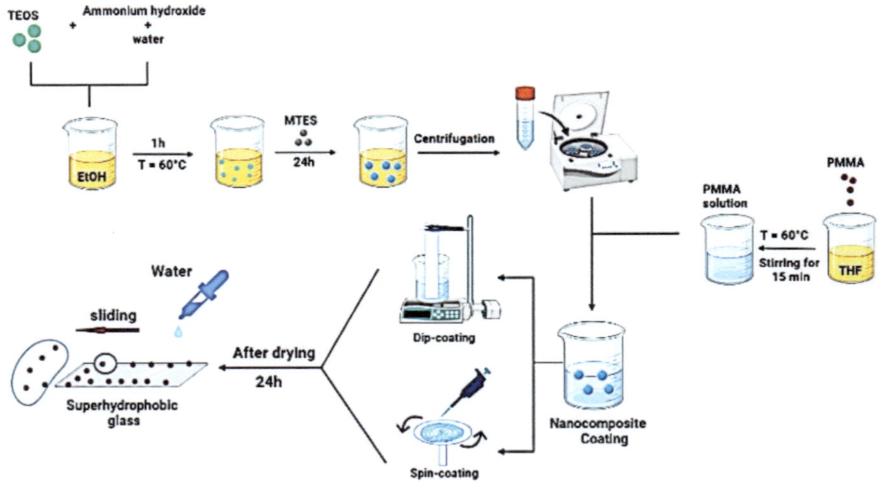

Fig. 1 Schematic illustration of the fabrication process of superhydrophobic coatings based on PMMA and SiO_2 hydrophobic

2.3 *Preparation of Superhydrophobic Surfaces*

The glass surfaces were first ultrasonically cleaned, in ethanol for 20 min, washed with purified water, and rinsed with acetone to eliminate any impurities. To prepare the coating solution, 0.2 g of PMMA was dissolved in 10 ml of THF at 60 °C, resulting in a clear, viscous solution. To enhance bonding between the PMMA resin and the nano-SiO_2, 0.6 g of hydrophobic SiO_2 was added to the solution, which was then stirred for an additional hour at 60 °C. The resulting suspension was sonicated for 60 min to create homogeneous dispersions and then spun-coated onto glass slides. This process was repeated several times, depending on the desired number of layers. The coated surfaces were dried overnight at 80 °C.

For deposition by immersion, a 0.5% by weight solution of PMMA was prepared in THF under the same conditions as the first formulation. Next, 1.4 wt.% of hydrophobic silica nanoparticles were dispersed in the PMMA solution, which was then stirred and sonicated for one hour. The glass slides were then immersed for one minute for several cycles, depending on the desired number of layers. The coated substrates were also dried at 80 °C for 24 h. (Refer to Fig. 1 for a visual representation of the process).

2.4 *Characterizations*

Chemical analysis of nanoparticle functionalization was studied by total analysis attenuated. Fourier transform infrared spectroscopy (ATR-FTIR) samples were examined using a Nicolet iS5 FTIR spectrometer. The thermogravimetric analysis (TGA) of the sample was measured by a thermogravimeter (NETZSCH TG209F1, Germany) to study the thermal stability and the grafting of MTES and PMMA on nano-SiO_2 under nitrogen from 25 to 700 °C with a heating rate of 10 °C min1. In addition, surface morphologies were observed using a scanning electron microscope (SEM, SUPRA55) at an accelerating voltage of 5.0 kV for SEM analysis. Samples are coated with a thin layer of gold before testing. The profilometry (Bruker XP-1 Model, Germany) was used to measure coating thickness. The surface transmittance was studied using a Cary 5000 UV-Vis spectrophotometer in the visible range. The distribution and size of SiO_2 nanoparticles were measured with the Zetasizer Nano ZS ZEN 3600. The contact angle and the slip angle of the superhydrophobic surfaces were measured by a goniometer with 1 L of deionized water droplets at room temperature extruded from a needle. If the water drop cannot leave the needle to adhere to the surface, the final average WCA and SA values were obtained by measuring at least 5 different points for each surface. To describe three-dimensional film imaging and measure surface roughness, atomic force microscopy (AFM) was used in non-contact mode using Veeco Dimension icon instrument, and AFM image analysis was performed by NanoScope-Analysis-v110R1 software.

3 Results and Discussions

3.1 Characterization of Hydrophobic SiO_2 and Nanocomposite Coating

The size and distribution of silica particles are widely recognized as critical factors in controlling both superhydrophobicity and coating transparency. Reducing the size of the nanoparticles results in increased surface transmittance, while larger particles have the opposite effect. To achieve a superhydrophobic surface with optimal transparency, nanoparticles smaller than 100 nm are recommended. In this study, strict control was exercised over several parameters, including temperature, preparation method, NH_4OH content, water temperature, and reaction time for SiO_2 nanoparticle synthesis.

SEM analysis (Fig. 2a) demonstrates that the optimized conditions yield non-condensed spheres with diameters ranging between 60 and 70 nm. The monodispersity of SiO_2 nanoparticles significantly influences the formation of superhydrophobic surfaces with nanoscale roughness. Additionally, the particle size distribution of silica particles (Fig. 2b), determined by a zeta-sizer, exhibits a reasonably uniform population with an average diameter of 69 nm. A comparison was made between the SiO_2 powder before and after functionalization using MTES. It was observed that when a drop of water was placed on the non-functionalized powder, the drop spread completely due to the presence of hydroxyl groups on the nano-SiO_2 surface. These groups could interact with the water droplets, leading to their propagation on the powder (Fig. 2c). In contrast, after the functionalization step, a drop deposited on the surface assumed a spherical shape, exhibiting a contact angle greater than 150°, and could freely slide on the powder (Fig. 2d). This visual observation confirms the successful execution of the functionalization process. To further validate this, the chemical composition of the nano-SiO_2 surface before and after functionalization was analyzed using FT-IR, as depicted in Fig. 2e. Both spectra of the two samples exhibit characteristic absorption peaks at 1100, 470, and 781 cm^{-1}, corresponding to the symmetrical stretching and bending vibrations of Si–O–Si. These peaks confirm the formation of a three-dimensional network of SiO_2. Additionally, an absorption peak at 950 cm^{-1} and a broad absorption peak at 3394 cm^{-1} indicate the stretching vibrations of -OH and Si–OH, respectively. However, after modifying the SiO_2 surface with MTES, the absorption of -OH and Si–OH stretching vibrations weakened, accompanied by the emergence of two new peaks in the modified curve located at 2975 and 1275 cm^{-1}. These peaks correspond to the vibrations of C-H and Si-CH_3 bonds, respectively. These findings provide preliminary evidence that the CH_3 groups partially replaced the OH groups on the surface of the silica nanoparticle through grafting. This replacement of Si-OH by Si-CH_3 rendered the nanoscale SiO_2 hydrophobic, elucidating the contrasting behavior observed between the two samples when exposed to a water droplet.

Furthermore, TGA analyses were conducted on the SiO_2 samples before and after functionalization to validate the grafting of methyl groups and assess their quantity. The results are illustrated in Fig. 2f. For the unmodified nanoparticles,

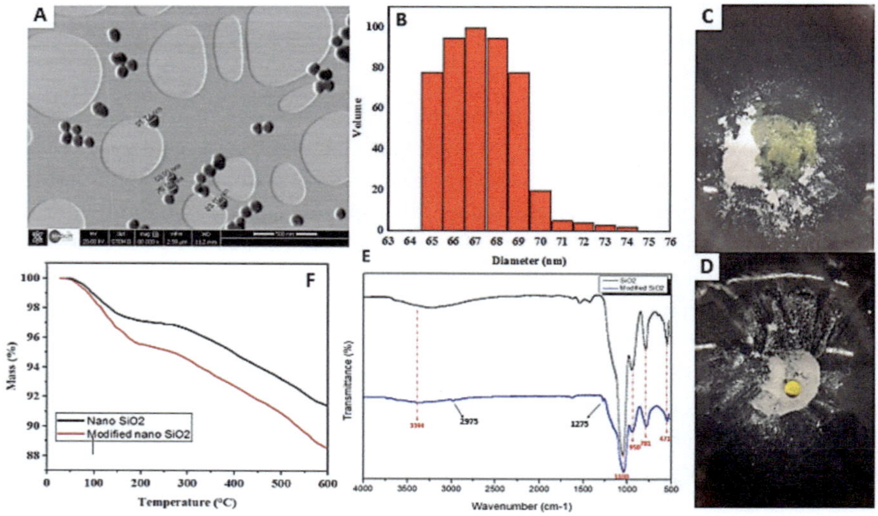

Fig. 2 (**a**) SEM image of prepared SiO_2 nanoparticles. (**b**) The size distribution of silica particles. (**c**) Image of a drop of water deposited on the SiO_2 powder before functionalization. (**d**) Image of a drop of water deposited on the SiO_2 powder after functionalization. (**e**) FT-IR infrared spectra of the nanoparticles before and after modification. (**f**) TGA curve of SiO_2 nanoparticles before and after modification

a weight loss was observed within the temperature range of 100–200 °C, corresponding to the volatilization of water absorbed by the hydroxyl groups present on the surface of the nano-SiO_2. Between 200 and 300 °C, the nanoparticles exhibited stability, but a subsequent weight loss occurred when the temperature surpassed 300 °C, attributed to the loss of hydration resulting from the condensation reaction between the OH groups of SiO_2 nanoparticles. In contrast, after the functionalization of nano-SiO_2 with MTES, the thermal weight loss between 100 and 200 °C became more pronounced compared to that of the unmodified nanoparticles. This can be attributed to the decomposition of the CH_3 groups, and the volatilization of water absorbed on the surface of the modified hydroxyl groups. As the temperature exceeded 200 °C up to 600 °C, the thermal decomposition tendencies of both the modified and unmodified samples were nearly identical. At 600 °C, the weight loss of the modified and unmodified SiO_2 samples was approximately 12% and 8%, respectively, indicating an approximate 4% incorporation of MTES on the SiO_2 surface. Based on the percentage of residual weight at 600 °C for the modified and unmodified SiO_2, although the thermal stability of the modified nano-SiO_2 is lower than that of the unmodified nano-SiO_2, it still satisfies the requirements for the preparation of superhydrophobic coatings.

3.2 *Superhydrophobic Surfaces Prepared by Spin-Coating*

3.2.1 Coating Wettability Study

In order to fabricate a high-quality superhydrophobic nanocomposite coating, it is crucial to achieve a balance between the resin and nanoparticles. The nanoparticles should contribute to the formation of a hierarchical surface with sufficient roughness to repel water droplets, while the resin should provide the necessary adhesion to enhance the durability of the coating. In this study, superhydrophobic films were developed by incorporating MTES-modified nano SiO_2 into PMMA films using the spin coating technique. It should be noted that low concentrations of nanoparticles, such as 0.7, 1, and 1.5%, failed to attain a contact angle exceeding 150°. This can be attributed to the challenges associated with achieving proper adherence of the PMMA/SiO_2 combination on the glass surface, which often resulted in uncovered areas where water droplets could accumulate. Consequently, a slightly higher nanoparticle concentration was required.

The effects of several factors, including the number of deposited layers, resin concentration, deposited water pH, and spin-coating rate, on surface wettability were investigated and presented in Fig. 3. The results demonstrated that increasing the number of layers led to an augmented contact angle and reduced sliding angle. For instance, after depositing a single layer of the coating, the contact angle increased from 40° to 130°, with complete absence of sliding even at a substrate inclination angle of 90° (Fig. 3a). This phenomenon can be attributed to the insufficient amount of deposited nanoparticles, which failed to cover the entire surface and left uncovered areas prone to interactions with water droplets, hindering sliding. Additionally, a single layer deposition was inadequate to establish an effective hierarchical and rough structure. As more layers were deposited, the hydrophobic nanoparticles progressively covered the surface, resulting in the gradual formation of a rough structure. This led to an increase in the contact angle with water. A superhydrophobic surface was achieved only after the deposition of five layers of the coating. Following the deposition of five layers, the surface contact angle reached 154°, accompanied by a sliding angle of 6°, enabling water droplets to freely roll on the surface (refer to Fig. 4).

Furthermore, the 5-layers coating not only achieved superhydrophobicity with water at neutral pH but also demonstrated superhydrophobic properties with solutions at different pH levels (Fig. 3b). For solutions with pH values ranging from 4 to 13, the surface exhibited a contact angle greater than 150° and a sliding angle below 10°. Only the solution with a pH of 2 resulted in a contact angle of 144° with a sliding angle of 20°. These findings are of great significance in validating the efficacy of the coatings for specific applications, such as in solar energy fields where the coating may come

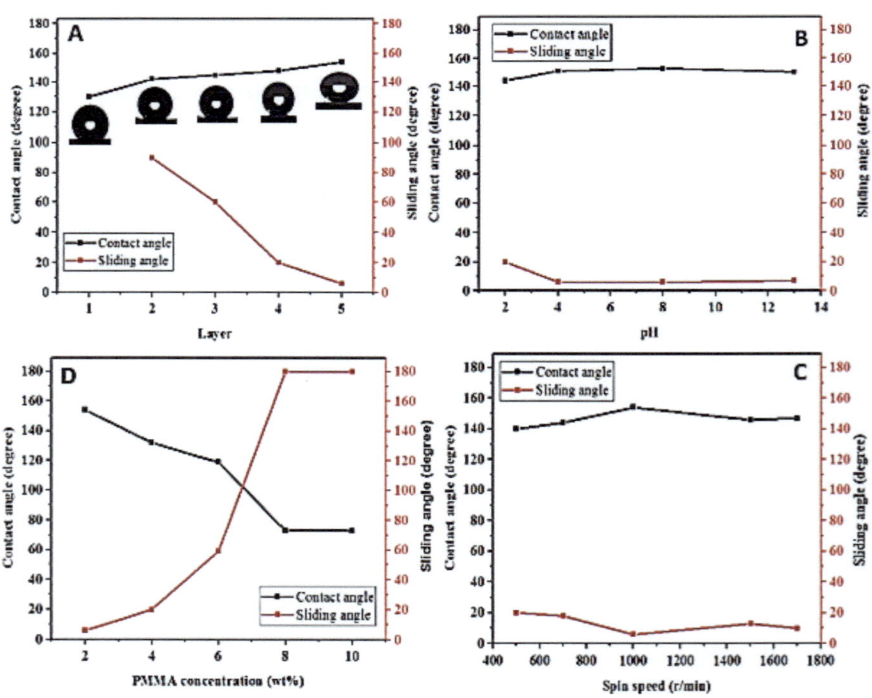

Fig. 3 (**a**) The variation of the CA and SA of the surface as a function of the number of layers of coating deposited. (**b**) Relationship between the pH of the water droplet and the AC and SA of the prepared superhydrophobic surface. (**c**) The effect of rotational speed on the contact angle and the slip angle of the prepared superhydrophobic surface. (**d**) The variation of the CA and SA of the surface as a function of PMMA concentration

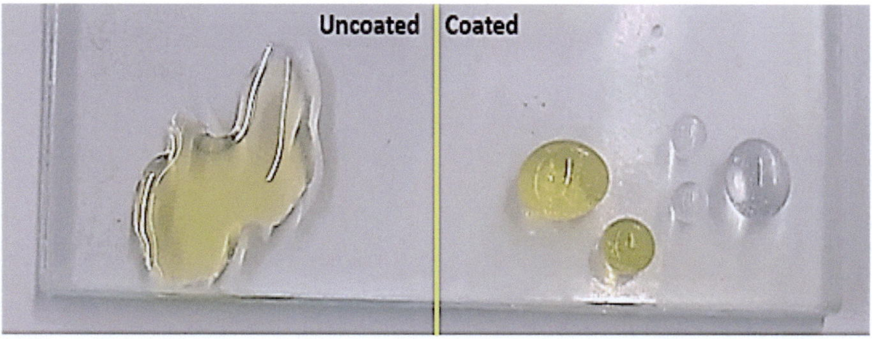

Fig. 4 Water droplets on uncoated glass surface and a coated surface (5 layers)

into contact with acidic or basic rainwater. To the best of our knowledge, transparent films based on PMMA prepared using the spin coating method have not previously been reported to exhibit superhydrophobic properties across various pH ranges. The rotational speed is a well-known influential parameter for controlling the thickness of the deposited coating and, consequently, its superhydrophobic properties. The observed effect of the rotational speed used for depositing the 5 layers of the coating on the surface contact angle is depicted in Fig. 3c. At low rotational speeds, the contact angle did not exceed 150°, and the same trend was observed when increasing the rotational speed. Only the intermediate speed of 1000 rpm yielded the desired outcome. This can be attributed to the fact that the rotational speed needs to be sufficiently high to reduce nano-agglomeration of SiO_2 particles without being too fast, which would result in fewer particles covering the surface. It should also be emphasized that the quantity of resin utilized plays a crucial role in controlling the surface properties, particularly in terms of coating durability. Increasing the resin content can enhance the fixation of nanoparticles onto the substrate. However, it is essential to avoid exceeding a certain concentration to maintain superhydrophobicity. In our case, we found that the optimal concentration to achieve a balance between superhydrophobicity and good adhesion of the coating is 2% wt. of PMMA resin. Below 2%, the coating is easily removed from the substrate due to insufficient resin to anchor the nanoparticles. On the other hand, exceeding 2% results in a reduction of the contact angle and an increase in the sliding angle. Changing the PMMA resin content from 2 to 8% by weight decreased the contact angle from 154° to 73°. This can be attributed to two factors. (a): The hydrophilicity of PMMA increases with the resin content, leading to a higher surface energy and a consequent decrease in the contact angle. (b): An increase in resin content affects the surface roughness by causing the nanoparticles to become immersed in the resin, thereby compromising the hierarchical nanostructure of the surface (Fig. 5).

Morphological analysis. Figure 6 showcases SEM images of the coatings prepared with varying numbers of layers, ranging from 1 to 5. The coating deposited in a single layer exhibited limited surface coverage and non-uniform deposition, resulting in interconnections between porous structures. This inadequate single-layer

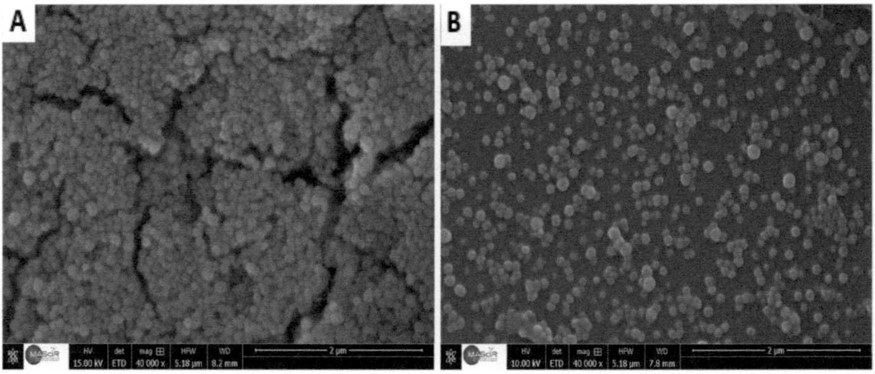

Fig. 5 (**a**) Hierarchical morphology of surfaces after depositing 5 layers of the coating with a concentration of 2% of PMMA. (**b**) Morphology of the surfaces after depositing 5 layers of the coating with a concentration of 8% wt of PMMA

deposition failed to achieve a superhydrophobic surface. With the deposition of a second layer, the coating began to spread further across the substrate, filling the voids between surfaces. The third layer provided nearly complete coverage with hydrophobic nanoparticles, but the structure appeared flat, lacking the required high roughness for superhydrophobicity. The fourth layer initiated the formation of a porous structure on the surface, capable of trapping water in its nanostructures. However, the height of these nanopores was insufficient to repel water droplets effectively. By depositing the fifth layer, the pores became more pronounced, resulting in a rough, porous, interconnected solid layer with numerous voids. This transformation enabled the transition from Wenzel to Cassie Baxter state, crucial for achieving superhydrophobicity. To further analyze the impact of the number of layers on surface properties, AFM was employed to examine the surface topography of the PMMA/SiO_2-based coating, as shown in Fig. 7. The AFM images revealed that the single-layer deposit displayed a relatively smooth surface with scattered small bumps, unevenly distributed across the surface, possibly indicating SiO_2 nanoparticle agglomerates that remained after the initial layer deposition. This observation aligns with the SEM results. Upon adding the second layer, new bumps emerged throughout the surface, although lighter compared to the single-layer deposition. The deposition of the third and fourth layers rendered a well-defined hierarchical surface, characterized by protrusions spread across the entire surface. With the fifth layer deposition, the variable protrusions became more prominent, effectively trapping the area between their microstructures and resulting in a superhydrophobic surface. Additionally, the roughness values presented in Fig. 8 demonstrate that the deposition of 5 layers of the coating increased the mean surface roughness (Ra) to 62.1 nm compared to 15.7 nm for the single-layer deposition. This confirms that the PMMA-SiO_2 coating achieved a nanoscale roughness of <100 nm, a highly recommended requirement for preparing a transparent superhydrophobic surface. Figure 8b depicts the thickness variation of the coating as a function of the number of layers, measured by profilometry. The coating thickness increases proportionally with the number of layers. The surface exhibiting superhydrophobicity demonstrated a coating thickness of 2.1 μm, consistent with the thickness of 2 μm obtained from the SEM analysis (refer to Fig. 7).

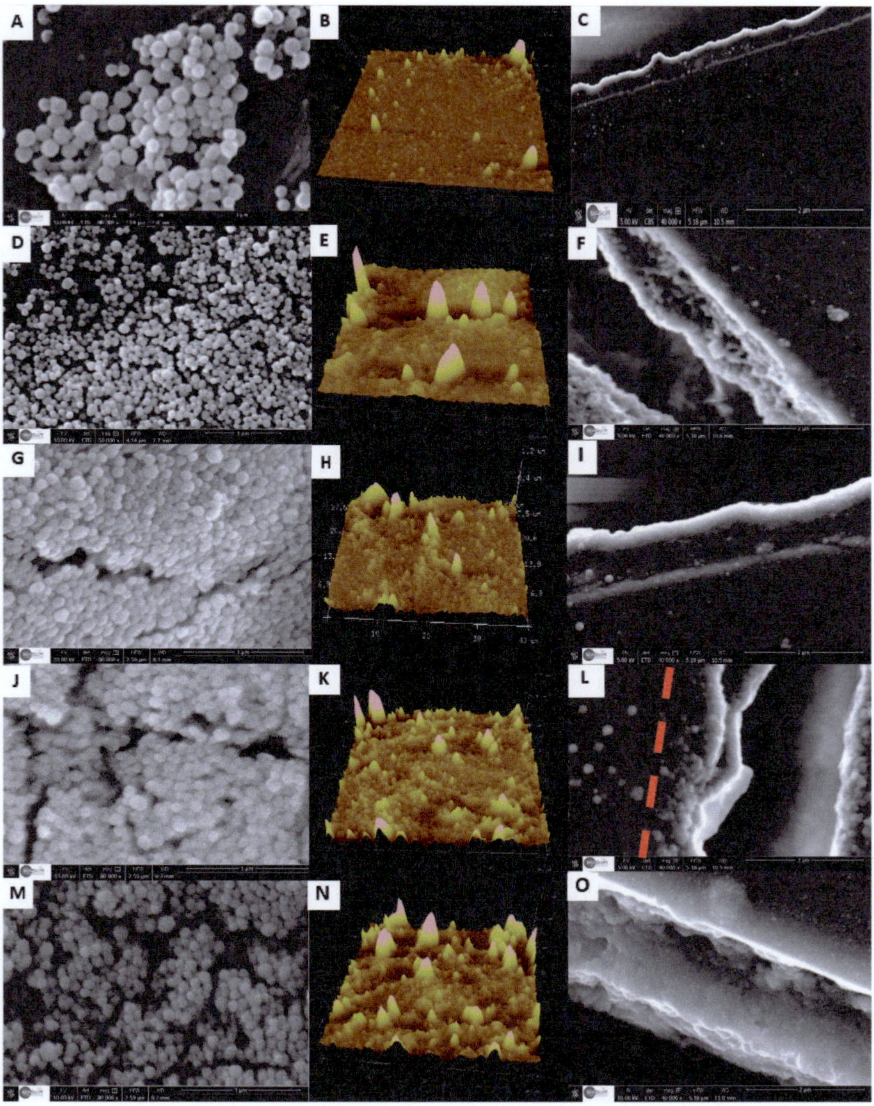

Fig. 6 Surface morphology by SEM, 3D morphology by AFM, and coating thickness by SEM: (**a**, **b**, **c**) one layer—(**d**, **e**, **f**) two layers—(**g**, **h**, **i**) three layers—(**j**, **k**, **l**) four layers—(**m**, **n**, **o**) five layers

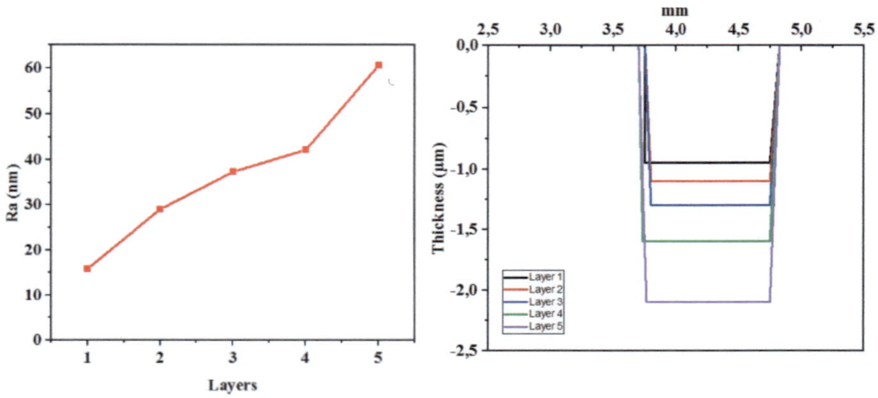

Fig. 7 (**a**) Roughness values Ra as a function of a number of layers. (**b**) The thickness of the coatings according to the number of layers deposited measured by profilometry

However, the phenomenon of superhydrophobicity is not solely determined by the thickness and roughness values, but also by the profile and structure of the texture peaks, their heights, and the distance between the peaks. It should be noted that smaller peak heights and wider spacing between the peaks facilitate water droplet penetration into the grooves and stronger adhesion to the surface. Therefore, a detailed analysis of the 2D profiles was conducted using AFM for the surfaces under investigation, as illustrated in Fig. 8. Based on the obtained profiles, it is evident that the surface exhibiting superhydrophobicity possesses closely spaced roughness peaks, similar to those observed on the other surface coatings. However, in terms of peak heights, the increase in the number of layers results in the creation of significantly deeper pores with a height of 300 nm, enabling the entrapment of an ample amount of air to repel water droplets, in contrast to the other surfaces. The peak heights on these surfaces do not exceed 120 nm, which elucidates the observed contact angle values.

3.3 Self-Cleaning Property

The self-cleaning capacity of our coating was tested using three different methods on both bare and coated glass substrates. In the first test, both substrates were exposed to the exterior at an inclination angle of almost 60° for 8 h. Figure 9a clearly shows that the coated substrate remained almost perfectly clean, while a large number of dust particles were deposited on the bare glass substrate. This indicates that the coating prevented dust deposition on the substrate. The replacement of hydroxyl groups with methyl groups in the coating prevented the formation of bonds with dust particles, reducing their deposition on the surface. This property is particularly useful in applications such as solar energy, where dust deposition on PV panels can reduce

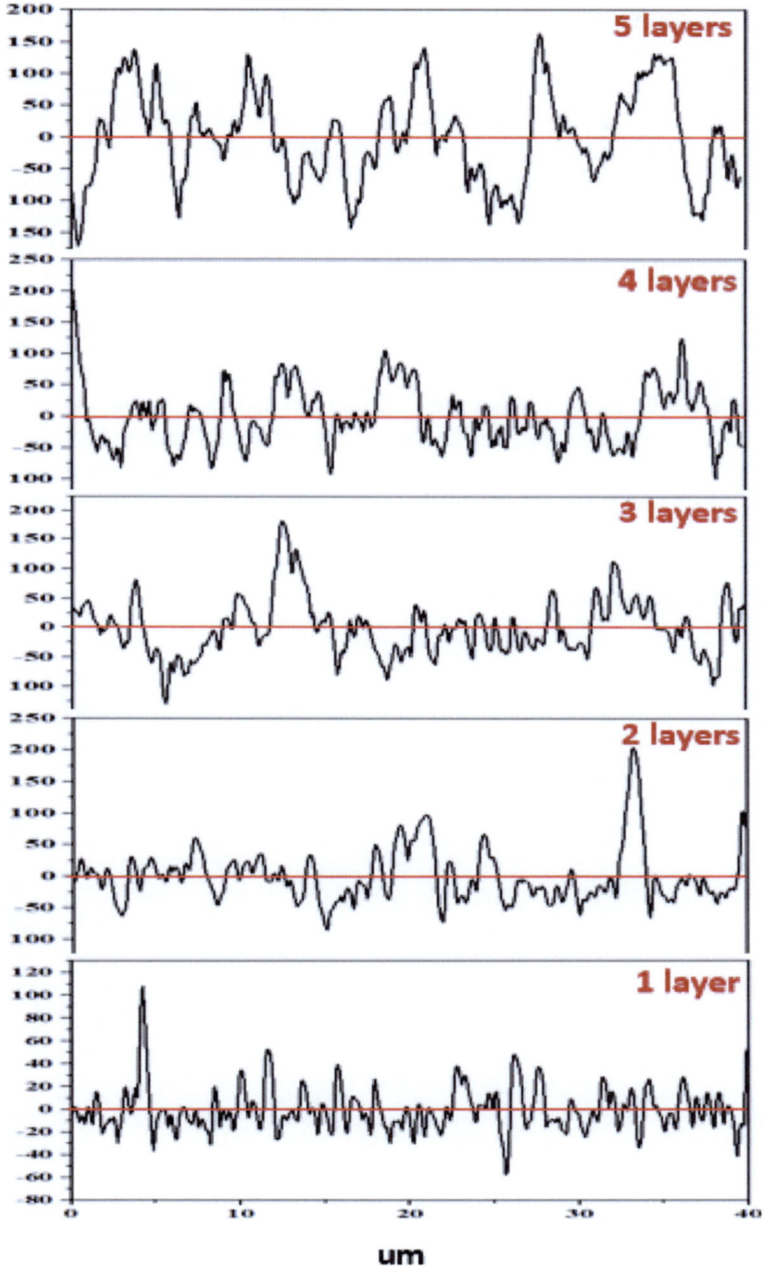

Fig. 8 2D roughness profiles of the glass surface coated with $PMMA/SiO_2$-based coating

their performance by up to 40%. In another self-cleaning test, sand was spread on both surfaces, and water droplets were drained. Figure 9b shows that when the water drop fell on the uncoated surface, it spread on the surface due to its hydrophilicity without any cleaning effect. However, when the drop was deposited on the coated surface, it assumed a spherical shape and quickly slid down the surface by gravity, bringing back with it the dust particles and leaving a clean path. This demonstrates the coating's ability to facilitate the self-cleaning process. We deposited 10 mg of dust on the coated substrate and then deposited a single drop on the surface. After weighing the remaining dust, we found that a single drop of water could eliminate 5.27 mg of dust due to the self-cleaning properties of the coating. This can be explained by the Cassie-Baxter model, where the hierarchical nanostructure of the surface reduces the contact area between the surface and the water droplet, allowing it to roll freely on the surface. Figure 9c shows another self-cleaning test that was applied to our coating, where both substrates were immersed for one minute in a beaker filled with dirty water. After drying, it was noticed that a film of dust had formed on the uncoated glass surface, which was not easily removed even with strong finger rubbing. However, the coated surface remained clean, with no film of dirt appearing, due to the coating's anti-wetting properties, which prevented salt water from adhering to the substrate even when immersed in dirty water.

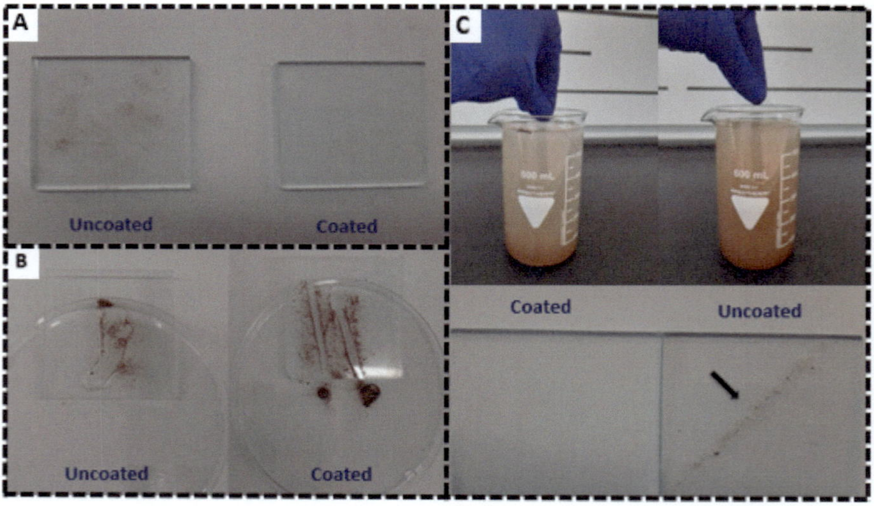

Fig. 9 (**a**) Coated and uncoated substrates after their exposure for 8 h outside. (**b**) Cleaning process of a coated surface and an uncoated surface. (**c**) Immersion of coated and uncoated substrate in salt water

3.4 Anti-Icing Application

To assess the performance of our superhydrophobic coating in preventing the formation of ice, we conducted an experiment in which a 3 ml water droplet was deposited on both the coated and uncoated glass substrates under similar conditions at a temperature of 25 °C. The substrates were then placed horizontally in a freezer at −1 °C for 15 min. After recovery, we captured images of the final shape of the water droplets on both types of substrates using a contact angle camera, as shown in Fig. 10. Our findings showed that the water droplet deposited on the uncoated glass substrate was completely frozen after 15 min at −1 °C, as evidenced by the total disappearance of the clear zone in the middle of the droplet. However, the water droplet deposited on the coated surface had a reduced contact angle after 15 min at −1 °C, but the clear zone in the middle was still observable, indicating that the coating delayed the formation of ice significantly. This delay in ice formation on the coated surface can be attributed to the low contact zone between the water droplet and the superhydrophobic surface, which has a hierarchical morphology. The contact between the deposited droplet and this surface is primarily on the heads of the bumps, while the remaining contact zone is between the droplet and the area trapped between the nanostructures. This area is a good thermal insulator, which slows down the energy exchange between the coated surface and the water droplet, making the heat transfer via the contact between the droplet and the superhydrophobic surface less efficient than that between the droplet and the uncoated surface.

3.5 Transparency Study

Transparency is a critical property for many applications, including window glass and solar cells. However, to achieve both superhydrophobicity and transparency, it is crucial to control and optimize the surface roughness. Generally, the roughness must be lower than 100 nm, which is less than $\lambda/4$ of visible light, to minimize Mie scattering, as was achieved in our coating. Figure 11a, b depict the transparency of the substrates after coating deposition by spin-coating and dip-coating, respectively. The coating does not influence the substrate behind it, as can be seen in Fig. 11c. This figure displays the percentage of optical transmission of the two coated substrates in the visible range (330–800 nm). The surface prepared by spin coating exhibited a transmittance between 74 and 79% in the visible range, while the surface prepared by dip coating showed a transmittance between 72 and 75%. The small shift in the transmittance can be attributed to the effect of roughness, which is well-known to decrease transmittance with an increase in roughness. The surface prepared by spin coating exhibited lower roughness than the surface prepared by dip coating, leading to the transmission of the surface prepared by spin coating to be superior to that of the surface prepared by dip coating.

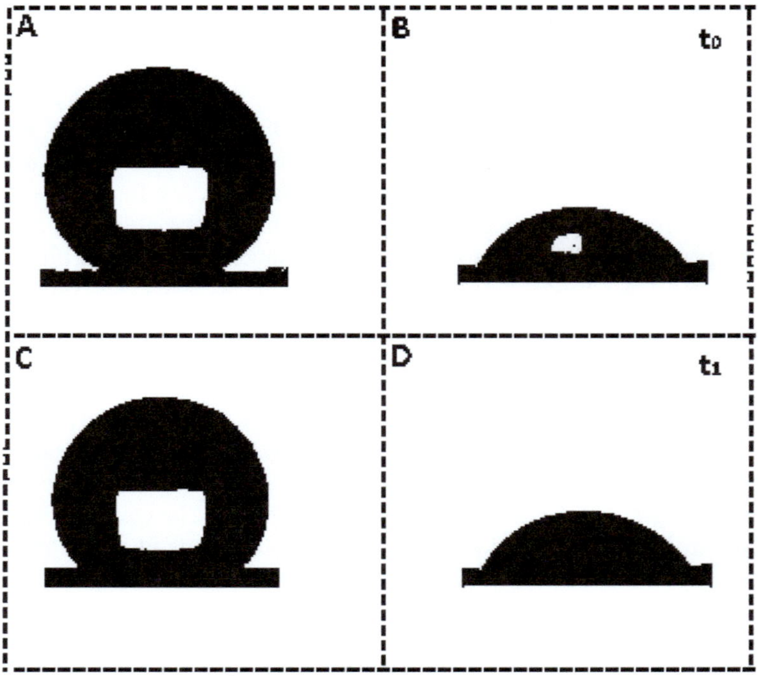

Fig. 10 Water drops freezing on coated glass (left) and uncoated coated glass (right). (**a** and **b**) Water drops were placed on surfaces at time t0 at 25 °C. (**c**) The drop of water spreads a little on the coated surface at −1 °C after t1 = 15 min. (**d**) Ice formation starts at −1°C on the uncoated surface after t1 = 15 min

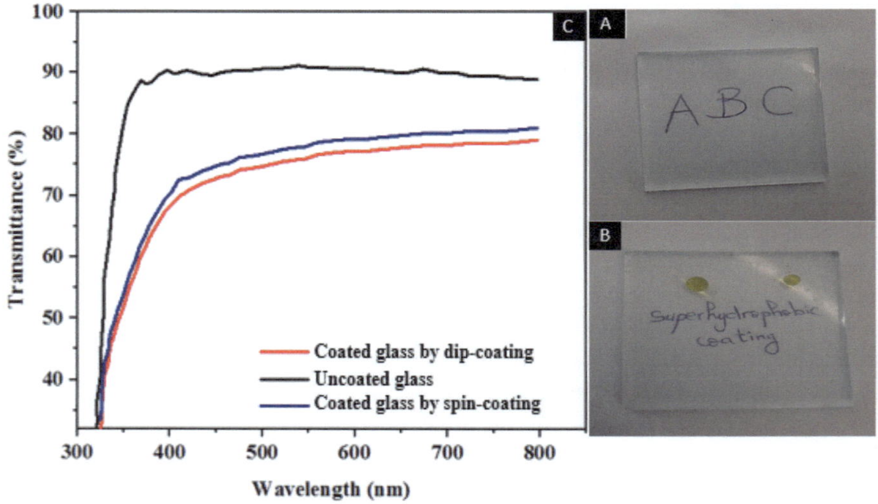

Fig. 11 (**a**) Transparent substrate of superhydrophobic glass prepared by spin-coating. (**b**) Superhydrophobic glass substrate prepared by dip-coating. (**c**) Coating transmittance spectrum

3.6 Durability Test

To validate the successful application of superhydrophobic surfaces in demanding conditions, it is crucial to assess the thermal stability and properties of the coatings. The thermal stability of superhydrophobic coatings was evaluated using the annealing method. Coated glass samples were subjected to different temperatures ranging from 0 to 280 °C for 1 h in an oven. After annealing, the samples were allowed to cool at room temperature for 1 h, and the contact angles and slip angles were measured. Figure 12 illustrates the contact angles obtained for the spin-coating and dip-coating samples at various annealing temperatures. It was observed that the contact angle (CA) and slip angle (SA) remained unchanged up to an annealing temperature of 200 °C for both samples. Even at this high temperature, the transparency of the coatings remained unaffected, and the water droplets maintained a spherical shape, exhibiting free rolling on the surface. This preservation of superhydrophobic behavior is promising for self-cleaning applications. However, at a temperature of 250 °C, the samples lost their superhydrophobic characteristics. Notably, the reduction in contact angle was more pronounced for the spin-coated surface compared to the dip-coated surface. The spin-coated surface became hydrophilic with a contact angle of 60°, whereas the dip-coated surface retained hydrophobicity with an angle of 113°. To understand this behavior, an energy-dispersive X-ray (EDX) analysis was conducted to examine any compositional changes in the surface at 250 °C. Table 1 presents the data, indicating a progressive decrease in the mass percentage of carbon at 250 °C. This reduction can be attributed to the degradation of the carbon chains in the PMMA component. Furthermore, SEM morphology analysis revealed the appearance of large, uncovered areas at 250 °C (Fig. 13), confirming the degradation of the PMMA chains that hold the hydrophobic SiO_2 nanoparticles. Consequently, the release of nanoparticles resulted in exposed areas, reducing surface roughness and the contact angle.

To assess the UV resistance of our coating, we subjected the coated surfaces to UV radiation for over 24 h. Results indicated that both the contact angle and slip angle remained unchanged (as depicted in Fig. 14). Furthermore, we conducted a long-term

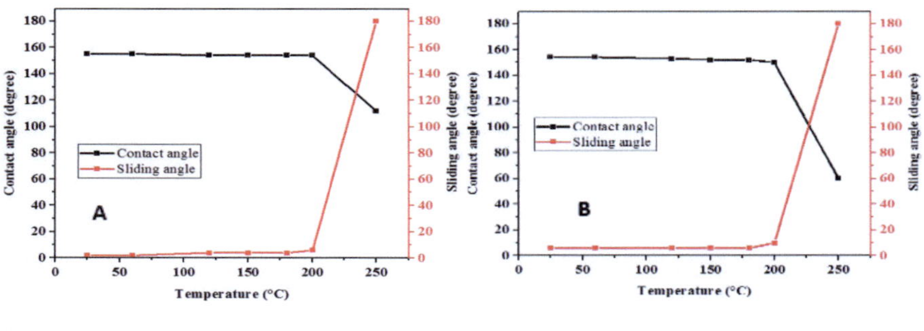

Fig. 12 WCA and SA of the two surfaces after calcination at different temperatures: (**a**) prepared by dip-coating and (**b**) prepared by spin coating

Table 1 Analysis by energy dispersive X-rays (EDX) of the surface coated by spin coating

Elements	Before the thermal stability test (atomic %)	After the thermal stability test (atomic %)
Carbon	39,86	17,32
Oxygen	38,44	30,81
Silicon	21,7	51,87

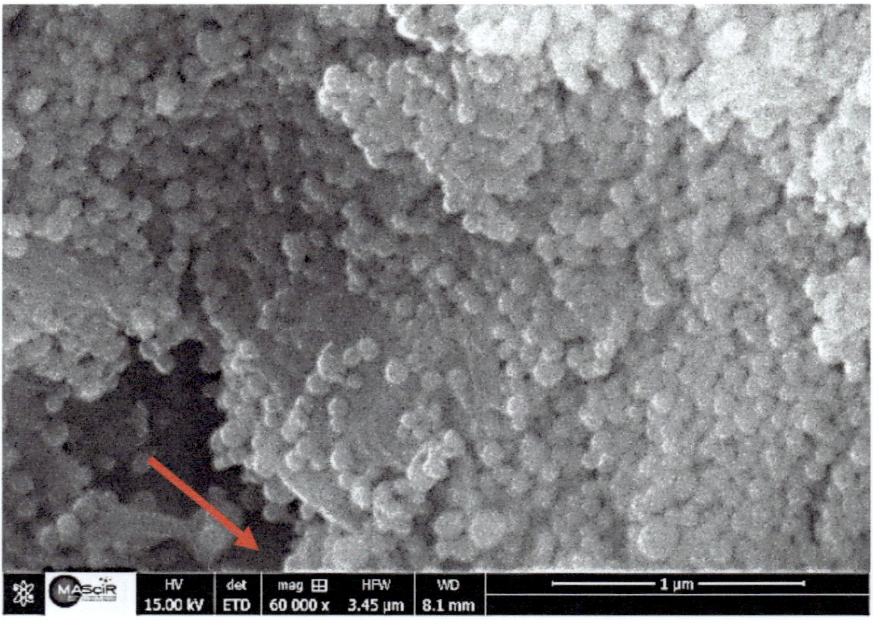

Fig. 13 SEM image of the glass surface coated by spin coating after calcination at 250 °C

study of the coated surfaces under ambient conditions for a duration exceeding one month and found that there was no impact on their superhydrophobicity.

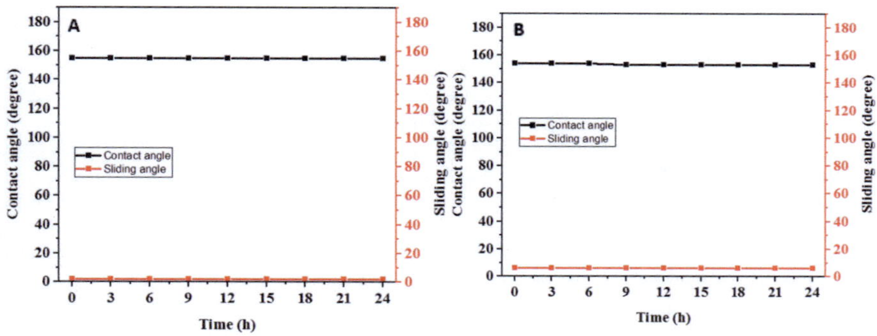

Fig. 14 WCA and SA of surfaces after exposure to radiation for 24 h: (**a**) prepared by dip-coating and (**b**) prepared by spin coating

4 Conclusion

In summary, we have successfully developed superhydrophobic and transparent glass surfaces through the deposition of a nanocomposite coating comprising PMMA and hydrophobic SiO_2. Two straightforward methods, spin-coating and dip-coating were employed for the coating process. Both surfaces exhibited a contact angle exceeding 150° and a slip angle less than 10°. Furthermore, the porous nature of the surfaces, with roughness below 100 nm, ensured superhydrophobicity while maintaining high transmittance. The spin-coated and dip-coated coatings demonstrated transmittance values of 78% and 75%, respectively. Additionally, thorough thermal and UV stability tests were conducted on the coatings. Remarkably, the coatings exhibited excellent resistance to UV radiation and retained their superhydrophobic properties even when exposed to temperatures up to 200 °C. These findings highlight the potential of PMMA resin as a promising alternative for the development of transparent superhydrophobic coatings. It should be noted, however, that the PMMA-based coating in conjunction with SiO_2 posed challenges in achieving full coverage of the glass surface, necessitating the deposition of multiple layers. The spin-coating technique required fewer layers compared to the dip-coating technique, resulting in improved transparency. Conversely, the dip-coating technique yielded a higher contact angle compared to spin-coating, even with a nanoparticle concentration of less than 1.4%, while spin-coating necessitated 6% nanoparticle concentration.

Acknowledgements This work benefited from the support of the Chair "Sustainable Energy" led by Mohammed VI Polytechnic University, sponsored by OCP.

References

1. Tan, X. *et al.* A simple fabrication of superhydrophobic PVDF/SiO2 coatings and their anti-icing properties. *J. Mater. Res.* **36**, 637–645 (2021).
2. Wu, B. *et al.* A superhydrophobic coating harvesting mechanical robustness, passive anti-icing and active de-icing performances. *J. Colloid Interface Sci.* **590**, 301–310 (2021).
3. Huang, J. *et al.* A facile preparation of superhydrophobic L-CNC-coated meshes for oil-water separation. *RSC Adv.* **11**, 13992–13999 (2021).
4. Shi, Y. *et al.* Magnetic, superelastic and superhydrophobic porous thermoplastic polyurethane monolith with nano-Fe3O4 coating for highly selective and easy-recycling oil/water separation. *Appl. Surf. Sci.* **535**, 147690 (2021).
5. Ogunsona, E. O., Muthuraj, R., Ojogbo, E., Valerio, O. & Mekonnen, T. H. Engineered nanomaterials for antimicrobial applications: A review. *Appl. Mater. Today* **18**, 100473 (2020).
6. Qing, Y. *et al.* Facile fabrication of superhydrophobic surfaces with corrosion resistance by nanocomposite coating of TiO2 and polydimethylsiloxane. *Colloids Surfaces A Physicochem. Eng. Asp.* **484**, 471–477 (2015).
7. Lou, C., Zhang, R., Lu, X., Zhou, C. & Xin, Z. Facile fabrication of epoxy/polybenzoxazine based superhydrophobic coating with enhanced corrosion resistance and high thermal stability. *Colloids Surfaces A Physicochem. Eng. Asp.* **562**, 8–15 (2019).
8. Liravi, M., Pakzad, H., Moosavi, A. & Nouri-Borujerdi, A. A comprehensive review on recent advances in superhydrophobic surfaces and their applications for drag reduction. *Prog. Org. Coatings* **140**, 105537 (2020).
9. Mohseni, M., Far, H. S., Hasanzadeh, M. & Golovin, K. Non-fluorinated sprayable fabric finish for durable and comfortable superhydrophobic textiles. *Prog. Org. Coatings* **157**, 106319 (2021).
10. Shen, L., Hu, H., Wang, S. & Fu, H. Preparation of super hydrophobic mMoS2/PDMS coating for fabrics. *React. Funct. Polym.* **143**, 104315 (2019).
11. Wu, Y. *et al.* A review of self-cleaning technology to reduce dust and ice accumulation in photovoltaic power generation using superhydrophobic coating. *Renew. Energy* **185**, 1034–1061 (2022).
12. Liu, H., Li, X., Lv, L., Liu, Z. & Chen, J. Fast fabrication of silicone-modified polyurethane/SiO_2composite superhydrophobic coating with excellent anti-icing and self-cleaning behaviour. *Mater. Res. Express* **7**, (2020).
13. Li, W., Zhan, Y. & Yu, S. Applications of superhydrophobic coatings in anti-icing: Theory, mechanisms, impact factors, challenges and perspectives. *Prog. Org. Coatings* **152**, 106117 (2021).
14. Karunakaran, R. G., Lu, C. H., Zhang, Z. & Yang, S. Highly transparent superhydrophobic surfaces from the coassembly of nanoparticles ($\leq$ 100 nm). *Langmuir* **27**, 4594–4602 (2011).
15. Aoki, Y. & Naganuma, R. One-step fabrication of composites with superhydrophobic surfaces using spin-coating with polydimethylsiloxane/silica/solvent suspension. *Mol. Cryst. Liq. Cryst.* **0**, 1–8 (2021).
16. Li, Y. *et al.* Phase Separation of Silicon-Containing Polymer/Polystyrene Blends in Spin-Coated Films. *Langmuir* **32**, 3670–3678 (2016).
17. Chen, S., Lin, D., Li, H., Lai, X. & Zeng, X. Facile fabrication of superhydrophobic, flame-retardant and conductive polyurethane sponge via dip-coating. *Mater. Lett.* **287**, 129307 (2021).
18. Siddiqui, A. R., Maurya, R., Katiyar, P. K. & Balani, K. Superhydrophobic, self-cleaning carbon nanofiber CVD coating for corrosion protection of AISI 1020 steel and AZ31 magnesium alloys. *Surf. Coatings Technol.* **404**, 126421 (2020).
19. Mirzadeh, S., Asefnejad, A., Khonakdar, H. A. & Jafari, S. H. Improved surface properties in spray-coated PU/TiO2/graphene hybrid nanocomposites through nonsolvent-induced phase separation. *Surf. Coatings Technol.* **405**, 126507 (2021).
20. Yang, Y., He, H., Li, Y. & Qiu, J. Using Nanoimprint Lithography to Create Robust, Buoyant, Superhydrophobic PVB/SiO2 Coatings on wood Surfaces Inspired by Red roses petal. *Sci. Rep.* **9**, 1–9 (2019).

21. Ke, C., Zhang, C., Wu, X. & Jiang, Y. Highly transparent and robust superhydrophobic coatings fabricated via a facile sol-gel process. *Thin Solid Films* **723**, 138583 (2021).
22. Adak, D., Bhattacharyya, R., Saha, H. & Maiti, P. S. Sol-gel processed silica based highly transparent self-cleaning coatings for solar glass covers. *Mater. Today Proc.* **33**, 2429–2433 (2019).
23. Pardo-Figuerez, M., López-Córdoba, A., Torres-Giner, S. & Lagaron, J. M. Superhydrophobic Bio-Coating made by co-continuous electrospinning and electrospraying polyethylene terephthalate films proposed as easy emptying transparent food packaging. *Coatings* **8**, (2018).
24. Nguyen, B. Q. H., Shanmugasundaram, A., Hou, T. F., Park, J. & Lee, D. W. Realizing the flexible and transparent highly-hydrophobic film through siloxane functionalized polyurethane-acrylate micro-pattern. *Chem. Eng. J.* **373**, 68–77 (2019).
25. Yoon, J. *et al.* Wet-Style Superhydrophobic Antifogging Coatings for Optical Sensors. *Adv. Mater.* **32**, 1–7 (2020).
26. Chen, L. *et al.* Robust and elastic superhydrophobic breathable fibrous membrane with in situ grown hierarchical structures. *J. Memb. Sci.* **547**, 93–98 (2018).
27. Liu, X., Wang, Y., Chen, Z., Ben, K. & Guan, Z. A self-modification approach toward transparent superhydrophobic glass for rainproofing and superhydrophobic fiberglass mesh for oil-water separation. *Appl. Surf. Sci.* **360**, 789–797 (2016).
28. Sutar, R. S. *et al.* Superhydrophobic Nanocomposite Coatings of Hydrophobic Silica NPs and Poly(methyl methacrylate) with Notable Self-Cleaning Ability. *Macromol. Symp.* **393**, 1–5 (2020).
29. Latthe, S. S., Terashima, C., Nakata, K., Sakai, M. & Fujishima, A. Development of sol-gel processed semi-transparent and self-cleaning superhydrophobic coatings. *J. Mater. Chem. A* **2**, 5548–5553 (2014).
30. Yilgör, E., Söz, C. K. & Yilgör, I. Wetting behavior of superhydrophobic poly(methyl methacrylate). *Prog. Org. Coatings* **125**, 530–536 (2018).
31. Meena, M. K., Sinhamahapatra, A. & Kumar, A. Superhydrophobic polymer composite coating on glass via spin coating technique. *Colloid Polym. Sci.* **297**, 1499–1505 (2019).
32. Pan, S., Wang, N., Xiong, D., Deng, Y. & Shi, Y. Fabrication of superhydrophobic coating via spraying method and its applications in anti-icing and anti-corrosion. *Appl. Surf. Sci.* **389**, 547–553 (2016).
33. Wang, Z. *et al.* A veil-over-sprout micro-nano PMMA/SiO2 superhydrophobic coating with impressive abrasion, icing, and corrosion resistance. *Colloids Surfaces A Physicochem. Eng. Asp.* **601**, 124998 (2020).

Enhancement of Room Temperature Magnetoelectric Coupling in $Na_{0.5}Bi_{0.5}TiO_3$ –$BaFe_{12}O_{19}$ Based Flexible Polymer Film

Khuraijam Jyotsna and Sumitra Phanjoubam

Abstract This study reports the enhancement of magnetoelectric coupling in biphasic and ternary composites of $BaFe_{12}O_{19}$ ceramics and flexible films. Three systems were fabricated using different methods: single-phase $BaFe_{12}O_{19}$ via sol-gel, biphasic $Na_{0.5}Bi_{0.5}TiO_3$–$BaFe_{12}O_{19}$ via solid-state, and ternary PVDF – $Na_{0.5}Bi_{0.5}TiO_3$–$BaFe_{12}O_{19}$ via solution casting. The structural and morphological studies were performed using an X-ray Diffractometer and Scanning Electron Microscope which showed successful formation of the phases in the three systems. Dielectric properties investigated at room temperature showed the dominance of interfacial polarization at lower frequencies. The dielectric constant was found to be highest in the biphasic composite resulting from the large difference in the resistivity of the constituent phases, while the lowest was observed in the ternary system due to the dominance of the PVDF polymer. The room temperature multiferroic property of the studied systems was confirmed from the P-E and M-H loop measurements. Furthermore, magnetoelectric interaction was analyzed which showed the single phase $BaFe_{12}O_{19}$ to have a small ME coupling value of ~0.439 mV/cmOe. In comparison, the biphasic composite had a higher value of ~26.12 mV/cmOe. This was found to be greatly enhanced up to ~562 mV/cmOe in the ternary polymer film which can be attributed to better connectivity between the different phases. The polymer film was subjected to a finger-tapping force to investigate the energy harvesting property, and an open circuit voltage of about 11.3 V could be observed. With further engineering and better design, the fabricated composite material appears to have a bright future in magnetoelectric sensors, pressure sensors, and mechanical energy harvesters.

Keywords Magnetoelectric · Biphasic · Ternary · PVDF · Flexible

---

K. Jyotsna · S. Phanjoubam (✉)
Department of Physics, Manipur University, Canchipur, Manipur, India
e-mail: sumitraphanjoubam@gmail.com

A. Vaseashta et al. (eds.), *Proceedings of the Eighth International Symposium on Dielectric Materials and Applications (ISyDMA'8)*, Springer Proceedings in Physics 349, https://doi.org/10.1007/978-3-031-84809-4_8

1 Introduction

With the ever-increasing demand for multifunctional devices, the quest for novel materials with multiple functionalities has been on the rise. In this regard, multiferroic materials with the ability to exhibit more than one ferroic order simultaneously have gained significant attention [1]. These materials with electric and magnetic coupling, also known as magnetoelectric (ME) effect, offer a wide variety of applications in new generation memory devices, multifunctional devices, spintronics, sensors, etc. allowing miniaturization of electronic devices [2, 3]. However, most single-phase multiferroic materials like rare earth manganites operate below room temperature. $BiFeO_3$, being the only known room-temperature multiferroic material, has been extensively studied for this reason. This material, however, has certain drawbacks like high leakage current and low ME effect [4]. Therefore, the focus has been shifted to the exploration of other materials that can produce robust ME coupling at room temperature. Recently, certain types of hexagonal ferrites such as substituted M-type, U-type, and Y-type hexagonal ferrites have been characterized as room-temperature multiferroic material [5]. Among these, the simplest one is the Barium hexaferrite (BaM), an M-type hexaferrite, with the formula $BaFe_{12}O_{19}$. It possesses hard magnetic behavior with a large saturation magnetization and large coercivity finding its place in applications like permanent magnets, magnetic recording media, and high-frequency devices for electromagnetic wave absorbing technology [6]. Various studies have been performed on BaM by substituting ions of different valencies in the Ba or Fe site to tune its magnetic and electric properties [7–9]. Besides, it has good chemical stability at high temperatures which makes it suitable for fabricating composites with other materials without compromising its purity. However, BaM shows weak electric polarization and this can be overcome by mixing it with an intrinsically ferroelectric material to form a composite such as Sodium Bismuth Titanate with the chemical formula $Na_{0.5}Bi_{0.5}TiO_3$ (NBT) which has been chosen for the present study. It has emerged as the most viable candidate as it is lead-free and has good ferroelectric and piezoelectric properties in addition to having a high Curie temperature [10]. Apart from being non-toxic, it has a lower sintering temperature as compared to the mostly studied lead-free $BaTiO_3$ and is less expensive as compared to $K_{0.5}Na_{0.5}NbO_3$. Owing to its excellent properties, NBT-based ceramics have been studied by various groups for applications in actuators, sensors, and capacitors [11–13].

Although bulk composite systems exhibit a higher magnetoelectric (ME) effect, their hard and brittle nature makes them unsuitable for the miniaturization of electronic devices [14, 15]. This issue can be addressed by incorporating the multiferroic ceramic composite into a polymer matrix. This approach opens up new opportunities for the developed system to be applied in flexible wearable electronics. Recently, research on flexible electronics has become significantly important due to its versatile nature and potential applications in various fields, including wearable technology, medical devices, and next-generation displays [16]. Also, the increasing energy demands have led to the depletion of conventional energy sources from fossil fuels and the quest for alternative energy sources is more critical than ever. While other renewable sources of energy are being exploited, the ability to convert small mechanical or vibrational energy into usable electrical energy would be a great advantage. Flexible magnetoelectric materials are promising in this regard since they possess piezoelectric properties that can effectively convert ambient vibrations or waste mechanical energy into electrical energy, thereby contributing to sustainable energy solutions. In this work, Polyvinylidene difluoride (PVDF) has been chosen as the flexible polymer matrix because of its good dielectric property, easy processibility, good mechanical strength, thermal stability, and chemical resistance to environmental exposure, and most importantly, its intrinsic piezoelectric nature [17, 18]. The present work demonstrates the development of a ternary composite system starting with single-phase BaM and shows the enhancement of ME coupling through the addition of different phases as in biphasic $Na_{0.5}Bi_{0.5}TiO_3$–$BaFe_{12}O_{19}$ (NBT-BaM) and PVDF–$Na_{0.5}Bi_{0.5}TiO_3$–$BaFe_{12}O_{19}$ (PVDF-NBT-BaM) composites. Additionally, the multifunctional properties of the PVDF-NBT-BaM system are highlighted, indicating its potential use for future advancements in magnetic sensing and energy harvesting.

2 Experimental

Stoichiometric amounts of analytical grade precursors, such as barium acetate and ferric nitrate nonahydrate, were used for the sol-gel synthesis of the hexaferrite phase. The synthesis of NBT was carried out adopting the same method using sodium acetate, bismuth nitrate pentahydrate, and titanium isopropoxide. The detailed synthesis procedures have been reported in our previous works [19–21]. The precursors were initially dissolved in distilled water with an appropriate amount of citric acid and the solution was continuously stirred for 2 h at 80 °C to obtain a homogeneous solution. Ethylene glycol was then added to the solution and the temperature was raised to 150 °C after which combustion initiated. This resulted in the formation of a black ash which was then ground and calcined. The fabrication of the biphasic composite system was carried out through a solid-state route. Calcined powders of BaM and NBT were taken in the ratio 1:3, mixed, ground, and pressed into pellets, which were then sintered at 1100 °C to obtain the biphasic composite.

The formation of the ternary flexible composite system was achieved through a simple solution casting method. PVDF powder was first dissolved in dimethyl formamide solution after which 5% wt. of the biphasic composite was added. The solution was then continuously stirred to attain homogeneity and subsequently cast onto a petri dish. The petri dish was then kept in an oven for evaporation for 30 minutes after which a free-standing thick film was obtained.

The phase purity of the synthesized materials was characterized by XPert Pro Panalytical X-Ray Diffractometer (XRD) at room temperature using Cu-Kα radiation. The morphological studies were performed from the JEOL JSM-6010LA W-Scanning Electron Microscopy (SEM) images. Impedance analyses were performed by using an Agilent 4284A LCR meter at room temperature. The electric polarization studies were performed using a Marine India P-E loop tracer while for the room temperature magnetic studies; a Vibrating Sample Magnetometer (Model-Lakeshore 7400) was employed. A SR865A 4 MHz DSP lock-in amplifier was used for the magnetoelectric measurement. Finally, the energy harvesting measurements were carried out using a Keysight B2901A a source/measure unit.

3 Results and Discussions

3.1 Structural and Morphological Properties

The structural properties of the synthesized systems can be studied from the XRD pattern presented in Fig. 1a–c. The formation of single-phase BaM was confirmed by the existence of peaks corresponding to the P63/mmc hexagonal structure in Fig. 1a. In Figure. 1b, peaks corresponding to both P63/mmc phase and R3c rhombohedral phase were observed without any impurity peaks which confirmed the successful formation of the biphasic composite. Rietveld analysis was performed on the XRD patterns to refine the structural parameters, and the obtained parameters are presented in Table 1. In Figure. 1c, the presence of the peaks corresponding to α- and β-crystalline phases of the PVDF polymer along with all the prominent peaks of NBT and BaM show the successful formation of the ternary system.

The SEM images of the three systems are presented in Fig. 2. The image shown in Fig. 2a, displays uniformly sized grains with well-defined grain boundaries for the single-phase BaM. In contrast, Fig. 2b shows differently sized grains for the NBT-BaM composite which can be attributed to the presence of two different phases with dissimilar grain sizes in the system. In Fig. 2c, the SEM micrographs show the NBT-BaM particles dispersed in the PVDF matrix. Some particle agglomeration can be observed in the ternary system which may be attributed to the magnetic nature of the constituent BaM phase.

3.2 Room Temperature Dielectric Studies

The room temperature dielectric properties of the synthesized systems are presented in Fig. 3a–c. The frequency-dependent behavior of the dielectric constant can be understood by taking into account the Maxwell-Wagner two-layer model [22, 23]. According to this model, a dielectric material consists of conductive grains surrounded by highly resistive grain boundaries. At low frequencies, charge carriers can hop through the grain and begin to accumulate at the highly resistive grain boundaries, thereby increasing space charge polarization and hence leading to a high dielectric constant. At higher frequencies, the charge carriers alter their direction of

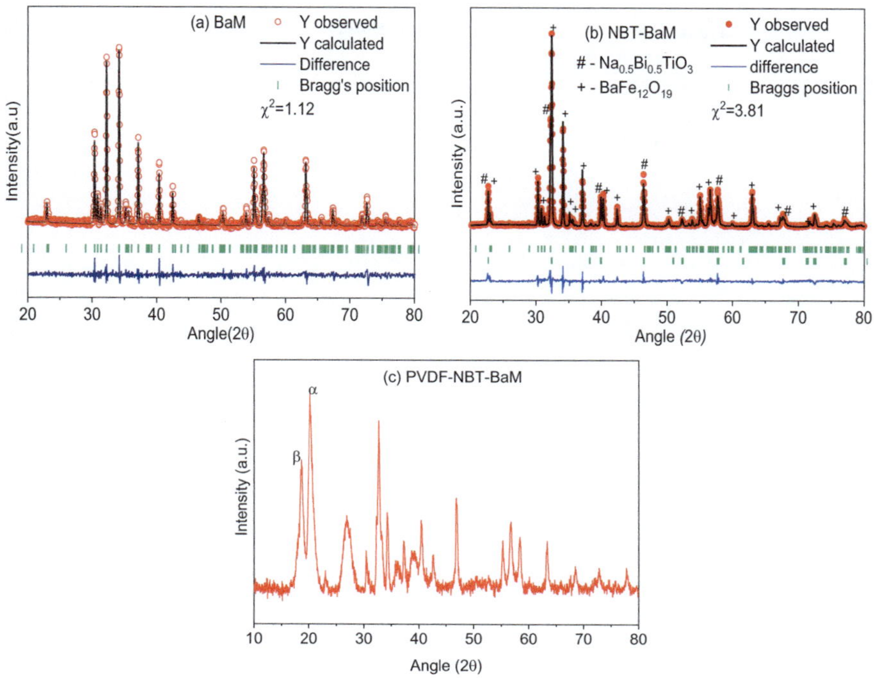

Fig. 1 XRD pattern of the (**a**) single phase BaM [20] (**b**) biphasic NBT-BaM composite [21] (**c**) ternary PVDF-NBT-BaM composite

Table 1 Structural parameters obtained from Rietveld refinement for the ceramics, BaM and NBT-BaM composite

Sample	Phase	Crystal structure	Space group	a = b (Å)	c (Å)
Single phase	BaM	Hexagonal	P63/mmc	5.8893	23.1986
Biphasic composite	NBT	Rhombohedral	R3c	5.5295	13.5707
	BaM	Hexagonal	P63/mmc	5.9001	23.2147

motion more frequently and their probability of accumulating at the grain boundaries reduces which results in the decrease of the dielectric constant. The tangent loss (tan δ) curves also follow a similar behavior. With the addition of NBT in the BaM material, the accumulation of space charges increases at the interface of NBT/BaM due to their difference in conductivities, and the dielectric constant is greatly enhanced in the biphasic composite system [24]. This enhancement of dielectric property suggests the potential usage of the composite material in capacitive and energy storage devices [25]. In contrast, the dielectric constant is drastically reduced in the ternary PVDF-containing composite which may be attributed to the dominance of the PVDF polymer which constitutes 95% of the ternary system and has a comparatively lower dielectric value than ceramic materials.

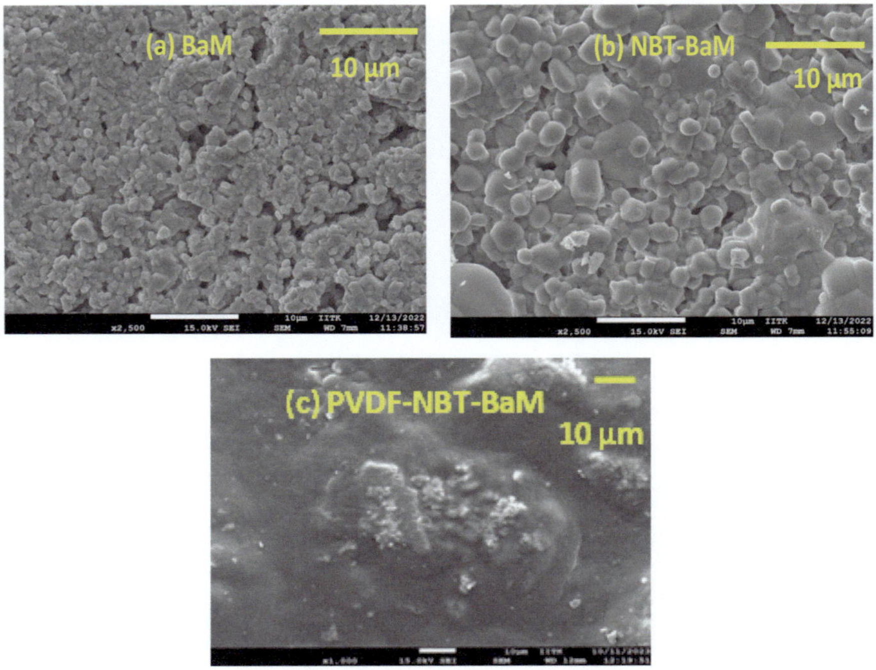

Fig. 2 SEM images of the (**a**) single phase BaM (**b**) biphasic NBT-BaM composite (**c**) ternary PVDF-NBT-BaM composite

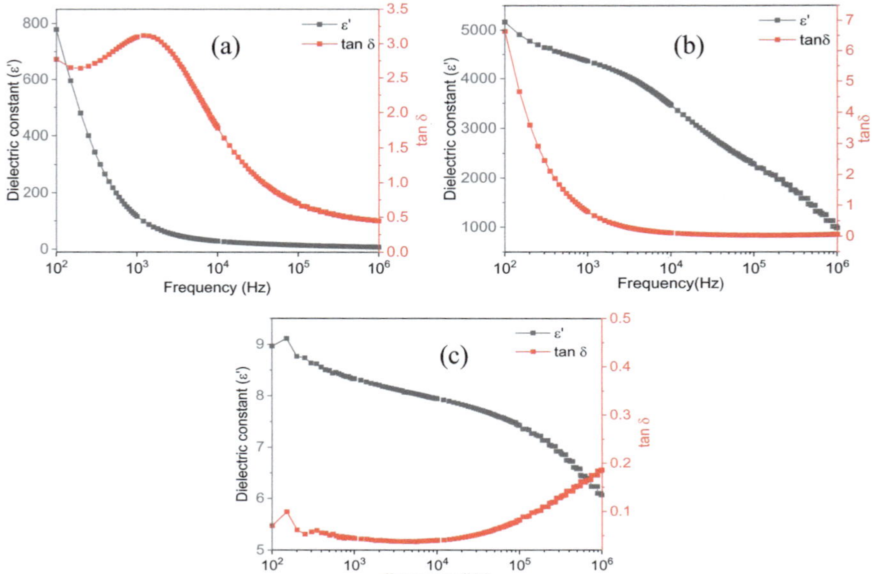

Fig. 3 Frequency-dependent dielectric behavior of (**a**) single phase BaM (**b**) biphasic NBT-BaM composite (**c**) ternary PVDF-NBT-BaM composite

3.3 Multiferroic and Magnetoelectric Properties

The electric polarization studies of the ceramic systems can be analyzed from the P-E hysteresis loops. While barium hexaferrites have been reported to possess ferroelectric properties in some studies, the P-E loop for the presently synthesized BaM could not be obtained supposedly due to its leaky nature because of the presence of Fe^{3+}/Fe^{2+} ions [26]. The P-E hysteresis curve of the NBT-BaM composite ceramic taken at room temperature is exhibited in Fig. 4a. The values of maximum polarization (P_{max}), and remnant polarization (P_r) for the studied systems are tabulated in Table 2. It can be observed that NBT-BaM composite possesses good P_{max} and P_r values as compared to other multiferroic bulk composites, indicating their utility in piezoelectric energy harvesting devices [27–29]. Figure. 4b represents the P-E loop of the ternary composite. The low values of P_{max} and P_r can be attributed to the dominance of PVDF polymer which has a comparatively lower ferroelectric property as compared to ceramics.

The room temperature magnetic M-H loops of the synthesized ceramics are presented in Fig. 5a, b while that of the ternary composite is presented in Fig. 5c and the magnetic parameters obtained are tabulated in Table 3. It is to be noted that the magnetic property in the studied systems originate from the presence of $Fe^{3+}–O^{2-}–Fe^{3+}$ exchange interactions in the BaM structure where the Fe ions are located at 5 crystallographic sites namely, 2a, 2b, 12k, $4f_1$ and $4f_2$ [26]. It can be observed that the addition of the non-magnetic phases in BaM reduces the saturation magnetization in the biphasic and ternary composites. A decrease in the coercivity values is also observed which indicates the disturbance of magnetic spins in the BaM structure. It is known that pure BaM has large coercivity and magnetic anisotropy due to the alignment of all its magnetic spins in the easy c-axis. Disturbance in the spin

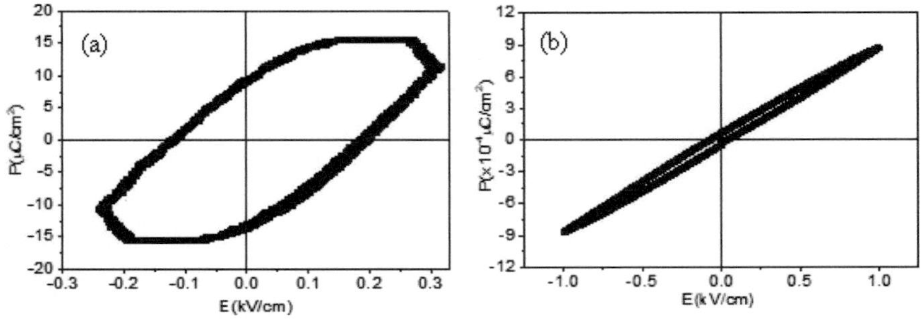

Fig. 4 Room temperature P-E loop of the (**a**) NBT-BaM (**b**) PVDF-NBT-BaM composite

Table 2 Parameters obtained from P-E loop analysis

Sample	Pmax (μC/cm²)	Pr (μC/cm²)
NBT-BaM	15.53	11.38
PVDF-NBT-BaM	8.62×10^{-4}	0.58×10^{-4}

alignment leads to the development of non-collinear cycloidal magnetic spin and the emergence of spin–orbit current/Dzyaloshinskii-Moriya (DM) interaction within the BaM structure due to which ME interaction arises [31, 32]. Therefore, the decrease in coercivity may indicate an increased DM interaction and hence an enhanced ME coupling.

In order to test the magnetoelectric interaction in the composite systems and to assess its usability in device applications, their magnetoelectric coupling property was studied using the dynamic method where the multiferroic specimen is subjected to a DC magnetic field varying from (0–1000) Oe along with a modulating AC magnetic field of 1 Oe at a frequency of 1 kHz. The voltage generated is measured using a lock-in amplifier and the magnetoelectric coefficient (α) is calculated using the formula given by Rivera [33]:

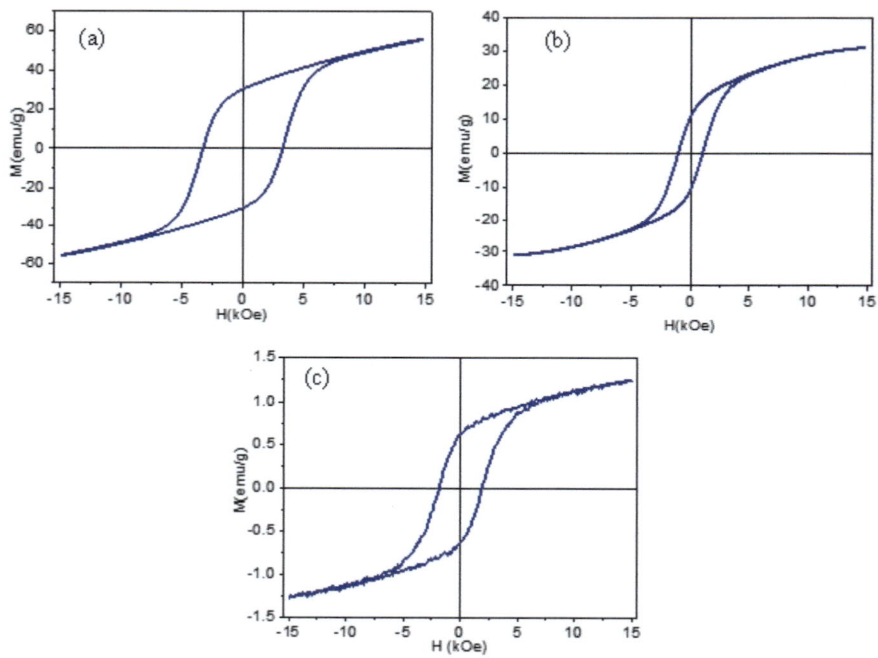

Fig. 5 Room temperature M-H loops of the (**a**) single phase BaM (**b**) biphasic NBT-BaM composite ceramics

Table 3 Magnetic parameters were obtained from the M-H curves

Sample	Ms (emu/g)	Hc (Oe)	α (mV/cmOe)
BaM	55.678	3257	0.439
NBT-BaM	30.99	1100	26.12
PVDF-NBT-BaM	1.24	1856	561.9

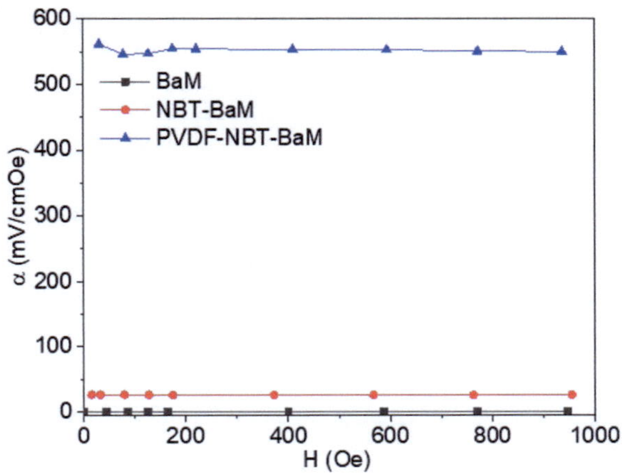

Fig. 6 Magnetoelectric coupling constant versus DC magnetic field of the three systems

$$\alpha = \frac{V_0}{H_{ac}t},$$

where V_0 is the voltage induced in the samples, H_{ac} the applied AC magnetic field, and t the thickness of the sample. The ME coefficient vs DC magnetic field of the synthesized systems is presented in Fig. 6 and tabulated in Table 3. The maximum ME coefficient values were found to be 0.439 mV/cmOe for the single phase BaM system and 26.12 mV/cmOe for biphasic NBT-BaM composite, while for the ternary PVDF-NBT-BaM system, the value of α was greatly enhanced to 561.9 mV/cmOe. ME coupling in hexaferrites has been reported to originate from spin-orbit current or Dzyaloshinskii-Moriya interaction [34]. However, this current is small in pure BaM due to the presence of minimal spin canting and high leakage current in the BaM material due to charge transfer between Fe^{4+}/Fe^{3+} ions [35]. The observed enhancement in ME coupling in the biphasic NBT-BaM composite can be understood as due to the mechanical interaction between the two phases [36]. The dramatic enhancement in α on the addition of PVDF to the NBT-BaM system can be attributed to the improved interfacial adhesion between the phases and superior displacement transfer properties of the PVDF matrix besides the mechanical interaction between the phases [37, 38]. The exceptional ME coupling value observed in the ternary flexible film makes it ideal for magnetic energy harvesting devices.

3.4 Energy Harvesting Property

The mechanical energy harvesting property of the fabricated flexible film is examined by subjecting the device to a repetitive finger-tapping force in the absence and presence of a small external DC magnetic field (100 Oe). Figure. 7a, b respectively show the rectified output voltage and current generated by the device. The maximum output voltage and current generated are found to be 11.3 V and 89 nA respectively

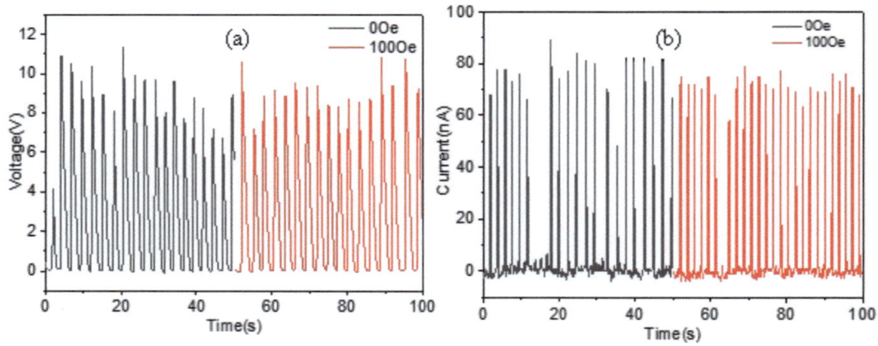

Fig. 7 Voltage and current output of the PVDF-NBT-BaM composite

at 0 Oe. This indicates the energy harvesting ability of the fabricated polymer film from small mechanical vibrations. The generation of current with applied mechanical force can be attributed to the poling effect caused by stress. When a vertical force is applied, dipoles that were initially oriented randomly align themselves along the direction of the applied force due to poling induced by the stress applied. As a result, a voltage is generated across the surface of the nanogenerator. In the presence of a magnetic field, the maximum voltage and current outputs were reduced to 10.8 V and 79 nA respectively. The reduction in output voltage and current observed with the application of a small DC magnetic field can be ascribed to the magnetostrictive behavior of the BaM filler. The magnetostrictive stress induced by the application of the magnetic field may lock the neighboring dipoles in a particular direction which causes the dipoles to become less responsive to the applied mechanical force [39]. This observation shows the magnetically tunable energy harvesting property of the developed nanogenerator. With further modification and better design, the fabricated device has a promising future in the development of a multifunctional device that encompasses mechanical energy harvesting properties along with stray magnetic field detection [40].

4 Conclusion

Single-phase BaM, biphasic NBT-BaM composite, and ternary PVDF-NBT-BaM composite were synthesized. The successful formation of the three different systems was confirmed by the XRD analysis where peaks corresponding to their respective constituent phases were observed in each system without any impurity peak. SEM image analysis confirmed the presence of grains and their distribution in the different systems. The materials showed typical dielectric dispersion behavior with frequency giving enhanced dielectric constant in the biphasic system and reduced dielectric constant in the ternary system. The room temperature PE and MH loops indicated the multiferroic nature of the studied systems. Magnetoelectric interactions in the studied systems were analyzed at room temperature and the coupling coefficient was found to be greatly enhanced from ~0.439 mV/cmOe in BaM to ~26.12 mV/cmOe with the addition of the NBT phase in the BaM system and further up to ~ 561.9 mV/cmOe in the ternary flexible film, which points to its suitability towards the fabrication of magnetoelectric sensors. The produced film was subjected to a finger-tapping force to investigate the energy harvesting property, and this produced an open circuit voltage of about 11.3 V. With further engineering and better design, the fabricated composite material appears to have a bright future in magnetoelectric sensors, pressure sensors, and mechanical energy harvesters.

Acknowledgements The authors are thankful to Prof. Monika Tomar, Department of Physics, Miranda House, Delhi University for providing measurement facilities. One of the authors, Khuraijam Jyotsna is thankful to the Ministry of Social Justice and Empowerment for providing financial assistance in the form of NFOBC-SRF.

References

1. Eerenstein W., Mathur N.D., Scott J.F.: Multiferroic and magnetoelectric materials. Nature **442**, 759 (2006).
2. Spaldin N.A., Ramesh R.: Advances in magnetoelectric multiferroics. Nature Mater **18**, 203 (2019).
3. Liang X., Chen H., and Sun N. X.: Magnetoelectric materials and devices. APL Materials **9**, 041114 (2021).
4. Vashisth B. K., Bangruwa J. S., Beniwal A., Gairola S. P., Kumar A., Singh N., and Verma V.: Modified ferroelectric/magnetic and leakage current density properties of Co and Sm co-doped bismuth ferrites. Journal of Alloys and Compounds **698**, 699 (2017).
5. Pullar R. C.: Multiferroic and Magnetoelectric Hexagonal Ferrites. Mesoscopic Phenomena in Multifunctional Materials: Synthesis, Characterization, Modeling, and Applications, In: Saxena A. and Planes A. pp. 159–200, Springer, Berlin, Heidelberg (2014).
6. Shepherd P, Mallick KK, Green RJ: Magnetic and structural properties of M-type barium hexaferrite prepared by co-precipitation. Journal of Magnetism and Magnetic Materials. Journal of Magnetism and Magnetic Materials **311**, 683 (2007).
7. Lu S., Liu Y., Yin Q., Chen J., Li J., Wu J.: Effects of Ce-Zn co-substitution on the structural and magnetic properties of M-type barium hexaferrites. Journal of Magnetism and Magnetic Materials **564**, 170068 (2022).
8. Gupta S, Deshpande SK, Sathe VG, Siruguri V. Effect of scandium substitution on magnetic and transport properties of the M-type barium hexaferrites. Journal of Alloys and Compounds **815**, 152467 (2020).
9. Yasmin N, Yasmin S, Zahid M, Gillani SF, Islam MU, Altaf M, et al. Impact of Ho-Ni substitution on structural, morphological, and dielectric characteristics of BaFe12O19 M-type hexagonal ferrite. Physica B: Condensed Matter **581**, 411950 (2020).
10. Li Z, Yang Q, Wang C, Zhang J, Wang Z, Gao B, et al. A Brief Review of Sodium Bismuth Titanate-Based Lead-Free Materials for Energy Storage: Solid Solution Modification, Metal/metallic Oxide Doping, Defect Engineering and Process Optimizing. Crystals **13**, 295 (2023).
11. Ji H, Wang D, Bao W, Lu Z, Wang G, Yang H, et al.: Ultrahigh energy density in short-range tilted NBT-based lead-free multilayer ceramic capacitors by nanodomain percolation. Energy Storage Materials **38**, 113 (2021).

12. Fan P, Liu K, Ma W, Tan H, Zhang Q, Zhang L, et al.: Progress and perspective of high strain NBT-based lead-free piezoceramics and multilayer actuators. Journal of Materiomics **7**, 508 (2021).
13. Kanuru SR, Baskar K, Dhanasekaran R.: Synthesis, structural, morphological and electrical properties of NBT–BT ceramics for piezoelectric applications. Ceramics International **42**, 6054 (2016).
14. Mao Q, Wu J, Hu Z, Xu Y, Du Y, Hao Y, et al.: Magnetoelectric devices based on magnetoelectric bulk composites. J. Mater. Chem. C **9**, 5594 (2021).
15. Sezer N, Koç M.: A comprehensive review on the state-of-the-art of piezoelectric energy harvesting. Nano Energy **80**, 105567 (2021).
16. Hassan M, Abbas G, Li N, Afzal A, Haider Z, Ahmed S, et al.: Significance of Flexible Substrates for Wearable and Implantable Devices: Recent Advances and Perspectives. Advanced Materials Technologies **7**, 2100773 (2022).
17. Yan J, Liu M, Jeong YG, Kang W, Li L, Zhao Y, et al.: Performance enhancements in poly(vinylidene fluoride)-based piezoelectric nanogenerators for efficient energy harvesting. Nano Energy **56**, 662 (2019).
18. Xie L, Wang G, Jiang C, Yu F, Zhao X.: Properties and Applications of Flexible Poly(Vinylidene Fluoride)-Based Piezoelectric Materials. Crystals **11**, 644 (2021).
19. Jyotsna K., Chanu L. P., and Sumitra P., Effect of Fe substitution on the structural, microstructural, and dielectric properties of sodium bismuth titanate. Materials Today: Proceedings **65**, 2781 (2022).
20. Jyotsna K. and Sumitra P.: Improved magnetoelectric coupling of Co–Ti substituted barium hexaferrite at room temperature and related electrical investigations. Materials Chemistry and Physics **317**, 129186 (2024).
21. Jyotsna K., Monika T., Sumitra P.: Strong magnetoelectric coupling in novel Na0.5Bi0.5TiO3–BaFe12–2xCoxTixO19 multiferroic composite., Ceramics International 50, 33189 (2024).
22. Maxwell J. C., *A Treatise on Electricity and Magnetism* (Clarendon Press, 1873).
23. Koops CG.: On the Dispersion of Resistivity and Dielectric Constant of Some Semiconductors at Audiofrequencies. Phys. Rev. **83**, 121 (1951).
24. Wang, SB, Wang, CB, Shen Q, Zhang LM.: Enhanced dielectric properties in Ba0.85Ca0.15Zr0.10Ti0.90O3/La0.67Ca0.33MnO3 laminated composite. Scripta Materialia **144**, 40 (2018).
25. Wu Y, Fan Y, Liu N, Peng P, Zhou M, Yan S, et al.: Enhanced energy storage properties in sodium bismuth titanate-based ceramics for dielectric capacitor applications. J. Mater. Chem. C **7**, 6222 (2019).
26. Zeng Z, Lan M, Zhang Q, Wu H, He J, Ao H, et al.: Effect of solution concentration on magnetoelectric properties of barium ferrite ceramics. Appl. Phys. A **127**, 946 (2021).
27. Rather M ud D, Ganie AN, Mala NA, Fatema M, Ansari DA, Lone AG, et al.: Unravelling the linear and biquadratic magnetoelectric coupling in Ba0.95Sn0.05Ti0.95 Ga0.05O3 – CoFe1.8Ga0.2O4 particulate multiferroic composites. Journal of Alloys and Compounds **946**, 169266 (2023).
28. Rather M ud D, Samad R, Want B.: Improved magnetoelectric effect in ytterbium-doped BaTiO3 –CoFe2O4 particulate multiferroic composites. Journal of Alloys and Compounds **755**, 89 (2018).
29. Lekha CSC, Kumar AS, Vivek S, Saravanan KV, Anantharaman MR, Surendran KP, et al.: Room temperature magnetoelectric properties of lead-free alkaline niobate based particulate composites. Ceramics International **45**, 8115 (2019).
30. Baykal A, Sözeri H, Güngüneş H, Auwal I, Shirsath SE, Sertkol M, et al.: Synthesis and Structural and Magnetic Characterization of BaZnxFe12−xO19 Hexaferrite: Hyperfine Interactions. J Supercond Nov Magn **30**, 1585 (2017).
31. Shao Y, Huang F, Zhang J, Yan S, Xiao S, Lu X, et al.: Magnetoelectric Coupling Triggered by Noncollinear Magnetic Structure in M-Type Hexaferrite. Advanced Quantum Technologies **4**, 2000096 (2021).

32. Xu C, Zhang AL, Feng Z, Lu W, Kang B, Zhang J, et al.: Spin-orbit coupling in magnetoelectric Ba3(Zn1−xCox)2Fe24O41 hexaferrites. Phys. Chem. Chem. Phys. **21**, 25826 (2019).
33. Bueno-Baques D, Grossinger R, Schonhart M, Doung GV, Sato R, Corral-Flores V, et al.: The magnetoelectric effect in pulsed magnetic fields. Journal of Applied Physics **99**, 08D908 (2006).
34. Shao Y, Shu L, Huang F, Yan S, Lu X, Zhu J.: Enhanced magnetoelectric effect by spin-orbit coupling in M-type hexaferrite. Phys. Rev. B **106**, 144419 (2022).
35. Rai HM, Singh P, Saxena SK, Mishra V, Warshi MK, Kumar R, et al.: Room-Temperature Magneto-dielectric Effect in LaGa0.7Fe0.3O3+γ; Origin and Impact of Excess Oxygen.Inorg. Chem. **56**, 3809 (2017).
36. Palneedi H, Annapureddy V, Priya S, Ryu J.: Status and Perspectives of Multiferroic Magnetoelectric Composite Materials and Applications. Actuators **5**, 9 (2016).
37. Lu L, Ding W, Liu J, Yang B. Flexible PVDF-based piezoelectric nanogenerators. Nano Energy **78**, 105251 (2020).
38. Adhlakha N, Yadav KL, Truccato M, Manjusha, Rajak P, Battiato A, et al.: Multiferroic and magnetoelectric properties of BiFeO3-CoFe2O4-poly(vinylidene-fluoride) composite films. European Polymer Journal **91**, 100 (2017).
39. Sasmal A, Sen S, Chelvane JA, Arockiarajan A. PVDF-based flexible magnetoelectric composites for capacitive energy storage, hybrid mechanical energy harvesting and self-powered magnetic field detection. Polymer **281**, 126141 (2023).
40. Hajra S, Vivekananthan V, Sahu M, Khandelwal G, Joseph Raj NPM, Kim SJ: Triboelectric nanogenerator using multiferroic materials: An approach for energy harvesting and self-powered magnetic field detection. Nano Energy **85**, 105964 (2021).

Additive Manufactured Compact Microwave Absorber

Gregory Peter Le Sage and Muhammad Shumail

Abstract A high-performance microwave absorber was created using Fused Deposition Modeling (FDM) 3D printing. Both a narrow band and a broadband absorber were created. The narrow band absorber was designed at 4.9 GHz, mid-band in WR-187 waveguide. The broadband absorber tried to achieve the best attenuation across the entire 3.95–5.85 GHz band. Two types of carbon loaded polylactic acid (PLA) plastic and unloaded PLA were printed with variable percentages of air to create different values of dielectric constant and loss tangent. A stack of five or six rectangular pieces of plastic each with different characteristics were optimized through computer simulation in HFSS and analytic modeling in MATLAB to maximize attenuation. The stack progressed from the lowest loss and lowest dielectric constant to the highest at the shorting end. The final narrow band load had simulated return loss of 291 dB at 4.9 GHz with an analytic solution in MATLAB, and a corresponding result of 119 dB in HFSS. The measured return loss of the 3D printed attenuator was 73.254 dB at 4.929 GHz. The total length of the absorber is 2.44 inches. A commercial absorber for WR-187 with a return loss of 40 dB has a length of 13 inches. The experiment proves that an effective microwave absorber can be created using 3D printing.

Keywords Microwave load · Matched load · Waveguide · Attenuation · 3D printing

1 Introduction

Fused Deposition Modeling (FDM) 3D printing was used to create a narrow band microwave absorber with attenuation of 73 dB at 4.93 GHz with a length of 2.44 inches. This load works only at a single frequency. A broadband absorber had an average attenuation of 28.81 dB across the band and had a length of 2.18 inches.

G. P. Le Sage (✉) · M. Shumail
SLAC National Accelerator Laboratory, Menlo Park, CA, USA
e-mail: lesage@slac.stanford.edu

A. Vaseashta et al. (eds.), *Proceedings of the Eighth International Symposium on Dielectric Materials and Applications (ISyDMA'8)*, Springer Proceedings in Physics 349, https://doi.org/10.1007/978-3-031-84809-4_9

A commercial broadband load with an attenuation of 40 dB across the band is 13 inches long. For a single-frequency application, this 3D-printed load outperforms a commercial load in terms of attenuation and length. Additive manufacturing or 3D printing has been shown to have important applications to microwave technology. Microwave structures such as slot array antennas [1–5], waveguide filters [6–8], and lenses [9, 10] have been demonstrated. Both structural elements and custom dielectric elements are possible. Microwave lenses created with 3D printing utilize the fact that custom dielectric constants are possible by choosing the "fill factor" or percentage of plastic and air in a 3D print. While structural 3D printing can utilize Stereo Lithography (STL) 3D printing where the dielectric properties of the material are not important, especially if parts are metal plated, FDM 3D printing is more useful for creating custom dielectrics since most FDM plastics have much lower loss at microwave frequencies than most SLA resins. There is one laser-cured resin 3D printing process with a low-loss material [11]. Given that the purpose of this exercise was the creation of an absorber, loss of SLA resin is not a limiting factor. It is possible to use SLA resins to create microwave absorbers. For a desktop SLA printer, Formlabs PU Rigid 650 has a dielectric constant 4.7 at 1 MHz and a loss tangent of 0.088. In cases where small and intricate absorbers are needed, the higher resolution capability of SLA 3D printing compared to FDM could become important.

Conceptually, an effective microwave absorber will match a high-loss element to the impedance of free space so that reflection is minimized. In fact, the dielectric constant and loss tangent of the blocks in the attenuator stack shown here do increase from the front to the shorting plane as determined through optimization. Several groups have published related techniques based on 3D printing lossy plastic including multilayer structures [12–15]. Lossy material and 3D printing have also been used to create frequency selective surfaces.

2 Absorber Design

Taken from [5], dielectric constant and loss tangent scale from bulk parameters and fill factor with the following equations. The fill factor, or percentage of plastic is ff.

$$\varepsilon_{custom} = ff * \varepsilon_{bulk} + (1 - ff) * \varepsilon_{hole} = \varepsilon' - j\varepsilon'' \tag{1}$$

$$\tan\delta_{custom} = ff * \tan\delta_{bulk}\left(\frac{\varepsilon_{r,bulk}}{\varepsilon_{r,custom}}\right) + (1 - ff) * \tan\delta_{hole} * \left(\frac{\varepsilon_{r,hole}}{\varepsilon_{r,custom}}\right) \tag{2}$$

Assuming $\varepsilon_{r,hole} = 1$, and $\tan\delta_{hole} = 0$, the equations simplify to the following.

$$\varepsilon_{custom} = ff * \varepsilon_{bulk} + (1 - ff) \tag{3}$$

$$\tan \delta_{custom} = ff * \tan \delta_{bulk} \left(\frac{\varepsilon_{r,bulk}}{\varepsilon_{r,custom}} \right) \quad (4)$$

Experimentally, these relationships hold well for fill factors above about 30% but start to diverge for smaller fill factors. The HFSS optimizer finally produced good return loss after thousands of iterations, but as soon as the MATLAB analytic model was proven, that was used going forward since its result takes seconds. Application of other plastics with intermediate loss and dielectric constants will also improve broadband performance.

The carbon loaded FDM filaments were Proto Pasta [16] conductive and Sunlu carbon loaded [17]. The unloaded PLA plastic was FlashForge brand [18]. The success of the optimized design depends entirely on knowing the correct bulk values of ε_r and tan δ for the three plastics. All the plastic samples were measured using multiple techniques to characterize dielectric constant and loss tangent. These included cavity perturbation, coaxial transmission, external coaxial probing, and waveguide transmission. Not all techniques worked for all plastics, and the results varied by more than a factor of two.

A new waveguide measurement technique proved the most accurate since its predicted values led to the proven optimized device. The dielectric measurement technique uses a two-reflection method and an S parameter method. In the two-reflection method, the dielectric sample is positioned at the input side of a shorted waveguide. A second measurement positions the dielectric sample at the shorting plane. These arrangements are shown in Figs. 1 and 2.

The propagation constant γ is related to the reflection coefficients Γ_1 and Γ_2 by the following equation.

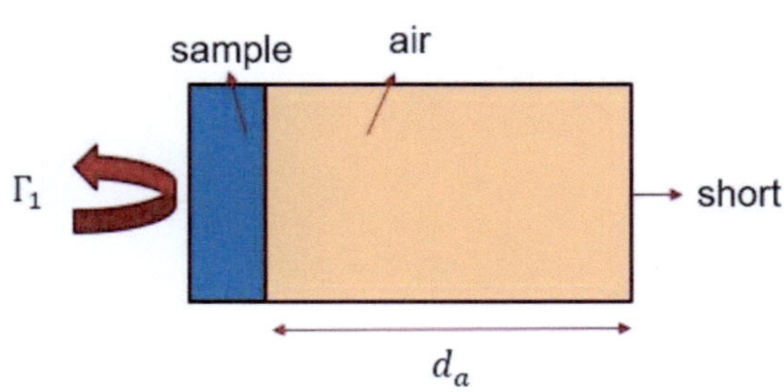

Fig. 1 First reflection measurement

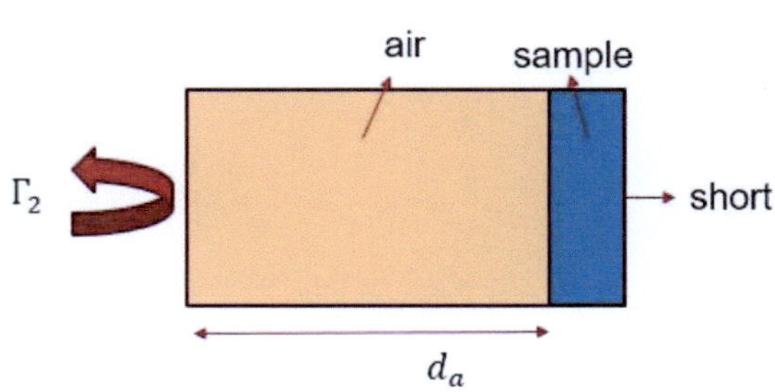

Fig. 2 Second reflection measurement

$$\gamma = j\beta_a\left(-\frac{j}{\sin\beta_a d_a(1+\Gamma_1)}\left((1-\Gamma_1)\cos\beta_a d_a - \left(e^{-j\beta_a d_a} + e^{j\beta_a d_a}\Gamma_1\right)\frac{1-\Gamma_2 e^{j2\beta_a d_a}}{1+\Gamma_2 e^{j2\beta_a d_a}}\right)\right)^{1/2} \tag{5}$$

where β_a is 2π divided by the waveguide wavelength in air and d_a is the waveguide length minus the sample length as shown in Figs. 1 and 2.

From γ, the dielectric constant and loss tangent are calculated as follows.

$$\varepsilon_r = \frac{\left(\frac{\pi}{a}\right)^2 - \mathrm{Re}\left[\gamma^2\right]}{\left(\frac{2\pi}{\lambda_0}\right)^2}$$

$$\tan\delta = \frac{\mathrm{Im}\left[\gamma^2\right]}{\left(\frac{2\pi}{\lambda_0}\right)^2\varepsilon_r} \tag{6}$$

where λ_0 is the free space wavelength, a is the waveguide width.

For the S parameter method, the dielectric sample is placed at an intermediate point in the waveguide and two port measurements are conducted. The arrangement is shown in Fig. 3.

The following parameters are calculated and measured.

$$\gamma_a = i2\pi\left(\frac{1}{\lambda_0^2} - \frac{1}{4a^2}\right)^{1/2}$$

$$\Delta S = S_{11}S_{22} - S_{12}S_{21}$$

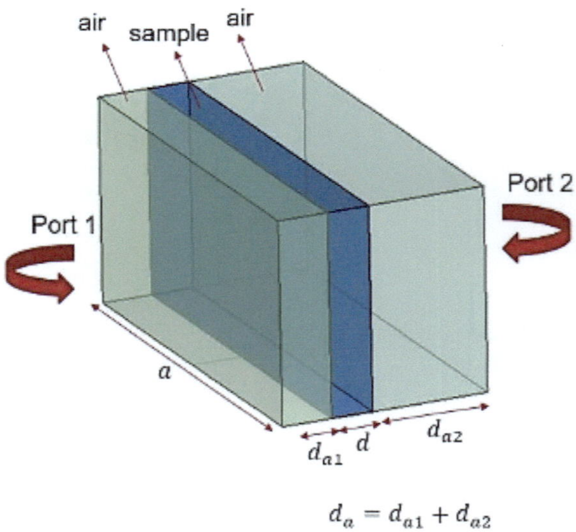

Fig. 3 S-parameter dielectric measurement

$$K = \frac{1 + \Delta S e^{2\gamma_a d_a}}{\left(\left(1 - \Delta S e^{2\gamma_a d_a}\right)^2 - 4S_{21}^2 e^{2\gamma_a d_a}\right)^{1/2}}$$

$$\gamma = \alpha + i\beta = \left(K + \sqrt{K^2 - 1}\right) \tag{7}$$

From these results, the dielectric constant and loss tangent are calculated using Eq. 6. This technique also determines the electrical length of the sample using the following equations.

$$\frac{S_{11}}{S_{22}} = e^{-2\gamma_a(d-d_s)} \tag{8}$$

$$d_s = d + \frac{Arg\left(\frac{S_{11}}{S_{22}}\right) + n \cdot 2\pi}{2\beta} \tag{9}$$

where d is the waveguide length and d_s is the sample length. The argument of S_{11}/S_{22} is the angular component of the complex value. One adds 2π as appropriate to find a physical length. The optimized 3D printed absorber simulated (HFSS) model is shown in the next figure. Each dielectric block is represented with a different color. Block 1 is nearest the air-filled input side of the waveguide (Fig. 4 and Table 1).

The magnitude and complex magnitude of the electric field is shown in the next figure. Note how the wavelength shortens as the dielectric constant increases (Fig. 5).

The optimized parameters for the model are shown in the next figure. The fill factors are 70.3, 80.2, 32.1, 64.0, 57.8, and 51.7%. These are all easy to print. Block lengths are 13.4, 7.35, 3.11, 7.80, 12.1, and 14.0 mm. The calculation of ε_r and tan δ is shown as a function of the fill factor. The optimized model swept over the waveguide band is shown in the next figure. This absorber only works at a single frequency (Fig. 6).

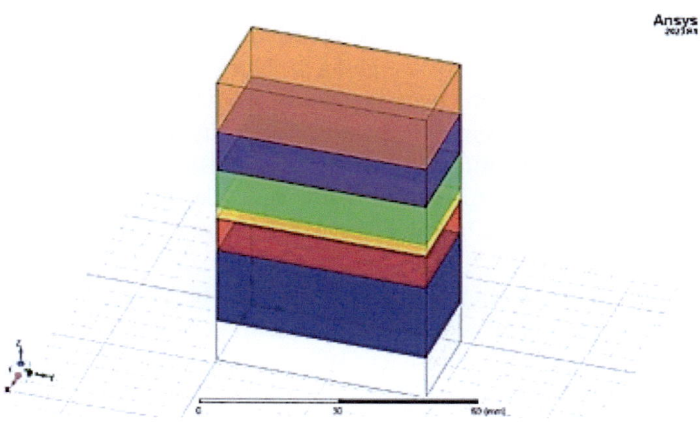

Fig. 4 Simulation of optimized narrow band 3D printed absorber

Table 1 Dielectric color codes and properties

Dielectric	Color	Length [mm]	ε_r	tan δ
1	Blue	16.851	2.246	0.007
2	Red	8.186	1.711	0.005
3	Yellow	3.900	3.154	0.011
4	Green	4.200	3.327	0.011
5	Purple	14.490	11.444	1.166
6	Orange	14.370	7.182	1.100

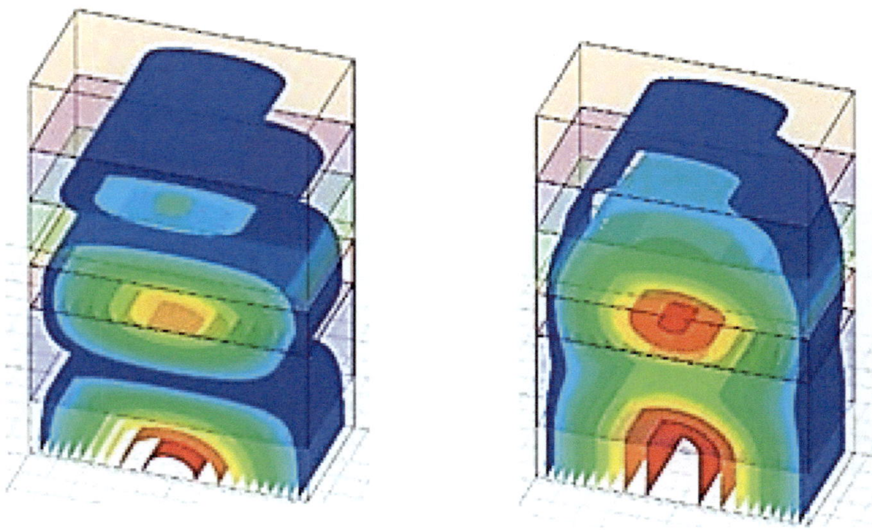

Fig. 5 Simulated electric field magnitude and complex magnitude

The attenuator components were 3D printed and assembled. The actual fill factor of each piece can be determined by weighing the block and comparing it to the solid design volume multiplied by the material density. Both PLA and Proto Pasta have a density of 1.24 g/cm^3. The Sunlu filament has a density of 1.26 g/cm^3. The actual lengths of each block were different than the design also, so the volume calculated in the drawing program was multiplied by the actual length divided by the design length. Actual weight divided by calculated 100% dense weight gives the actual fill factor.

The following table shows the measured/designed block lengths, the measured block weights, and the calculated/designed block fill factors (Table 2).

The 3D printed blocks that fit inside of WR-187 waveguide are shown in the next figure. All filaments are based on PLA plastic, so reasonable 3D printing parameters are used. The nozzle temperature was 200 C for the unloaded PLA and 220 C for the

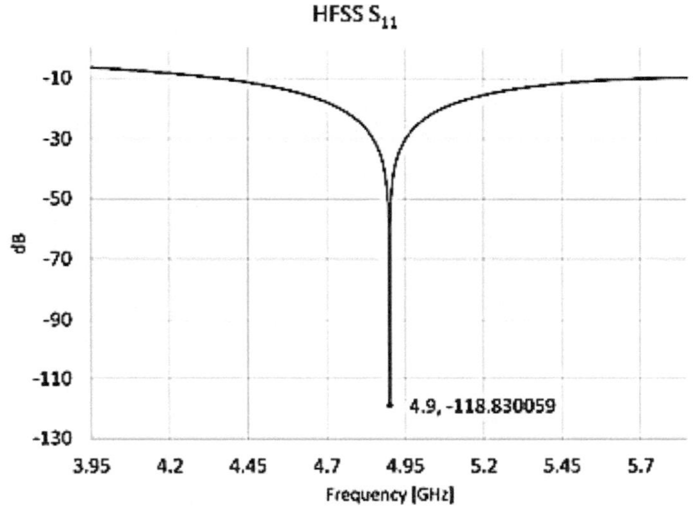

Fig. 6 Frequency sweep attenuation WR-187 band

Table 2 Design and measured absorber block properties

Block	Actual/design length	Calc/design fill factor
1	16.78/16.851	68.97/68.270
2	8.27/8.186	40.13/38.942
3	3.90/3.888	47.36/47.067
4	4.22/4.197	51.77/50.834
5	14.48/14.490	59.171/60.566
6	14.40/14.369	35.94/35.849

Block	Design ε	Measured ε	Design tan δ	Measured tan δ
1	2.246	2.1	0.007	0.005
2	1.711	1.6	0.005	0.005
3	3.154	3.2	0.011	0.012
4	3.327	3.4	0.011	0.012
5	11.444	11	1.166	1.3
6	7.182	5	1.100	1.5

carbon loaded PLA. The heated bed temperature was 55 C. A commercial FlashForge desktop FDM 3D printer was used to make all the parts (Fig. 7).

The experimental arrangement is shown in the next figure. A LibreVNA was used to perform the S11 measurement. This VNA was waveguide TRL calibrated with the pictured WR-187 launchers (Fig. 8).

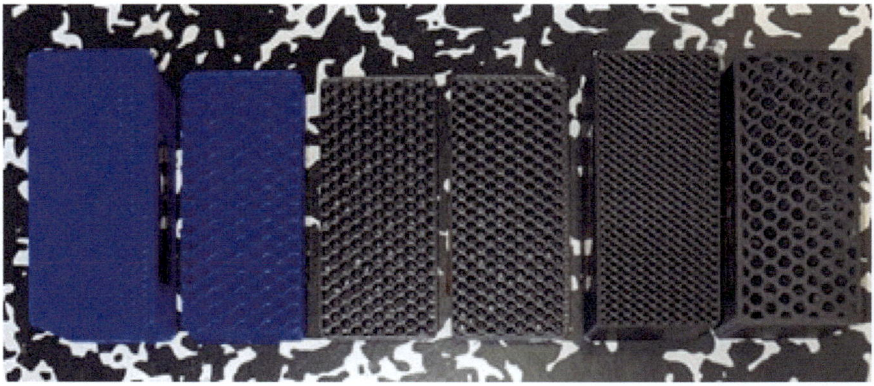

Fig. 7 3D printed absorber blocks

Fig. 8 Experimental arrangement

As shown in the next figure, the complete device had a minimum S_{11} of −73.254 dB at 4.929 GHz. The as-built device performs less well than the simulated, optimized design, though for practical purposes the performance is excellent (Fig. 9).

Though the absorber only works well at a single frequency, a 2.44-inch device has better attenuation (33 dB) than a 13-inch-long commercial load at that frequency. Clearly, custom dielectrics created by 3D printing can effectively attenuate a microwave signal. The optimizer using MATLAB was modified to find S_{11} values at ten frequencies, sum the result, and drop the lowest value. Dropping the lowest

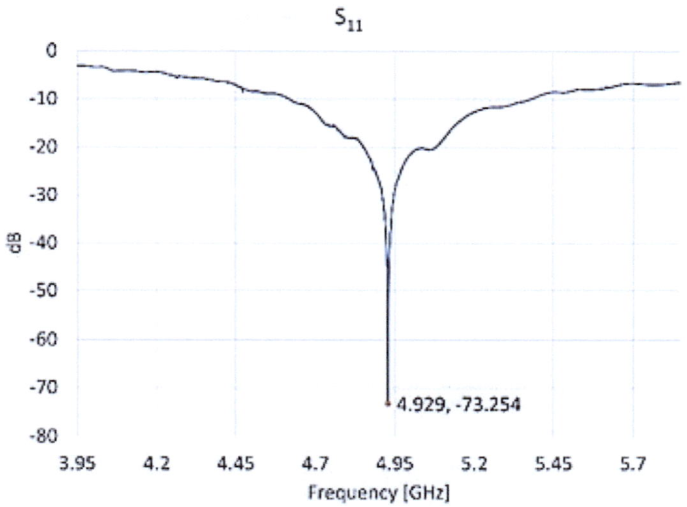

Fig. 9 Measured attenuation of 3D printed absorber

value was done to prevent the optimizer from finding a single optimum S_{11} value. The code then finds the optimal broadband solution. The MATLAB broadband result is shown in the following figure (Fig. 10).

The as-built broadband absorber did not achieve such a good result, but it was certainly much more broadband than the original load. The measured result is shown

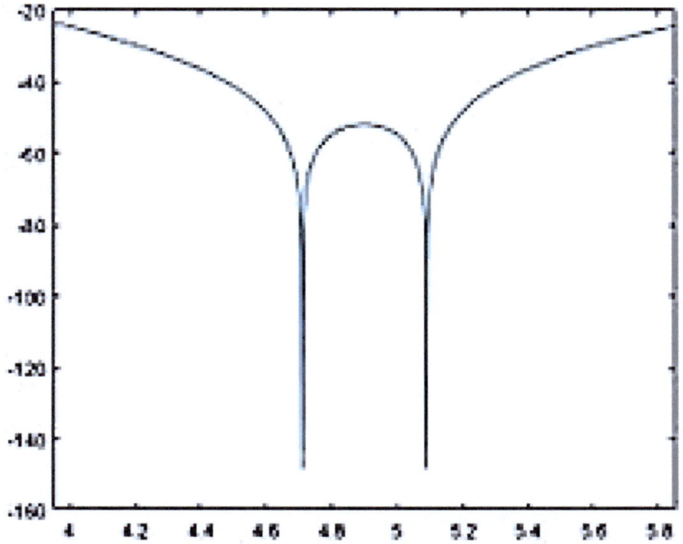

Fig. 10 Broadband absorber optimized simulation

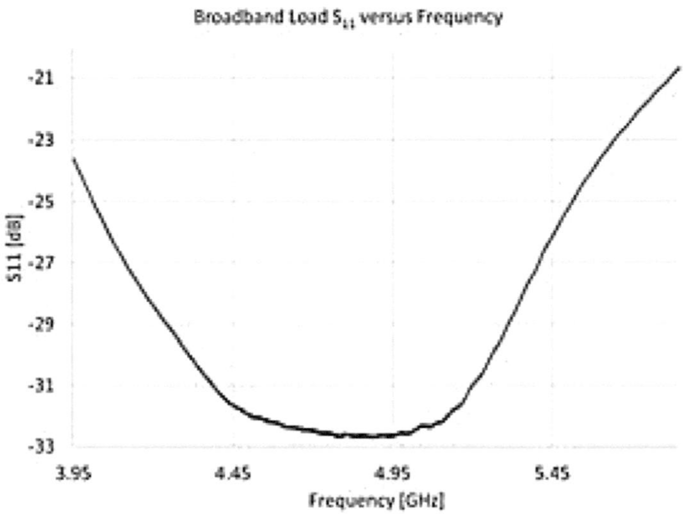

Fig. 11 Broadband absorber measurement

in the next figure. The average attenuation was −28.81 dB and the total load length was 2.18 inches (Fig. 11).

3 Conclusion

Using carbon-loaded PLA plastic and FDM 3D printing to create a very effective microwave attenuator has been proven. With 3D printing, the dielectric constant can be tailored in any desired three-dimensional configuration, so a great variety of problems can be solved. Computer optimization determines printed fill factors and dimensions. We will investigate using sintered 3D printed filaments to create UHV compatible, high-power loads and also investigate the use of commercial ceramic materials.

References

1. Le Sage, G.P., "3D printed waveguide slot array antennas, IEEE Access, vol. 4, pp. 1258–1265, (2016).
2. Tak, J., Kantemur, A., Sharma, Y., Xin, H., A 3-D-Printed W-Band Slotted Waveguide Array Antenna Optimized Using Machine Learning, in IEEE Antennas and Wireless Propagation Letters, vol. 17(11), pp. 2008–2012, Nov. (2018),
3. Le Sage, G.P., Thermal Frequency Drift of 3D Printed Microwave Components, Metals 2020, 10, 580; https://doi.org/10.3390/met10050580, 29 April (2020).

4. Le Sage, G.P., Dielectric Steering of a 3D Printed Microwave Slot Array, IEEE AWPL Special Cluster 2018, publication in November (2018)
5. Ghasemi, A., Laurin, J.-J., A Continuous Beam Steering Slotted Waveguide Antenna Using Rotating Dielectric Slabs, IEEE TRANSACTIONS ON ANTENNAS AND PROPAGATION, Submitted 25 Feb 2017, Revised 12 Mar (2018).
6. Le Sage, G.P., Stringfellow, R., 3D Printed, Self-Temperature Compensated Microwave Cavity Notch Filters," in IEEE Access, vol. 9, pp. 32969-32978, 2021, https://doi.org/10.1109/ACCESS.2021.3060637.
7. Tomassoni, C., Peverini, O.A., Venanzoni, G., Addamo, G., Paonessa, F., Virone, G., 3D Printing of Microwave and Millimeter-Wave Filters: Additive Manufacturing Technologies Applied in the Development of High-Performance Filters with Novel Topologies, in IEEE Microwave Magazine, vol. 21(6), pp. 24–45, June (2020)
8. Milan, S., 3-D Printed Microwave And Terahertz Passive Components, Ph.D. Thesis, University of Birmingham, June (2019). https://core.ac.uk/download/pdf/341793525.pdf
9. Monkevich, J.M., Le Sage, G.P., Design and Fabrication of a Custom-Dielectric Fresnel Multi-Zone Plate Lens Antenna Using Additive Manufacturing Techniques, IEEE Access, vol. 7, pp. 61452-61460, (2019).
10. Wang, C., Yu, W.S., Liu, Y.F., Peng, L., A 3D printed all dielectric Fresnel zone plate lens with high gain characteristic at the X-band, 2021 International Conference on Microwave and Millimeter Wave Technology (ICMMT), Nanjing, China, pp. 1–3, (2021)
11. https://www.rogerscorp.com/advanced-electronics-solutions/radix-printable-dielectric
12. Zhang, W., Mi, R., Khilkevich, V., 3D Printed Multilayer Microwave Absorber," 2022 IEEE International Symp. on Electromagnetic Compatibility & Signal/Power Integrity, Spokane, WA, USA, pp. 59–63, (2022) https://doi.org/10.1109/EMCSI39492.2022.9889603
13. Ping, G., Liang, H., Yan, L., Zheng, L., Wei, X., 3D-printed carbon fiber/polyamide-based flexible honeycomb structural absorber for multifunction broadband microwave absorption." Carbon 185, 272–281 (2021).
14. Dahyun, D.L., Jinwoo, P. Jaemin, L. Dowon, N., Jeongwoo, L., Jaeho, C., Wonjoon, C., Broadband mechanical metamaterial absorber enabled by fused filament fabrication 3D printing," Additive Manufacturing 55, 102856 (2022).
15. Kronberger, R., Soboll, P., New 3D printed microwave metamaterial absorbers with conductive printing materials," 2016 46th European Microwave Conference (EuMC), London, UK, pp. 596–599, (2016). https://doi.org/10.1109/EuMC.2016.7824413 Date accessed: 3/15/2024
16. https://www.proto-pasta.com/collections/all/products/carbon-fiber-pla Date accessed: 3/15/2024
17. https://www.sunlu.com/products/pla-carbon-fiber Date accessed: 3/15/2024
18. https://www.flashforgeshop.com/product/flashforge-pla Date accessed: 3/15/2024

The Effect of Pollution on High Voltage Power Lines Insulators

Messaoud Hamouda, A. Menasria, A. Laidi, M. A. Hartani, I. Hassani, O. Ouledali, A. Oudrane, and S. Laribi

Abstract This work relates to the study of the effect of pollution on the insulators of high-voltage overhead power lines. The study will be carried out using a new laboratory model for flashover studies. This model resembles a real outdoor insulator with respect to its shape and pollution. We carried out a series of experiments to measure the resistance of pollution between the electrode of mass and a static discharge in various positions of the discharge on the surface of the electrolyte as well as the current and critical voltage. The experimental results of this model and their analysis were performed in two groups: The first is the measure of pollution resistance between the ground electrode and a static discharge for many discharge positions. The second is the measure of critical current and critical voltage according to the position of the discharge starting point. A comparison with the case of a point-to-point gap reveals that the electric field in the air in the vicinity of the electrolyte surface is responsible for the discharge elongation, and the critical conditions of flashover are the weakest initial electric conditions necessary to create this field distribution. The results obtained verify and agree with those of other research works referring to devices considered as prototypes of work and allowing to validation of this model.

Keywords Discharge · Flashover · High voltage · Insulator · Pollution

1 Introduction

Flashover insulators on high-voltage lines interrupt the distribution of energy, often leading to load shedding on the line supported by the faulty insulator. It results in a short circuit between the part carrying the high voltage and the earthing system, created by an electric arc generally established on the surface of the polluted layer covering the insulator. The origin of this phenomenon is related to the presence of

M. Hamouda (✉) · A. Menasria · A. Laidi · M. A. Hartani · I. Hassani · O. Ouledali · A. Oudrane · S. Laribi
Faculty of Science and Technology, LDDI Laboratory, University of Adrar, Adrar, Algeria
e-mail: mes.hamouda@univ-adrar.edu.dz

A. Vaseashta et al. (eds.), *Proceedings of the Eighth International Symposium on Dielectric Materials and Applications (ISyDMA'8)*, Springer Proceedings in Physics 349, https://doi.org/10.1007/978-3-031-84809-4_10

an uneven polluting layer, covering the insulators of airlines installed by the sea, industrial regions, or in the desert. The combination of this with water brought by an atmospheric moisture supply forms a conductive film (electrolyte) on the surface of the insulator and contributes to the degradation of these dielectric properties. The processes of developing electrical discharge on the surface of polluted insulators are very complex. Among the parameters identified that make this process complex are in particular—the variability of surface resistance.

In the literature, several experimental models have been used, the aim of which is to find the criteria and mechanisms for the evolution of flashover discharge. The simplest and most used model is designed by Obenaus [1] the so-called one-directional model, on its basis Alston and Zoledziowski [2] affirmed the existence of a critical point of circumvention corresponding to the critical values of voltage, current, and critical length, beyond these values the circumvention is possible. Hampton [3] showed that critical conditions are achieved when the potential gradient in the electrolyte is higher than the potential Gradient within the positive column of the discharge. Hesketh [4] generalized the conditions of Hampton, stating that the change in the discharge is only possible if it causes an increase in the current consumed by the system. Wilkins [5] reformulated Hesketh's criterion as follows: the lengthening movement of the discharge occurs when the power supplied by the source increases during the discharge.

After listing the flashover criteria, we have noticed that the previous work has resulted in the same sizes that constitute the sizes of the same critical point, from which the discharge becomes unstable, and the flashover will take place. It is assumed that this agreement results due to the peculiarity of the single-turn model. Flazi [6] has proposed a new model called two-direction or the discharge can stretch between two directions at two different grooves. He developed his criterion, by making a distinction between the flashover criteria and the elongation criteria, by criticizing the first model criteria or the one-direction model. Based on these two models, we conducted experiments aimed at studying the effect of pollution represented here by electrolytic resistance on the development of polluted insulator flashover discharge.

2 Experimental Set-Up

We used our experiences on two models: One and two-directional models. In the first case, we conducted our experiments on a so-called one-directional model.

It consists of a plexiglass plate dug with a strain filled with the electrolyte (H_2O + NaCl) of resistivity ρ, representing the pollution. A high-voltage electrode from the capacitor discharge, overlooking the strain at a height "h". A low-voltage electrode is placed at one of the ends of the screw figure (see Fig. 1). In the second case, previous experiments were repeated on another so-called flashover model. It consists of an electrolyte-filled strain and has two mass electrodes at both ends. A high-voltage electrode above the crane at a height of approximately 2 mm above the electrolyte and a distance between the two mass electrodes (see Fig. 2).

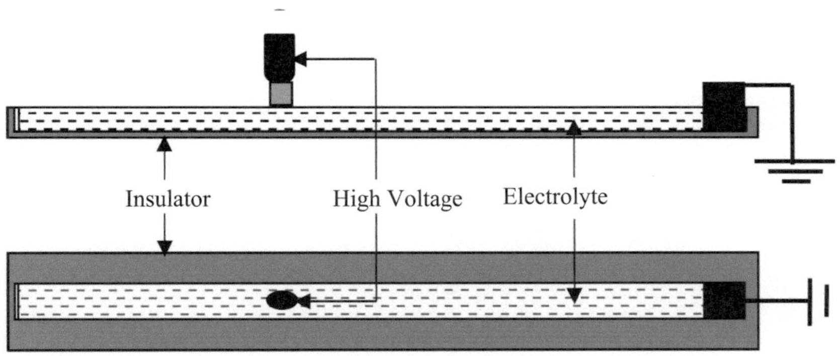

Fig. 1 One directional model

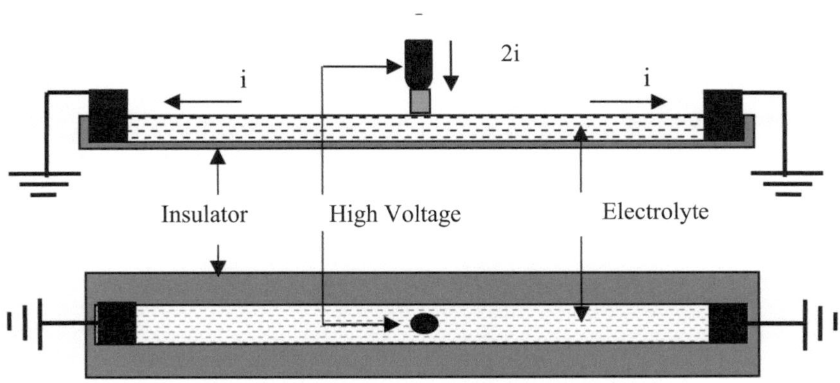

Fig. 2 Two directionals model

Our grooves (in the number of five), have the same length (L = 40 cm) and a width that differs from one groove to another, varies from 5.6 to 13.6 cm, and the depth of each of them of 3 mm. By sliding the high-voltage electrode along the hose at different positions ranging from 11 to 19 cm. At each position, the critical voltage and current are taken. The same manipulation is repeated with the five grooves.

3 Results and Discussion

3.1 Results on the One-Direction Model

The test results are illustrated graphically in Figs. 3, 4, 5 and 6 respectively.

Influence of the critical voltage. Figure 3 shows the variation of the critical voltage in positive polarity depending on the electrolytic resistivity ρ = 2, 4, 6 and 8

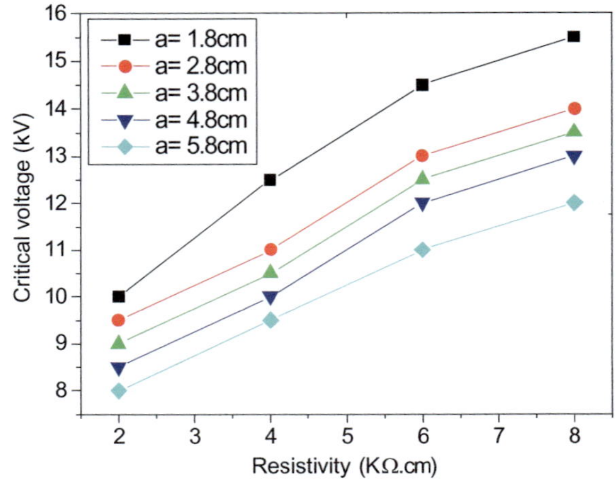

Fig. 3 Critical voltage as a function of resistivity for different widths a and thickness $e = 3$ mm, at position $x = 11$ cm

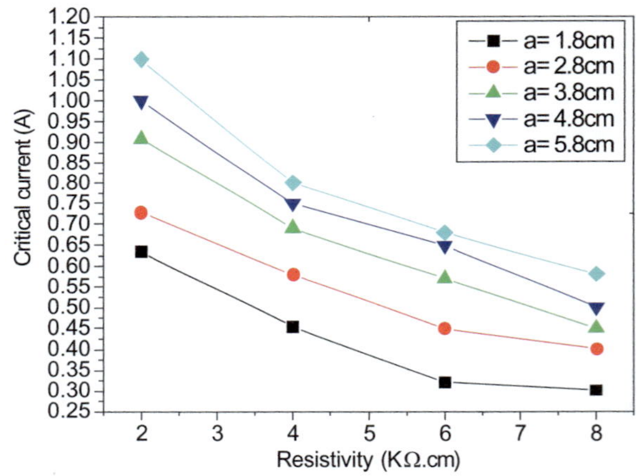

Fig. 4 Critical current as a function of resistivity for different widths a and thickness $e = 3$ mm, at position $x = 11$ cm

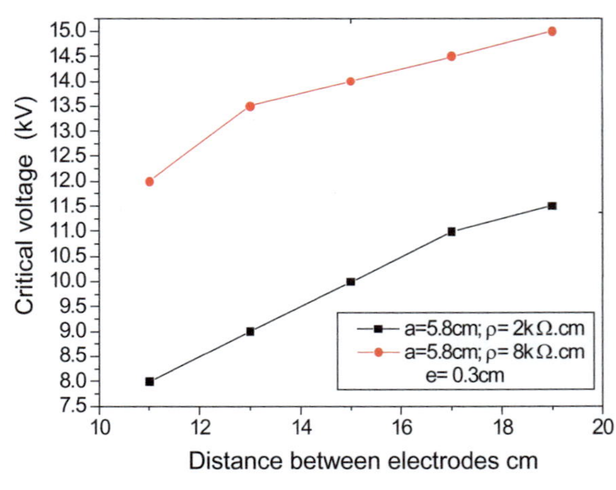

Fig. 5 Critical voltage as a function of the distance between electrodes

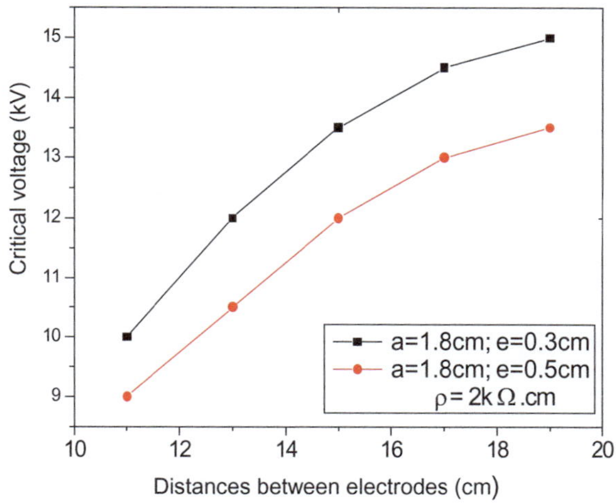

Fig. 6 Critical voltage as a function of the distance between electrodes

kΩ cm for different widths ranging from 1.8 to 5.8 cm, and thickness equal to 3 mm at the position x = 11 cm. The first observation to be made is that: (a): Critical voltage increases rapidly with the increase of the resistivity ρ for the low values of ρ, this increase is slower for the high values, so we find that the resistance is a parameter on which the tension depends; (b): the appearances of the lower curves correspond to the broad grooves and the upper appearance to the narrow; and (c): curves look almost the same as any thickness. Curves have almost the same appearance regardless of width. The interval between the two voltage curves is the same for small resistance values ρ = 2 kΩ cm. Critical tensions decrease from 5 kV from one width to another from a = 1.8 cm to a = 5.8 cm, for small resistivities ρ = 2 kΩ cm, and with any thickness.

Influence of the critical current. Figure 4 shows critical current characteristics based on electrolytic resistance that vary from ρ = 2, 4, 6 and 8 kΩ cm for thickness (e = 3 mm) and different widths that range from 1.8 cm to 5.8 cm. We observe that the variation of the critical current is inversely proportional to the value of the resistivity. They decrease quickly with small values and more slowly with large values. We note that all curves have almost the same appearance for any thickness. The lower lines correspond to the narrow channels of a = 1.8 cm width and the upper lines to the wide channels of a = 5.8 cm wide. The lines of the curves are the same for any width.

Influence of cell length on critical voltage. The influence of the length of the cell on the critical voltage. We measured critical voltage values corresponding to inter-electrode distances along the length of cell L. These distances vary from 11 to 19 cm between the position of the high voltage electrode and the mass electrodes.

Influence of cell length on critical current. The variations in critical current depending on the distance between electrodes are shown in Figs. 7 and 8, which represent the influence of resistance and thickness, respectively. These curves are croissant and slightly linear and have the same appearance. When the resistivity ρ or thickness e is kept constant, the current increases with the distance between electrodes.

Influence of linear resistance. Figure 9 represents the variation of the critical voltage in linear resistance for channel widths ranging from a = 1.8 cm to a = 5.8 cm and an electrolyte thickness of e = 3 mm. We notice that the critical voltage increases rapidly with the increase of the linear resistance for low values. This increase is slower for higher resistance values. These results agree with several authors [7–11]. For resistance values greater than 49.5 kΩ, the critical voltage is expected to become constant with a value of approximately 15.5 kV. The curves have the same appearance, and they are identical for any value of r. For a resistance r = 1 kΩ cm we have the same voltage U = 10 kV, even though the widths are different.

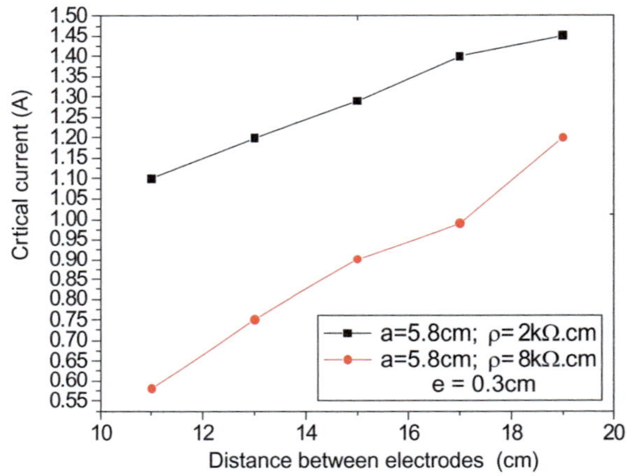

Fig. 7 Critical current as a function of the distance between electrodes

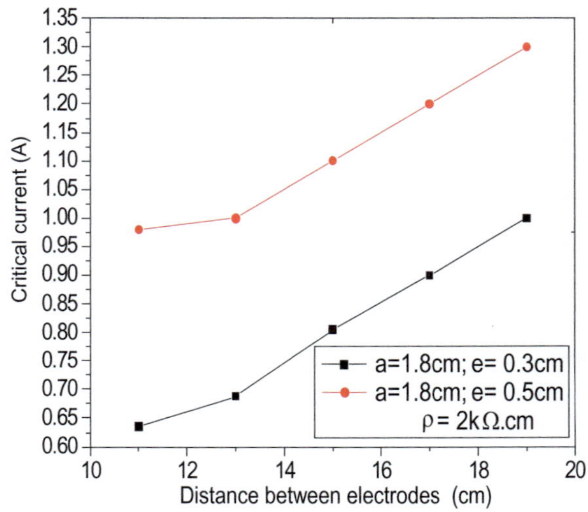

Fig. 8 Critical current as a function of the distance between electrodes

Fig. 9 Critical voltage as a function of resistance per unit length for different widths a and thickness $e =$ 3 mm, at position $x = 11$ cm

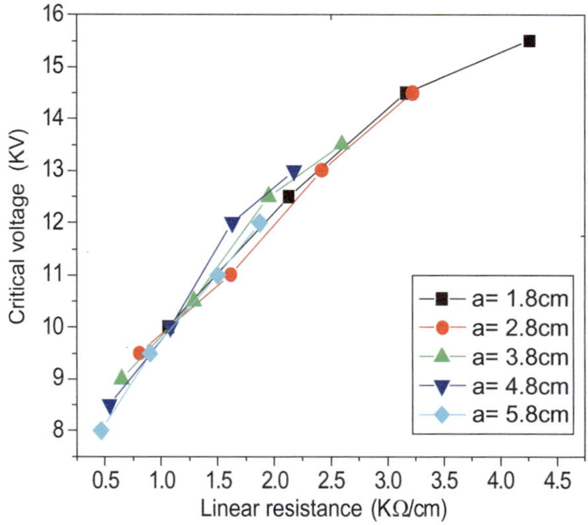

Figures 9 and 10, respectively, represent the variation of the critical voltage and the critical current according to the resistance per unit of length for different channel widths ranging from 1.8 to 5.8 cm and the electrolyte thicknesses ranging between 0.3 cm. We find that the critical voltage for the low resistors increases rapidly with the increase of the resistance. Then this increase is slower for large resistors. So, we can say that the resistance per unit of length is a parameter on which the voltage depends. This result was found by other authors [12, 13]. This is explained by the Flazi hypothesis [6], which shows that the value of the resistance influences the ionization process and the distribution of field lines in size and direction. Thus, for the high resistance values, the increase of the r value of the product therefore, a decrease in the initial diameter of the discharge, an increase in the length of the foot especially in the part of low current density, and a tendency of the current lines to become more and more parallel to the surface of the electrolyte.

This allows us to say that from a certain value of r the current lines become almost parallel to the surface of the electrolyte, especially after the first moments of elongation, and the penetration area of the current then covers the entire surface of the electrolyte. Some current lines thus reach the electrode [14]. We repeated the same experiments on the second model, called the two-directional model. After analyzing the results, we observed that the figures for the one-directional model were almost identical to those for the two-directional model. We, therefore, thought it useful to compare the values of the critical voltages and currents of the two models for two grooves: one narrow with a width of a = 1.8 cm and the other side with a width of a = 5.8cm and a thickness of = 3 mm (see Figs. 11 and 12).

In both cases, we can see that electrolytic resistivity has a very significant influence on the critical value of the voltage. This resistivity increases as the critical voltage increases. The current decreases as the resistivity increases. We can also see that the curves are almost the same for both single and dual circuits. Hence the difference

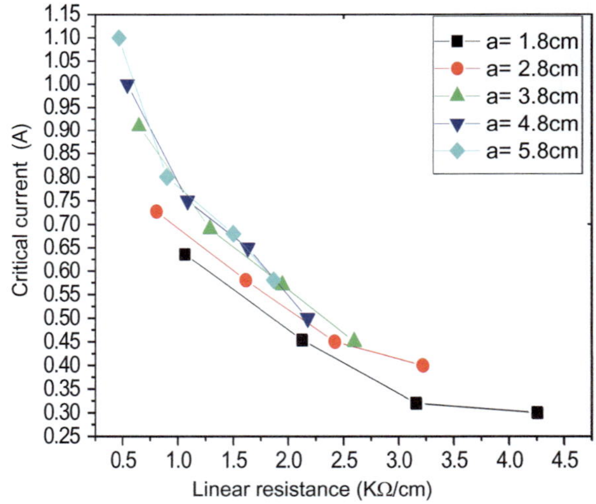

Fig. 10 Critical current as a function of resistance per unit length for different widths a and thickness e = 3 mm, at position x = 11 cm

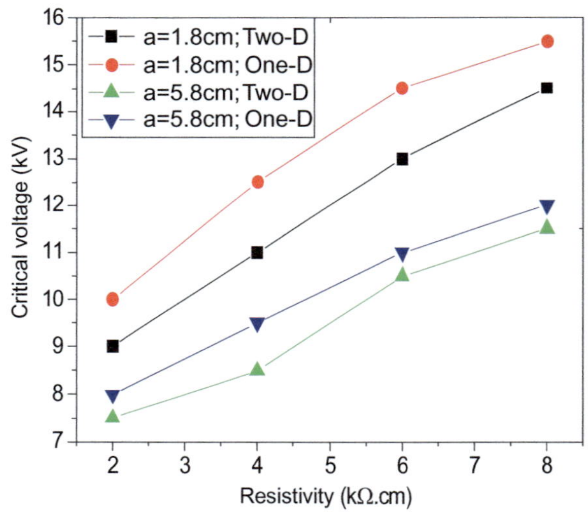

Fig. 11 Critical voltage as a function of the resistivity of two widths at one and two directional channels

between the curves which is due to the position of the H.T electrode (single-ended model X = 11cm, on the other hand for the double-ended model X = 10 cm). Therefore, these results lead us to say that the conditions are also the same and that the parallel current in the two-directional model has no influence on the critical conditions on the side where bypassing does not occur for the one-directional model. The figures above show that the voltage is proportional to the electrolytic resistivity, and the current is inversely proportional, and these results agree with the majority of the authors [15–18].

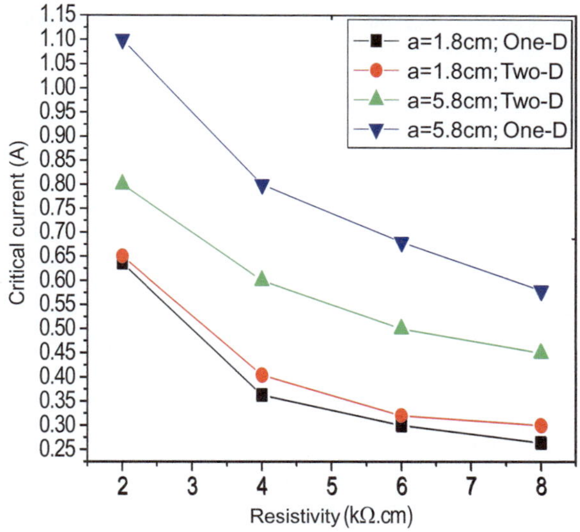

Fig. 12 Critical current as a function of the resistivity of two widths. One-D: one directional, Two-D: two directionals

4 Conclusion

Our rectangular model allowed us to refine the experiments for the determination of critical circumvention conditions and the effect of pollution on the insulators of high voltage. We have pointed out that in both cases: one or two directionals, the critical conditions are almost the same. from the results obtained on the influence of various parameters, we found that: neither, the dimensions of the grooves, nor the electrolytic resistance, taken separately, can influence the critical flashover conditions, but this is the case if these parameters are taken together and are connected by the ratio of the surface resistance ($r = \rho/a.e$). Hence the main parameter on which critical sizes depend is the linear resistance. This result is consistent with those found by other researchers [19, 20].

References

1. Obenaus, F., Dieubershlogspunnung verschutzler isolatoren. ETZ, **vol. 56**, pp 368 – 370, (1935).
2. Alston, L.l., Zoledziowski, S., Growth of discharges on polluted insulators, Proceedings, vol.110, n°7 , pp.1260–1266 (1963).
3. Hampton. B.F., Flasheover mechanism of polluted insulation. Proc I.E.E., Vol III, n°5, p.p. 985–90, (1964).
4. Hesketh. S., General criterion for the prediction of pollution flashover. IEE, Proceeding, vol.114, pp 985-990, (1967).
5. Wilkins, R., Flashover voltage of high voltage insulators with uniform surface pollution films. Proc. IEE, vol. 116, n°3, p.p. 457-65. (1969).

6. Flazi, S., Etude du contournement électrique des isolateurs H.T. pollués. Critère d'élongation de la décharge et dynamique du phénomène. Thèse de doctorat Es-science, Université Paul Sabatier, Toulouse (1987).
7. Mahi, D., Observation de l'allongement d'une décharge électrique sur des surfaces faiblement conductrices. Thèse de doctorat d'état, Université Djillali Liabès, Sidi Bel Abes, (2002).
8. Adjabi, M., Mise au point d'un critère de contournement à partir des modèles électrolytiques. C.N.H.T' 97. Tizi-Ouzou, 17 - 19 Nov. (1997).
9. Gorur, R.S., Flashover of contaminated nonceramic outdoor insulators in a wet atmosphere. IEEE, trans on dielectric and electrical insulation **vol. 5**, **N° 6**, Dec. (1998).
10. Flazi, S., Hadi, H., Rabah, K.L., .Hamouda, M., Boudjella. A., Dynamic Aspect of the Direct Current Flashover on Rectangular and New Disk Model. IEEE Conference on Electrical Insulation and Dielectric Phenomena (CEIDEP), Canada.17–19 Nov (2001).
11. Hamouda, M., Flazi, S., Critical Condition of Flashover on a Disk Model. CNHT'2003, 01-, USTMB Oran, 03dec (2003).
12. Flazi, S., Ouis, A., Hamouda, M., Hadi, H., Dynamic feature of the discharge in the equivalent electric circuit of flashover. IET Proc. Generation, Transmission & Distribution Vol. 1, Issue1, 2007. ISSN 1751–8687, (2007).
13. Hamouda, M., Flazi, S., Critical condition of flashover on a new model of polluted insulators. Research Journal of Applied Sciences ISSN: 1815–932X, Vol 2 N° 4 pp 471–475 September (2007)
14. Flazi, S., Hamouda, M., Study of the pollution resistance and current distribution in front of the distribution discharge. International Symposium on H.V Engineering ISH'2003 25–29, Delft Netherland, (2003).
15. Zhou, H., Chen, Y., Prediction of Insulator Pollution Flashover Voltage Based on Data Mining Technology. IOP Conf. Series: Earth and Environmental Science 692 (2021) 022070 IOP Publishing https://doi.org/10.1088/1755-1315/692/2/022070
16. Slama, M.E., Contribution à l'Etude de l'Influence de la Non-Uniformité de la Distribution de la Pollution basée sur la Méthode de la DDSE pour le Dimensionnement des Isolateurs des Lignes THT à Courant Alternatif. Mémoire de Magister, Dép. d'électrotechnique, faculté de génie électrique, USTO-MB, Juin 2002.
17. Derkaoui, A., Etude de la Validité d'un Nouveau Modèle Expérimental du Contournement des Isolateurs Haute Tension Pollués. Mémoire de magister, USTO-MB, Juin 15 juin 2004.
18. Bouhmama, M., Modélisation du contournement sur isolateur réel pollué. Thèse de doctorat, Université Sidi Belabas, (2018).
19. Salem, A.A., et al., Pollution Flashover Under Different Contamination Profiles on High Voltage Insulator. Numerical and Experiment Investigation: IEE Access; Vol 9; (2021) https://doi.org/10.1109/ACCESS.2021.3063201 (2021).
20. Salem, A.A., et al.: Pollution Flashover Characteristics of High Voltage Outdoor Insulators. Arabian Journal for Sciences and Engineering; Vol 47; pp 2711-27299; (2022).

Effects of Change of Chirality of Green Dielectric Gold Nanoparticles on Antimicrobial Activity

Karen M. Soto, Angelica Gódinez-Oviedo, Adriana Romo-Pérez, Sandra Mendoza, José Mauricio López-Romero, Jorge Pineda-Piñón, Luis M. Apatiga-Castro, and Alejandro Manzano-Ramírez

Abstract Gold nanoparticles (AuNPS) are widely used metallic nanoparticles with unique surface plasmon characteristics. They have applications in biomedical and therapeutic fields. The present study aims to study the effect of AuNPS chirality surface on the dielectric constant, electrical conductivity, and antimicrobial activity. AuNPs were prepared using green synthesis with Pirul's aqueous extract. The modification of chirality is carried out by adding L or D-cysteine to the surface and corroborating it through a chirality analysis. TEM characterized the AuNPs to determine their size and morphology, and the particle size dispersion was evaluated by DLS, Z potential, and electrical parameters were obtained with a precision component analyzer. The disk technique and the MIC and MBC against microorganisms with medical importance were evaluated for microbiological activity. AuNPs with different morphologies (spherical, triangular, and hexagonal) and diameters between 10 and 20 nm were obtained. The diameter increased by 3–4 nm after surface functionalization, while the morphologies remained similar. The functionalization of the

K. M. Soto (✉) · J. M. López-Romero · A. Manzano-Ramírez
Centro de Investigacion y de Estudios Avanzados del I.P.N., Unidad Querétaro, Libramiento Norponiente 2000, Querétaro, Mexico
e-mail: karen.soto@cinvestav.mx

A. Gódinez-Oviedo · S. Mendoza
Departamento de Investigación y Posgrado en Alimentos, Facultad de Química, Universidad Aut´Onoma de Querétaro, Querétaro, Qro., Mexico

A. Romo-Pérez
Instituto de Química, Universidad Nacional Autónoma de México, Circuito Exterior S/N, Ciudad Universitaria, C.P., Coyoacán, Ciudad de México, México

J. Pineda-Piñón
Centro de Investigación en Ciencia Aplicada y Tecnología Avanzada, Instituto Politécnico Nacional, Unidad Querétaro, Cerro Blanco No. 141, Colinas del Cimatario, Querétaro, Qro, Mexico

L. M. Apatiga-Castro
Centro de Física Aplicada y Tecnología Avanzada (CFATA), Universidad Nacional Autónoma de México, Querétaro, México

A. Vaseashta et al. (eds.), *Proceedings of the Eighth International Symposium on Dielectric Materials and Applications (ISyDMA'8)*, Springer Proceedings in Physics 349, https://doi.org/10.1007/978-3-031-84809-4_11

surface was verified through a chirality study, observing that the cysteine adhered to the surface, giving them L or D chirality. It was also confirmed that the functional groups of the cysteine and the extract are present on the surface with the help of an IR analysis. AuNPs present strong dielectric dispersion corresponding to the alpha relaxation region, and the conductivity values are dependent on the size of the particles; this parameter can be related to a major accumulation and better activity in cells. It was found that the change in chirality affects the antimicrobial properties of the particles since when modifying to a chirality D, the activity decreases. At the same time, L increases by up to 30% and allows antifungal activity against C. albicans. The functionalization of green gold nanoparticles was achieved by changing their chirality by adding the cysteine in its L or D conformation; these changes impact the electrical properties and antimicrobial activity, increasing or decreasing it according to the chirality obtained.

Keywords Dielectric materials · Gold nanoparticles · Biomedical application · Antimicrobial activity

1 Introduction

Chirality is a physical–chemical phenomenon found at various hierarchical levels, ranging from the atomic and molecular to the supramolecular and macroscopic scales. This property refers to two enantiomers of a chiroptical compound that are characterized by the capability to rotate in the opposite direction; this feature confers them peculiar optical activities, such as circular dichroism, circularly polarized luminescence, and non-linear optics [1, 2]. Taking inspiration from this aspect, in recent years, researchers have focused on developing chiral inorganic nanostructures to combine the outstanding (bio)technological applications of nanoparticles (NPs) with the intriguing and multifaceted potentials of chiral systems with different biological and biomedical applications, such as imaging, diagnostics, biosensing, and therapeutics [3, 4]. Gold nanoparticles are particles with different morphologies, including nanospheres, nano bars, nanolayers, and nanocubes, with small sizes ranging from 1 to 100 nm. The synthesis method is based on reducing Au^{3+} with chemical or green reduction agents. generally don't present intrinsic chiral properties, but the functionalization with different compounds can be used to modify their chiral properties; one of these compounds is the cysteine that can be combined with –SH groups through Au–S bonding; this change in chiral properties affects other properties like morphology, size, conductivity, dielectric and possible applications [5]. The present study aims to study the effect of AuNPS chirality surface on the dielectric constant, electrical conductivity, and antimicrobial activity. AuNPs were prepared using green synthesis with Pirul's aqueous extract.

2 Methodology

2.1 *Extract and Characterization*

Molle (*Schinus molle*) (World Checklist of Selected Plant Families, WCSP record in review: Sp. Pl.: 389.) fruits were collected in Hidalgo (Mexico) and deposited in the "Dr. Jerzy Rzedowski" (QMEX) herbarium of Universidad Autónoma de Querétaro (Juriquilla, Querétaro, Mexico). The selected fruits were washed with deionized water and dried in a convection oven at 60 °C for 6 h. The dried residues were then powdered to fine particles using a mixer grinder (KRUPS GX4100) and passed through a 60-mesh screen (250 μm particle size). For Ultrasonic-Assisted Extraction (UAE), 10 g of ground Pirul seed was placed in a 250 mL beaker and extracted with 100 mL distilled water using an ultrasonic probe. The sonicator tip was immersed 30 mm into the sample at an amplitude of 50% for 15 min. The ultrasonication extraction process was carried out in ice water. The extracts were filtered (Whatman Paper no. 40) and stored at 4 °C for further analysis).

2.2 *Green Synthesis of Gold Nanoparticles with Pirul Seed Extract*

For the green synthesis of AuNPs, 10 mL of an aqueous solution of $HAuCl_4$ (0.5 mM) was mixed with 100 μL of the aqueous extract of Pirul seeds (P-AuNPs). The reaction mixture was incubated at 60 °C for 5 min, the reduction of Au^{3+} ions was monitored using color changes, and the UV−vis spectrum was recorded from 300 to 800 nm with baseline correction using water as the blank on a Spectra Max Tunable Microplate Reader (Molecular Devices Co., Sunnyvale, CA, USA).

2.3 *Synthesis of L/D-Cys-AuNPs*

The L/D-Cys-AuNPs were prepared as follows: first, 0.1. mL of 1×10^{-4} M of L or D-cysteine solutions were added to 5 mL of the green AuNP solution, followed by 2 h of stirring, before centrifuging the solutions. After centrifugation, the supernatants were removed. The precipitates were dispersed in deionized water by sonication.

2.4 *AuNPs Characterization*

The morphology and diameters of AuNPs were determined using a scanning transmission electron microscope (STEM SU8230, Hitachi, Tokyo, Japan). The diameters of the AuNPs were obtained by measuring at least 50 particles with Image J software. Infrared absorption spectra of dried AuNPs and lyophilized extracts were recorded in Fourier-transformed infrared (FT-IR) equipment (Spectrometer Spectrum Two, PerkinElmer, Waltham, MA, US). The electrical parameters were measured in the frequency range of 20 Hz up to 1 MHz using a WAYNE KERR precision component analyzer, model 6440 B, with the adapted methodology of Anwar et al. 2011 [6].

2.5 Antimicrobial Activity

Well-Diffusion antimicrobial analysis

The stock cultures of the six strains were streaked on trypticase soy agar (TSA) and were incubated at 35 °C for 24 h. From TSA plates, colonies were harvested, cell suspensions were adjusted to 0.5 McFarland standard, and a sterile cotton swab was straked on the surface of Muller Hinton agar (MHA). Four wells of 8 mm were made in the plates, and inside of them was put 50 μL of different concentrations (10 mg/mL, 5 mg/mL, 2.5 mg/mL, 1.25 mg/mL, 1 mg/mL, 0.1 mg/mL) of the *Schinus molle* extract and the different nanoparticles of *Schinus molle* and cysteine. The plates were incubated at 35 °C/24h, and the inhibition zone was measured.

3 Results and Discussion

3.1 Green Synthesis of Gold Nanoparticles and Characterization

TEM images and DLS particle size of the P-AuNPs and L/D-Cys-AuNPs are presented in Fig. 1. The particles obtained have different morphologies and sizes, with triangular and spherical morphologies predominating. The sizes in TEM micrographs are 19.86 ± 1.96, 22.69 ± 1.23, and 28.63 ± 2.31 nm for P-AuNPs, L-Cys-AuNPs, and D-Cys-AuNPs, respectively. It can be observed that modifying the particles' surface chirality significantly increases their size, mainly with a compound of chirality D. The hydrodynamic diameter obtained by DLS analysis showed that the smallest diameters were the P-AuNPs (59.75), followed by modified L-Cys-AuNPs (91.14) and D-Cys-AuNPs (110.1). The results agree with the order observed with the STEM micrographs. However, diameters are bigger, a fact ascribed to compounds stuck on the surface of the particles and the formation of aggregates of the particles that increase their size.

Zeta potential measures the magnitude of electrostatic (or charge) repulsion or attraction between particles and is one of the fundamental parameters known to affect stability. Particles with a zeta potential value above +30 mV (or below −30 mV)

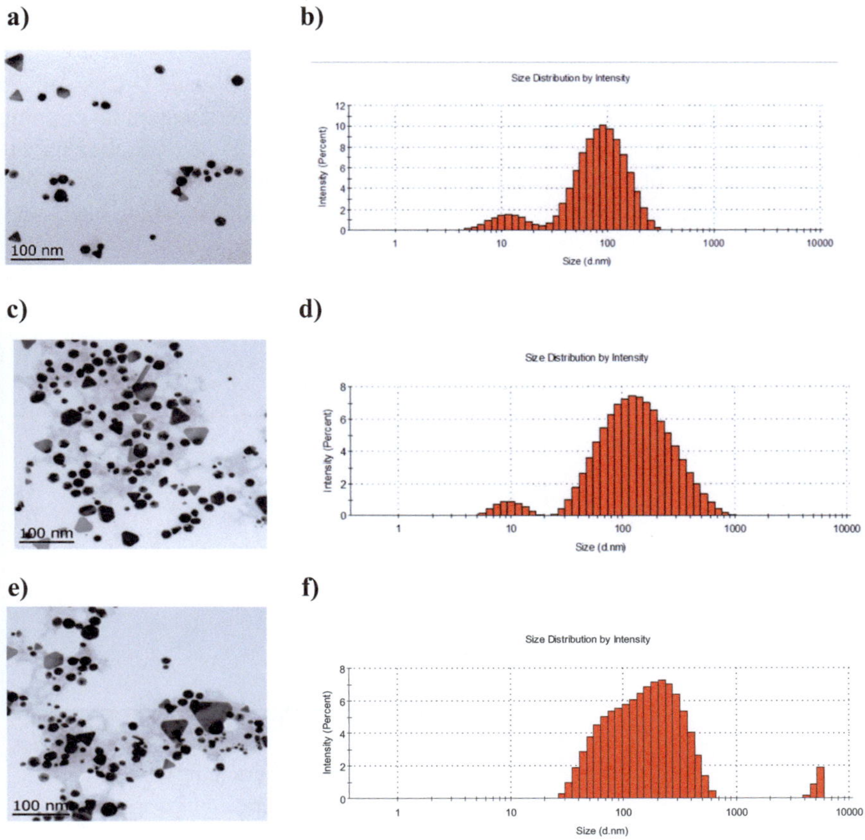

Fig. 1 TEM micrographics and DLS size of **a**, **b** P-AuNPs, **c**, **d** L-Cys-AuNPs and **e**, **f** D-Cys-AuNPs

are commonly considered stable. The particles present high stability and a negative charge of −4.88 mV, −2.85 mV, and −18.6 mV for P-AuNPs, L-Cys-AuNPs, and D-Cys-AuNPs, respectively. We can observe that the change in surface chirality affects the charge quantity but not the stability [7].

The IR spectra of the AuNPs are shown in Fig. 2, and these spectra were employed to detect the characteristic bands of the cysteine portion after P-AuNPs modification. Generally, amino acids exist as zwitterions (internal salts) and exhibit spectra characteristic of carboxylate and primary amine salts. The modified AuNPs present a characteristic band at 1600 and 1390 cm^{-1}, which corresponds to the asymmetric stretching of COO–, a band at 1532 cm^{-1} related to N single bond H bend, and a 3000–3500 cm^{-1} presents a very broad band which corresponds to of NH_3^+ stretch. We can't observe the absence of a band in modified nanoparticles approximately at 2500 cm^{-1}, related to the S–H cysteine group, confirming the S–Au interaction. A shift in the position of COO– and NH_3^+ stretching is likely due to a change in their dipole moment when cysteine binds on a metal surface with high electron density [8].

Circular electronic spectra (ECD) of P-AuNPs, L-Cys-AuNPs, and D-Cys-AuNPs (Fig. 3) show that the nanoparticles present the chirality of the compounds capping on the surface. An increase in the transition energy, about 225 nm, can indicate the synergetic interplay of the electrostatic interaction among the –SR side chains and the Au, aurophilic attraction in the Au(I)-Cys polymeric backbone, which likely may result in helicity in the nanostructures [9].

Fig. 2 IR spectra of the AuNPs and modified AuNPs

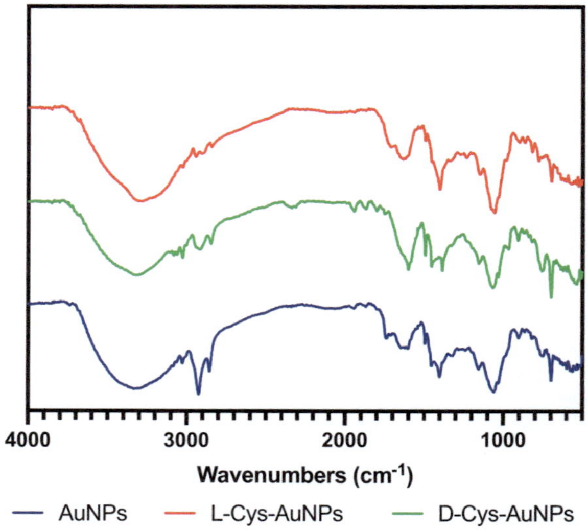

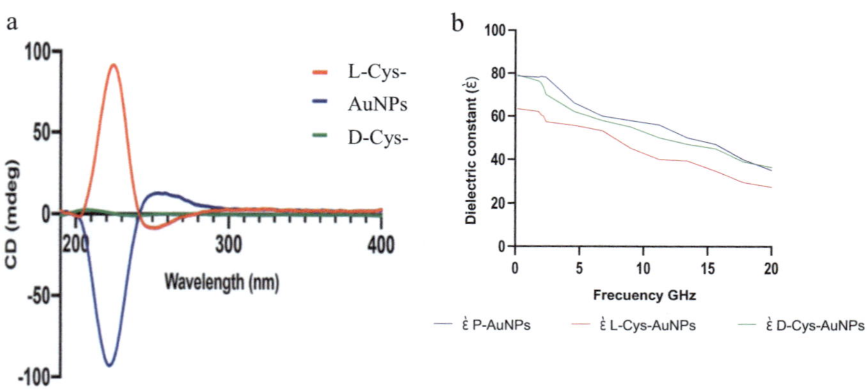

Fig. 3 **a** ECD spectra, and **b** dielectric constant of P-AuNPs, L-Cys-AuNPs, and D-Cys-AuNPs

AuNPs present strong dielectric dispersion corresponding to the alpha relaxation region. The ε′ in the selected range of 200 MHz–20 GHz was 78.72–35.11, 63.32–27.4, and 79.01 to 36.50 for P-AuNPs, L-Cys-AuNPs, and D-Cys-AuNPs, respectively. This parameter can be related to a major accumulation and better antimicrobial or antitumoral activity. The dielectric loss factor (e″) determines the ease with which an applied electromagnetic energy can heat a material. High values of ε″ lead to desirable volumetric heating, and low values of ε′ indicate the material's transparency to the incoming energy [10].

3.2 Antimicrobial Activity

The antibacterial activity of P-AuNPs and modified AuNPs against selected bacterial strains is displayed in Table 1. The results show that the antimicrobial activity is affected by functionalization; while the addition of L cysteine increases the activity, the addition in its D form decreases it in all strains; this can revert to the size changes that were observed when adding the surface and also to the changes in the dielectric constant of the particles, since the negative charges can interact in a better way with the positive charge of the membrane, achieving faster endocytosis and therefore greater effect. Greater activity could be observed against gram-positive bacteria compared to negative bacteria, and only particles with L conformation had an impact against the fungus *C. albicans*.

4 Conclusion

Changes in chirality affect all the properties of the gold nanoparticles, with the modification with chirality D having the greatest effect concerning the unmodified nanoparticles since the size and dielectric constant were affected, so the antimicrobial activity they presented decreased.

Table 1 Inhibition zone diameter of *Schinus molle* extract and nanoparticles of *Schinus molle* and cysteine modified AuNPs

Antimicrobial suspension		Microorganism					
		E. coli	*S. enterica*	*L. monocytogenes*	*S. aureus*	*S. epidermidis*	*C. albicans*
Schinus molle extract	10 mg/mL	11.7 ± 0.6	14.3 ± 0.6	17.0 ± 1.7	17.7 ± 1.5	17.7 ± 0.6	PI (18 ± 0.0)
	5 mg/mL	NI	NI	13.3 ± 1.5	13.7 ± 1.5	14.0 ± 1.0	PI (10.7 ± 0.6)
	2.5 mg/mL	NI	NI	12.3 ± 0.6	13.0 ± 1.0	12.0 ± 0.0	NI
	1.25 mg/mL	NI	NI	NI	9.7 ± 0.6	NI	NI
	0.1 mg/mL	NI	NI	NI	9.0 ± 0.0	NI	NI
P-AuNPs	10 mg/mL	21 ± 1.0	22.7 ± 1.5	19.7 ± 0.6	25.7 ± 1.5	26.7 ± 0.6	NI
	5 mg/mL	12.3 ± 1.5	16.7 ± 1.2	NI	13.0 ± 1.0	9.3 ± 0.6	NI
	2.5 mg/mL	NI	NI	NI	10 ± 0.0	NI	NI
	1.25 mg/mL	NI	NI	NI	NI	NI	NI
	0.1 mg/mL	NI	NI	NI	NI	NI	NI
D-Cys-AuNPs	10 mg/mL	15.7 ± 1.2	17.0 ± 1.7	14.7 ± 1.5	21.3 ± 1.7	21.0 ± 1.7	NI
	5 mg/mL	9.7 ± 0.6	10.7 ± 1.2	NI	15.7 ± 1.2	9.3 ± 0.6	NI
	2.5 mg/mL	NI	NI	NI	12.7 ± 0.6	NI	NI
	1.25 mg/mL	NI	NI	NI	NI	NI	NI
	0.1 mg/mL	NI	NI	NI	NI	NI	NI
L-Cys-AuNPs	10 mg/mL	24.0 ± 1.0	25.3 ± 0.6	26.3 ± 0.6	28.0 ± 1.0	32.0 ± 1.0	PI (23.0 ± 0.0)
	5 mg/mL	18.7 ± 1.5	22.0 ± 1.0	16.3 ± 0.6	20.3 ± 0.6	18.0 ± 0.0	PI (10.7 ± 0.6)
	2.5 mg/mL	13 ± 1.0	15.0 ± 1.0	NI	15.7 ± 0.6	13.3 ± 0.6	NI
	1.25 mg/mL	NI	11.0 ± 1.0	NI	13.7 ± 0.6	11.3 ± 0.6	NI
	0.1 mg/mL	NI	10.0 ± 1.0	NI	11.3 ± 0.6	NI	NI

NI: non inhibition, PI: Partial inhibition, some colonies growth inside inhibition halo

References

1. Hou, K.; Zhao, J.; Wang, H.; Li, B.; Li, K.; Shi, X.; Wan, K.; Ai, J.; Lv, J.; Wang, D.; et al. Chiral Gold Nanoparticles Enantioselectively Rescue Memory Deficits in a Mouse Model of Alzheimer's Disease. *Nat Commun, 11*, https://doi.org/10.1038/s41467-020-18525-2. (2020).
2. Cho, N.H.; Kim, Y.B.; Lee, Y.Y.; Im, S.W.; Kim, R.M.; Kim, J.W.; Namgung, S.D.; Lee, H.E.; Kim, H.; Han, J.H.; et al. Adenine Oligomer Directed Synthesis of Chiral Gold Nanoparticles. *Nat Commun, 13*, https://doi.org/10.1038/s41467-022-31513-y. (2022).
3. Soto, K.M.; Mendoza, S.; López-Romero, J.M.; Gasca-Tirado, J.R.; Manzano-Ramírez, A. Gold Nanoparticles: Synthesis, Application in Colon Cancer Therapy and New Approaches - Review. *Green Chem Lett Rev, 14*, 663–676. (2021).
4. Soto, K.M.; Luzardo-Ocampo, I.; López-Romero, J.M.; Mendoza, S.; Loarca-Piña, G.; Rivera-Muñoz, E.M.; Manzano-Ramírez, A. Gold Nanoparticles Synthesized with Common Mullein (Verbascum Thapsus) and Castor Bean (Ricinus Communis) Ethanolic Extracts Displayed Antiproliferative Effects and Induced Caspase 3 Activity in Human HT29 and SW480 Cancer Cells. *Pharmaceutics, 14*, 2069, https://doi.org/10.3390/pharmaceutics14102069. (2022).
5. Zhang, S.; Zhou, H.; Kong, N.; Wang, Z.; Fu, H.; Zhang, Y.; Xiao, Y.; Yang, W.; Yan, F. L-Cysteine-Modified Chiral Gold Nanoparticles Promote Periodontal Tissue Regeneration. *Bioact Mater, 6*, 3288–3299, https://doi.org/10.1016/j.bioactmat.2021.02.035. (2021).
6. Anwar, M.; Abdelhalim, K.; Mady, M.M.; Ghannam, M.M. Dielectric Constant, Electrical Conductivity and Relaxation Time Measurements of Different Gold Nanoparticle Sizes. *International Journal of the Physical Sciences, 6*, 5487–5491, https://doi.org/10.5897/IJPS11.1073. (2011).
7. Carone, A.; Emilsson, S.; Mariani, P.; Désert, A.; Parola, S. Gold Nanoparticle Shape Dependence of Colloidal Stability Domains. *Nanoscale Adv, 5*, 2017–2026, https://doi.org/10.1039/d2na00809b. (2023).
8. Devi, S.; Singh, B.; Paul, A.K.; Tyagi, S. Highly Sensitive and Selective Detection of Trinitrotoluene Using Cysteine-Capped Gold Nanoparticles. *Analytical Methods, 8*, 4398–4405, https://doi.org/10.1039/c6ay01036a. (2016).
9. Russier-Antoine, I.; Bertorelle, F.; Kulesza, A.; Soleilhac, A.; Bensalah-Ledoux, A.; Guy, S.; Dugourd, P.; Brevet, P.F.; Antoine, R. Chiral Supramolecular Gold-Cysteine Nanoparticles: Chiroptical and Nonlinear Optical Properties. *Progress in Natural Science: Materials International, 26*, 455–460, https://doi.org/10.1016/j.pnsc.2016.08.008. (2016).
10. Melo, A.A.; Rodrigues, E.P.; Vasconcelos, J.S.; Medeiros, E.S.; Oliveira, L.C.; Lima, A.M.N. Dielectric Function of Gold Nanoparticles Synthesized Using Camellia Sinensis Extract. *Plasmonics, 18*, 529–540, https://doi.org/10.1007/s11468-022-01776-8. (2023).

Meta GGA Calculations of the Optical and Electronic Properties of Pristine and N-doped Pentagraphene

Al Rey Villagracia and Hui Lin Ong

Abstract The effects of nitrogen doping on the optical and electronic properties of penta-graphene (PG) using first-principles calculations under the meta-generalized gradient approximation. Results showed that with less than 4% doping concentration of nitrogen with variation in locality, mid-gap states were added to the electronic state leading to a shift of the Fermi energy towards the valence band as shown in the band structure calculation. The peaks for Penta- graphene's quantum capacitance and excess charge are generally higher compared to graphene. The dielectric function, absorption spectrum, and energy loss function have been calculated for light polarization perpendicular to the plane of the PG sheets and compared with the properties of single-layer graphene. The calculated dielectric functions and energy-loss spectra are in good agreement with the previous literature for pure graphene. Peaks in the dielectric function of graphene can be found among the peaks of PG systems which have more peaks than graphene. It has been found that N doping does not significantly affect the peaks of the imaginary component of the dielectric function of PG systems. However, there is an increase in the absorption spectrum for the terahertz range of the PG systems when nitrogen is introduced. The results can be used to tailor the quantum capacitance of supercapacitors, and the optical properties of penta-graphene in the visible region. These materials can be used as organic light-emitting diodes, electrodes, and metamaterial.

Keywords Pentagraphene · Dielectrics · Metamaterial · Nitrogen · DFT

A. R. Villagracia (✉)
Department of Physics, De La Salle University, Manila, Philippines
e-mail: arcvillagracia@gmail.com

Advanced Nanomaterials Investigation and Molecular Simulations (ANIMoS) Research Unit, CENSER, De La Salle University, Manila, Philippines

H. L. Ong
Faculty of Chemical Engineering Technology, Universiti Malaysia Perlis (UniMAP), Jejawi, Perlis, Malaysia

Energy (WISE Centre), Universiti Malaysia Perlis (UniMAP), Jejawi, Perlis, Malaysia

A. Vaseashta et al. (eds.), *Proceedings of the Eighth International Symposium on Dielectric Materials and Applications (ISyDMA'8)*, Springer Proceedings in Physics 349, https://doi.org/10.1007/978-3-031-84809-4_12

1 Introduction

Dielectric materials play a crucial role in a wide array of technological applications, ranging from electronics and telecommunications to energy storage and medical devices. These materials, characterized by their ability to support electric fields while exhibiting minimal electrical conduction, have undergone significant advancements in recent years, driven by the demand for enhanced performance, efficiency, and miniaturization across various industries [1].

Current developments in dielectric materials encompass a broad spectrum of research endeavors aimed at addressing key challenges and unlocking new capabilities. One of the primary efforts of research is the quest for dielectric materials with high permittivity, low loss tangent, and stability over a wide range of frequencies and temperatures. Such materials hold the potential to revolutionize fields like capacitive energy storage, wireless communication, and dielectric resonator antennas [2]. Moreover, the emergence of novel fabrication techniques, including additive manufacturing, chemical vapor deposition, and self-assembly processes, has enabled the precise engineering of dielectric materials at the nanoscale. These advancements offer unprecedented control over material properties, paving the way for the development of customized dielectrics tailored to specific application requirements.

Furthermore, the integration of dielectric materials into emerging technologies such as flexible electronics, wearable devices, and Internet of Things (IoT) sensors presents new challenges and opportunities. Researchers are actively exploring flexible and stretchable dielectrics that can withstand mechanical deformation without compromising electrical performance, thus enabling the realization of next-generation flexible electronics and conformal sensor arrays [1].

Additionally, the pursuit of environmentally friendly and sustainable dielectric materials has gained traction in response to growing concerns regarding the ecological impact of electronic waste and hazardous materials. Bio-based polymers, organic–inorganic hybrids, and recyclable dielectric composites are among the eco-friendly alternatives under investigation, promising to reduce reliance on traditional petroleum-based dielectrics and mitigate environmental risks [3–9].

One of the recent materials being used as a dielectric material is graphene. Graphene-based dielectrics have been employed in various technological applications, including in energy storage devices such as graphene-based supercapacitors and lithium-ion batteries which exhibit enhanced energy storage performance due to the high surface area and conductivity of graphene dielectrics. Graphene-based dielectrics also enable the fabrication of flexible and stretchable electronic devices, including displays, sensors, and wearable electronics, by providing mechanical flexibility and robust electrical insulation. Lastly, it is being used in optoelectronic devices such as light-emitting diodes (LEDs), photodetectors, and solar cells, where they serve as transparent electrodes, barriers, and charge transport layers [10–12].

Despite the promising properties of graphene-based dielectrics, several challenges remain to be addressed, including the development of scalable fabrication methods, optimization of dielectric properties, and compatibility with existing manufacturing processes. An alternative material that can be used is penta-graphene (PG). Penta-graphene is a novel 2D carbon material proposed by Zhang et al. (2015). Unlike the hexagonal lattice of graphene, penta-graphene consists of pentagons, leading to a structure that combines sp2 and sp3 hybridized carbon atoms. This unique configuration endows penta-graphene with distinct properties, such as high mechanical strength, thermal stability, and an intrinsic band gap, which are not present in graphene. These characteristics make penta-graphene a potential candidate for dielectric applications in electronic devices [13–15].

Penta-graphene's structure comprises a planar arrangement of pentagons, which imparts unique electronic properties. The presence of an intrinsic band gap, as opposed to the zero-band gap in graphene, is particularly significant for dielectric applications. Zhang et al. demonstrated that penta-graphene exhibits a direct band gap of 3.25 eV, which can be beneficial for insulating applications. Penta-graphene is a promising candidate for dielectric applications due to its unique structural and electronic properties [13]. Its high dielectric constant, low dielectric loss, and mechanical flexibility make it suitable for various technological applications, including energy storage, flexible electronics, and optoelectronics. In this work, we investigated the electronic and optical properties of nitrogen-doped pentagraphene.

2 Methodology

In this work, we implemented the first principles of calculations [16, 17] for the characterization of the electronic properties of the different systems using the CAmbride Serial Total Energy Package (CASTEP) [18]. Figure 1 shows the different systems of interest in this study based on a 2 by 2-unit cell of pristine penta-graphene then doped with nitrogen atoms corresponding to less than 4% doping concentration and with a vacuum of 20 Å as a separation distance between layers. The exchange–correlation interaction was modeled using meta-generalized gradient approximation Regularized Strongly constrained and appropriately normed (RSCAN) [19] with Monkhorst k-point sampling of $21 \times 21 \times 1$, norm-conserving potentials, and an energy cut-off of 400 eV [20]. The formation energy (E_{form}) of the different systems was calculated using Eq. (1) using the pristine system as the reference energy and energies of isolated atoms for the dopants and replaced atoms. Equation (1) is given as:

$$E_{form} = E_{F\,aN} - (E_{pristine} + aE_N - aE_c) \tag{1}$$

where the $E_{F\,aN}$ represents the total energy of the systems doped with a number of nitrogen atoms, $E_{pristine}$ represents the energy of the pristine system, E_N is the energy of an isolated nitrogen atom, and E_C is the energy of an isolated carbon atom.

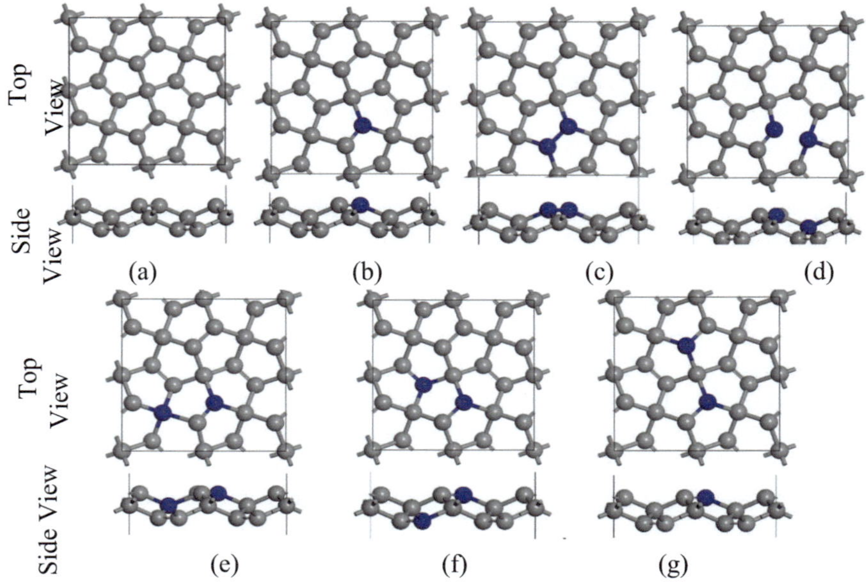

Fig. 1 Pentagraphene and N-doped PG system models **a** pristine PG **b** 1-Nitrogen-doped PG, **c–d** 2 adjacent nitrogen-doped PG systems, **e–g** 2 non-adjacent nitrogen-doped PG systems

3 Results and Discussion

In this section, we discussed the electronic and optical properties of pristine and N-doped pentagraphene systems.

3.1 Electronic Properties

The band structures of the different systems were plotted and shown in Fig. 2. Results showed that a pristine pentagraphene can be used as an organic light emitting diode (OLED) producing an ultraviolet (UV) light up to green light with a wavelength of 528 nm at a meta-GGA calculation while a local density approximation (LDA) and a GGA approximation produces 562–565 nm yellow light. Doping it with a nitrogen atom produces mid-gap states adjusting the band gap and producing a light with wavelengths corresponding to infrared (IR) light up to the red-light range. The wavelengths of light emitted from the de-excitation of electrons from the higher state to the ground state were computed based on the energy band gap (E_{gap}) using Eq. (2)

$$\lambda = \frac{hc}{E_{gap}} \approx \frac{1.235}{E_{gap}} \tag{2}$$

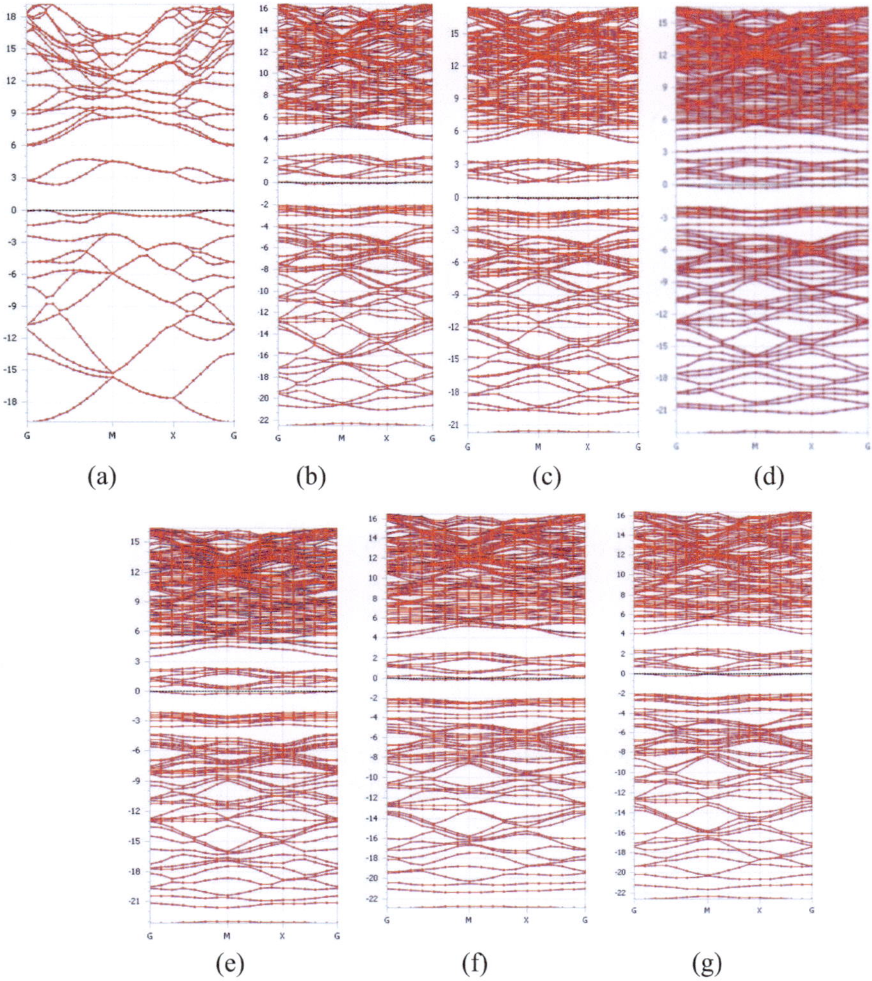

Fig. 2 Band structure of **a** pristine PG **b** 1-Nitrogen-doped PG (N1), **c–d** 2 adjacent nitrogen-doped PG systems [N2-0-1, N2-0-2], **e–g** 2 non-adjacent nitrogen doped PG systems [N2-1-1, N2-1-2, N2-1-3]

In terms of formation energy, e graphene is a preferred state than pentagraphene based on formation energy as validated by experiments. A single nitrogen-doped system has a lower formation energy than pristine by 0.48 eV if synthesized from pentagraphene. However, the nitrogen-doped systems have lower total energy, indicating a possible synthesis of the nitrogen-doped systems.

If the material is used as an electrode for a supercapacitor, the quantum capacitance (C_Q) and excess charge ($Q(\varphi)$ of the different systems can be computed from the density of states $D(\varepsilon)$ as a function of energy (ε) using Eqs. (3) and (4) [21, 22]:

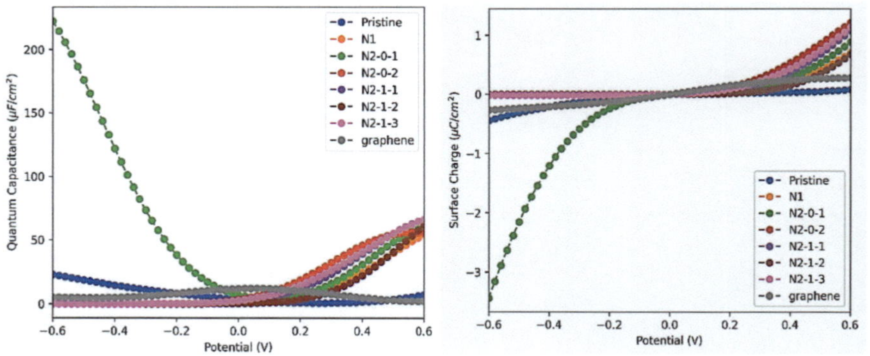

(a) (b)

Fig. 3 **a** Quantum capacitance and **b** excess charge of pristine PG and N-doped PG systems

$$C_Q = \frac{e^2}{4kT} \int_{-\infty}^{\infty} D(\varepsilon) \text{sech}^2 \frac{\varepsilon - \varphi}{2kT} d\varepsilon \tag{3}$$

$$Q(\varphi) = \int_{0}^{\varphi} C_Q(\varphi) d\varphi \tag{4}$$

where φ is the external potential in volts, T is the temperature in Kelvin (K), and k is the Boltzmann constant in eV. The quantum capacitance and excess charge of the different systems at 298 K were plotted and shown in Fig. 3.

Results showed that at the negative external potential, the 2 adjacent nitrogen-doped pentagraphene system (N2-0-1) yielded a higher quantum capacitance shown in Fig. 3a compared to the rest of the systems. In contrast, all nitrogen-doped systems have similar behavior with clustered values given a positive external potential. A similar trend for the excess charge is seen in Fig. 3a.

3.2 *Optical Properties*

In this section, the calculated optical properties of the different systems due to the electronic transitions using time-dependent perturbations of the ground state electronic state modeling the interaction between photon and electron. Three optical properties are presented in Fig. 4: dielectric function, absorption spectrum, and loss function.

The dielectric function would show the response of the material due to an external electric field. The absorption spectrum would indicate how much of the energy from

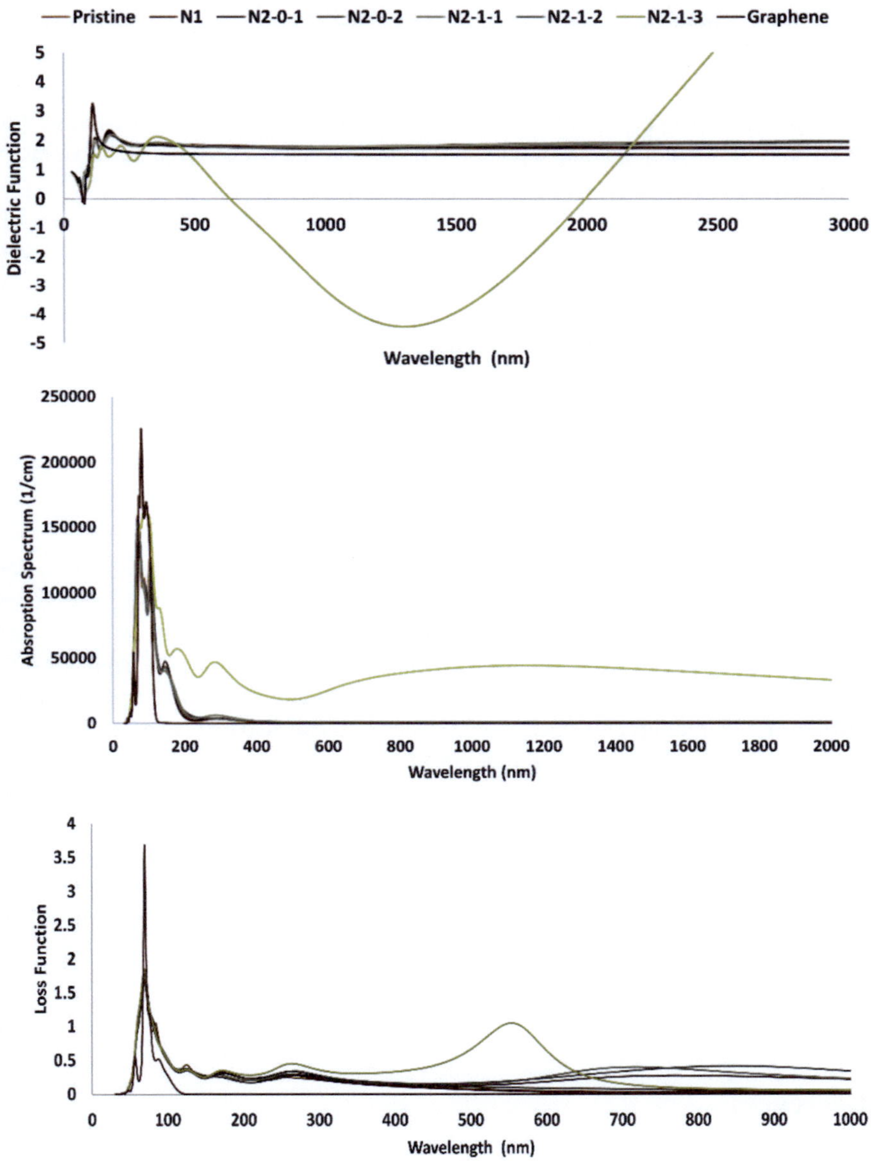

Fig. 4 Optical properties of pristine PG and N-doped PG systems

an external electric field can be absorbed. The loss function would demonstrate how much of the electron's energy is lost and converted to heat.

In these calculations, a polarized incident light normal to the surface of pentagraphene was used. Results have shown that the dielectric function of pentagraphene is more interesting than graphene. The peaks are almost the same for all pentagraphene systems with graphene. However, the dielectric functions of pentagraphene are generally higher above 100 nm with a distinct behavior for the 2 non-adjacent nitrogen-doped pentagraphene system N2-1-3. Its dielectric function from 600 to 2000 nm is negative indicating a potential meta material which has a negative index of refraction and can be operated with light rather than electricity.

The same thing can be said with its absorption spectrum where system N2-1-3 can absorb a wider range of light and a distinct peak at 550 nm of loss function and without energy loss above 800 nm. This means that this configuration can indeed work with light without energy loss above 800 nm. The common peaks near 100 nm indicate a large absorption and loss of energy in the UV range with graphene having a higher magnitude.

4 Conclusion and Recommendations

In this work, the electronic and optical properties of pristine and nitrogen-doped pentagraphene systems were calculated using density functional theory. Energetics showed that nitrogen-doped systems have lower total energy than graphene and pristine pentagraphene which indicates a possibility of being synthesized. Results also showed the potential of the material as an OLED emitting UV to green light for pristine, and IR/red light nitrogen-doped systems. It can also be used as an electrode for supercapacitors and a possible metamaterial that can be used for modern light-operated devices. Further investigations can be done by doping it with other atoms and checking its optoelectronic properties.

Acknowledgements The authors would like to thank the support of the Department of Physics, Office of the Vice President for Research and Innovation, and the Research and Grants Management Office of De La Salle University through the project 09 FR 1TAY23-1TAY24 entitled "Quantum Capacitance and Surface Charge of N-doped penta-graphene: A DFT Study".

References

1. Sharma, J.D., Sharma, M, Kumar, N., Ahluwalia, P.K., Computational study of the dielectric function and optical properties of a graphene nanostructure containing graphene quantum dot. *J Phys Conf Ser*, vol. 472(1), (2013), https://doi.org/10.1088/1742-6596/472/1/012010.
2. Phirouznia, A., Akbari, N., Lotfi, S., and Jamshidi-Ghaleh, K., Dielectric function of graphene-based quantum dot under the uniaxial stress A controllable red shift and band gap engineering by the stress process, *European Physical Journal B*, vol. 85, no. 8, (2012), https://doi.org/10.1140/epjb/e2012-30076-4.
3. Owi, W. T., Ong, H. L., Sam, S. T., Mern, C. K., Villagracia, A. R., Santos, G. N. C., Akil, H. M., A comparative study of green composites based on tapioca starch and celluloses, in *American Inst. of Physics Conf. Proc.*, (2017). https://doi.org/10.1063/1.4993361.

4. Ong, H. L., Villagracia, A. R., Owi, W. T., Sam, S. T., and Akil, H. M, Revealing the Water Resistance, Thermal and Biodegradation Properties of Citrus aurantifolia Crosslinked Tapioca Starch/Nanocellulose Bionanocomposites, *J Polym Environ*, (2020), https://doi.org/10.1007/s10924-020-01852-2.
5. Owi, W. T., Ong, H. L., Sam, S. T., Chia, C. H., Zakaria, S., Mohaiyiddin, M. S., Villagracia, A.R., and Md Akil, H., Comparative study of microcelluloses isolated from two different biomasses with commercial cellulose, *Bioresources*, vol. 11, no. 2, pp. 3453–3465, (2016), https://doi.org/10.15376/biores.11.2.3453-3465.
6. Mohaiyiddin, M. S., Ong, H. L., Akil, H. M., Yee, T. G., Adik, N. N. A. N., and Villagracia, A. R., Effects of polypropylene methyl polyhedral oligomeric silsesquioxanes and polypropylene-grafted maleic anhydride compatibilizers on the properties of palm kernel shell reinforced polypropylene biocomposites, *Polimeros*, vol. 26, no. 3, pp. 228–235, (2016), https://doi.org/10.1590/0104-1428.2038.
7. Anis, S. M. K., Ong, H. L., Akil, H. M., and Villagracia, A. R., Effect of compatibilizer on the accelerated weathering performance of polypropylene-silica nanocomposites, *Materials Research Innovations*, vol. 18, pp. S6–433-S6–438, (2014), https://doi.org/10.1179/1432891714Z.000000000993.
8. Ong, H. L., Toh, G. Y., Tsai, C. K., Arboleda, N., Culaba, A., Manzano, M. C., Villagracia, A.R., and David, M.., Effect of mesoporous nanoparticles from LCD glass panels waste toward polypropylene based hybrid composites, *2018 IEEE 10th Intl. Conf. on Humanoid, Nanotechnology, Information Tech., Comm., and Control, Envir. t and Manag., HNICEM 2018*, pp. 1–6, (2019), https://doi.org/10.1109/HNICEM.2018.8666336.
9. Ong, H. L., Yee, T. G., Nik Nur Azza, N. A., Muhammad Safwan, M., Villagracia, A. R., Mern, C. K., and Al Bakri, A. M.., Utilization of modified palm kernel shell for biocomposites production, *Key Eng Mater*, vol. 700, pp. 60–69, (2016), https://doi.org/10.4028/www.scientific.net/KEM.700.60.
10. Wang, M., and Li, C. M., Excitonic properties of graphene-based materials, *Nanoscale*, vol. 4, no. 4, pp. 1044–1050, (2012), https://doi.org/10.1039/c1nr10885a.
11. Ramadan, F. Z., Ouarrad, H., and Drissi, L. B., Tuning Optoelectronic Properties of the Graphene-Based Quantum Dots $C_{16-x}Si_xH$ Family, *Journal of Physical Chemistry A*, vol. 122, no. 22, pp. 5016–5025, (2018), https://doi.org/10.1021/acs.jpca.8b02704.
12. Farah, S. E., Ferhati, H., Dibi, Z., Djeffal, F., Performance analysis of broadband Mid-IR graphene-phototransistor using strained black phosphorus sensing gate: DFT-NEGF investigation, *Micro and Nanostructures*, vol. 163, (2022), https://doi.org/10.1016/j.spmi.2022.107187.
13. Zhang, S., Zhou, J., Wang, Q., Chen, X., Kawazoe, Y., and Jena, P., Penta-graphene: A new carbon allotrope, *Proc Natl Acad Sci U S A*, vol. 112, no. 8, pp. 2372–2377, (2015), https://doi.org/10.1073/pnas.1416591112.
14. Enriquez, J. I. G., Villagracia, A. R. C., Hydrogen adsorption on pristine, defected, and 3d-block transition metal-doped penta-graphene, *Int J Hydrogen Energy*, vol. 41, no. 28, pp. 12157–12166, (2016), https://doi.org/10.1016/j.ijhydene.2016.06.035.
15. Shahrokhi, M., Tuning the band gap and optical spectra of monolayer penta-graphene under in-plane biaxial strains, *Optik (Stuttg)*, vol. 136, pp. 205–214, (2017), https://doi.org/10.1016/j.ijleo.2017.02.033.
16. Kohn, W., Density-functional theory for excited states in a quasi-local-density approximation, *Phys Rev A (Coll Park)*, vol. 34(2), p. 737, (1986), https://doi.org/10.1103/PhysRevA.34.737.
17. Rojas, K. I., Villagracia, A. R., and Arboleda, N. B., H_2 adsorption on K decorated germanene surface: An ab-initio investigation, *Mater Res Express*, vol. 3, no. 11, p. 115015, (2016), https://doi.org/10.1088/2053-1591/3/11/115015.
18. Imai, Y., Mukaida, M., and Tsunoda, T., Calculation of electronic energy and density of state of iron-disilicides using a total-energy pseudopotential method, CASTEP, *Thin Solid Films*, vol. 381, no. 2, pp. 176–182, (2001), https://doi.org/10.1016/S0040-6090(00)01740-5.
19. Liu, R., Zheng, D., Liang, X., Ren, X., Chen, M., Li, W., Implementation of the meta-GGA exchange-correlation functional in numerical atomic orbital basis: With systematic testing on

SCAN, rSCAN, and r2SCAN functionals, *Journal of Chemical Physics*, vol. 159, no. 7, (2023), https://doi.org/10.1063/5.0160726/2907452.

20. Wisesa, P., McGill, K. A., Mueller, T., Efficient generation of generalized Monkhorst-Pack grids through the use of informatics, *Phys Rev B*, vol. 93, no. 15, p. 155109, (2016), https://doi.org/10.1103/PHYSREVB.93.155109/FIGURES/10/MEDIUM.
21. Bharti, B., Kumar, Y., Gupta, M., Sharma, S., Study of Quantum Capacitance of Pure and Functionalized Nb_2C and Ti_2C MXenes for Supercapacitor Applications, in *ECS Transactions*, (2022), pp. 1751–1760. https://doi.org/10.1149/10701.1751ecst.
22. Paz, K.A.C., Villagracia, A.R., David, M.Y., First-principles study on copper passivated armchair graphene nanoribbon for a supercapacitor electrode material, *Electrochim Acta*, vol. 462, p. 142753, (2023), https://doi.org/10.1016/J.ELECTACTA.2023.142753.

Laser-Power Dependence Effects on the Structural Stability of CeZr and ZrMn Catalysts Oxides Studied by Raman Spectroscopy

Antonio N. da Silva, Marcos Davi L. Gonçalves, Pierre B. N. Assis, and Alcineia C. Oliveira

Abstract The rapid growth in nanotechnology has driven the interest in studies on the structural properties of nanosized systems to expand their technological applications in catalysts. Structural properties of binary CeZr and ZrMn nanocomposite oxide catalysts were monitored towards the Laser Raman spectroscopy investigations providing new insights to control catalytic applications upon temperature ranges at which the laser power was varied. The vibrational properties were investigated by varying the incident laser power during Raman measurements from 0.017 to 4.0 mW. Previous studies on the local laser heating effects on samples have demonstrated that nanomaterials stability deteriorated due to thermal effects. This work applies laser power heating to observe its effects on the vibrational properties of binary nanocomposite oxide catalysts. Laser-induced local heating upon varying laser power values from 0.017 to 4.0 mW. Structural changes in nanocomposites were achieved upon increasing laser power, which induced local heating disorder causing the sintering of ZrMn nanocomposites. The oxygen vacancy defects over the CeZr catalyst were observed. ZrMn exhibited phase transitions from MnO_2 to α-Mn_2O_3 being the use of this nanocomposite limited to work at 1.1 mW.

Keywords Nanocomposite · Catalysts · Raman spectroscopy · Laser heating effects

A. N. da Silva (✉) · M. D. L. Gonçalves · P. B. N. Assis
Federal of Ceará University – Russas Campus, Russas, Ceará, Brazil
e-mail: neves@ufc.br

A. C. Oliveira
Físico-Química Department, Federal of Ceará University, Fortaleza, Ceará, Brazil

A. Vaseashta et al. (eds.), *Proceedings of the Eighth International Symposium on Dielectric Materials and Applications (ISyDMA'8)*, Springer Proceedings in Physics 349, https://doi.org/10.1007/978-3-031-84809-4_13

1 Introduction

The interest in studies on the structural properties of nanosized systems to expand their technological applications in catalysts has been growing [1–3]. The changes in the vibrational modes of the nanostructured oxide catalysts are assessed by laser Raman spectroscopy since the optical and acoustic phonons are confined within the particle of the nanosized oxides compared to that of the bulk [2, 4]. The effects of the anharmonicity and phonons confinement should be considered when studying local laser heating-inducing spectral feature changes in nanostructured oxides. Raman spectroscopy is a versatile technique in providing detailed information on nanostructures pressure- and temperature-induced transformations, surface-induced by laser effects, and phase transitions. The temperature-dependent anharmonic behavior of the phonon modes is a result of the anharmonic interactions providing changes in the phonon population with the consequent variations in the nanostructured oxide lattice parameters [2, 5]. More recently, several studies on the local laser heating effects have demonstrated that the stability of active sites of nanostructured oxides can deteriorate during the reactions due to the thermal effects, which implies the formation of undesired phases and leaching of the active sites, as well [6, 7]. Even though there are great advances in engineering for in situ catalytic systems operations in current years, these systems often do not enable monitoring the stability of the active sites under heating the solids during the reactions. This work deals with applying laser power heating to observe its effects on the vibrational properties of binary nanocomposite oxide catalysts.

2 Experimental

2.1 Synthesis of the Solids

The binary catalytic nanocomposite oxides namely CeO_2–ZrO_2 (CeZr) and ZrO_2–MnO_x (ZrMn) were synthesized, via a nanocasting route. A more detailed synthesis procedure and characterizations can be found in our previous reports [8, 9]. In brief, the mesoporous SBA-15 was used as a hard template along with an equimolar mixture of metal sources.

Synthesis of CeZr catalyst was accomplished using equimolar mixtures of ammonium cerium (IV) nitrate $(NH_4)_2Ce(NO_3)_6$ and zirconium chloride $ZrOCl_2{\cdot}8H_2O$ solutions and acetic acid, under stirring. The mesoporous SBA-15 hard template was subsequently added to the previous solutions. As formed, the nanocomposite was collected, and propylene oxide was added. The resulting solid was further filtered and submitted to steam of methanol being subsequently washed and dried under steam of methanol for 16 h. The SBA-15-supported CeZr oxide nanocomposite was obtained after calcination at 500 °C under airflow for 2 h. The hard template was finally removed from the nanocomposite with sodium hydroxide (NaOH) solution to obtain CeZr upon calcinating the solid under airflow at 600 °C [8, 9]. Similar procedures were adopted to obtain ZrMn nanostructured oxide by using $ZrOCl_2{\cdot}8H_2O$ and $Mn(NO_3)_2{\cdot}6H_2O$ as metal source precursors, respectively.

2.2 *Laser Raman Spectroscopy Studies*

The Raman spectroscopy technique was used to investigate the vibrational properties and their correlation with the structural properties of the nanostructured oxide catalysts. Raman spectra were obtained using a LabRam spectrometer possessing 1800 grooves/mm holographic grating, which was connected with a nitrogen-cooled charge-coupled-device detector. The spectra were collected with 10 data acquisitions per 180 s in a 100–2000 cm^{-1} spectral window using a 633 nm excitation wavelength. A silicon substrate was adopted as a reference for the measurements at 521 cm^{-1}. The experiments were performed from 0.017 to 4.0 mW on the samples. The intensity of the laser power was varied by switching filters.

The exciting light was focused on the sample with a 20 times microscope objective lens from an Olympus microscope apparatus. The numerical aperture (N.A.) was c.a. 0.4. It was found that there is a limit for the size area of the sample illuminated by the laser beam. The limit is determined by the diffraction of light and the minimum radius in the localized area is given by Eq. 1 [10]:

$$r = \frac{0.61\lambda}{N.A.} \tag{1}$$

The wavelength (λ) used here was of c.a. 633 nm, which gave a minimum radius of r = 0.97 μm. For the solids possessing the CeO_2 phase, the average crystallite sizes were estimated by using spectral data from Raman measurements, that is to say, the line linewidth of the Raman modes Γ, according to the following equation [11]:

$$\Gamma\left(cm^{-1}\right) = 5.48 + \frac{98.4}{D_{cal}(Raman)} \tag{2}$$

where D_{cal} designated the crystal size obtained from the Raman spectrum.

3 Results and Discussion

3.1 Laser Local Heating Effects

Spectral features of CeZr nanocomposite oxide catalyst by varying the laser power are shown in Fig. 1. The spectral features of CeZr have a typical major Raman mode assigned to T_{2g} optical phonon from cubic CeO_2, as highlighted in the square of Fig. 1a. Besides, the weaker Raman band at around 251 cm^{-1} (highlighted in the square of Fig. 1b) assigned to the CeO_2 vacancies are not clearly visible along with that of 188 cm^{-1} because of the huge absorption of light by the solid giving a high noise and the superimposition with the ZrO_2 modes, as well. The tetragonal ZrO_2 structure arises in the Raman spectra at around 275, 321, 523 and 655 cm^{-1} but only the Raman modes at 260 and 597 cm^{-1} appear as a result of the overlapping bands of CeO_2. Both of these modes can be attributable to the E_g vibrational mode of the tetragonal ZrO_2 phase and a non-degenerate longitudinal optical (LO) mode of ceria, respectively [12]. Interestingly, the Raman spectrum at 0.017 mW displays two weak and asymmetrical bands fitted to the D (1334 cm^{-1}) along with a shoulder at around 1617 cm^{-1} and G (1583 cm^{-1}), which arise from the residual graphitized carbon formed during the synthesis and calcination of the nanocomposite (Fig. 1b).

By increasing laser power from 0.017 to 4.0 mW, the Raman bands redshift and a broadening of the D and G band is observable whereas the T_{2g} mode of CeO_2 and their defect bands remain almost unchangeable, independently of the laser power applied (Fig. 1a). The intensity of the D and G bands ratios (I_D/I_G) gives an estimation of the degree of graphite layers crystallinity with the lowest values of I_D/I_G ca. 0.95 observed at 0.017 mW and the highest one of ca. 1.27 being achieved at 4.0 mW that is to say, more structural defects in the carbon species. Noteworthy, the band 984 cm^{-1} borders and increases its intensity (Fig. 2) compared with low laser power values and this is thought to originate more vacancies of oxygen due to the thermal effects. In agreement, the local heating induces the displacement of oxygen atoms from the fluorite lattice positions Ce–O with the consequent increase in oxygen vacancy concentrations [13]. The T_{2g} CeO_2 optical phonon exhibited a redshift of 7 cm^{-1} (Fig. 2a) and a FWHM narrowing of ca. 6.5 cm^{-1} is seen upon increment of the laser power from 0.017 to 1.1 mW, being unchangeable at higher laser power values (Fig. 2b). The thermal effect of particle growth with the consequent formation of small clusters of cerium oxide at CeZr surface is causing the shifts and line width narrowing [13].

The influence of laser power on the spectral features of ZrMn nanocomposite oxide catalyst is depicted in Fig. 3. The spectrum of ZrMn is dominated by a set of Raman bands scattered in the 1300–100 cm^{-1} spectral window from 0.017 to 4.0 mW (Fig. 3a). The unresolved bands seem to correspond to the optical phonons observed at 181 (E_g), 267 (E_g), 539 (E_g) and 653 (A_{1g}) cm^{-1} coming from ZrO_2 tetragonal phase at 0.017 mW, as seen in the magnification of this region in Fig. 3b. A better definition of the signals with remarkable changes is seen in the ZrMn spectrum taken at 0.58 mW (Fig. 3c), in which many bands split into multiplets. This should be related

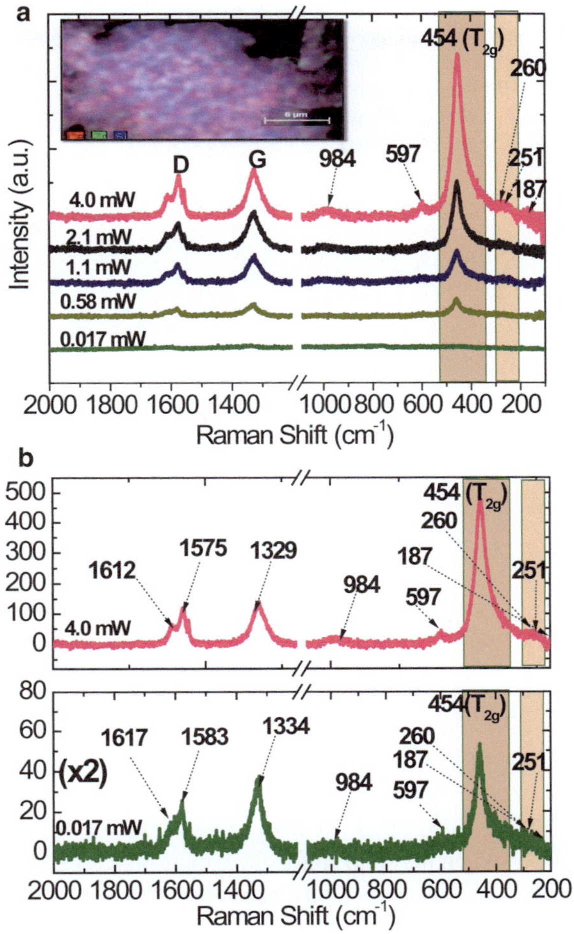

Fig. 1 **a** Raman spectra of CeZr collected at various laser power values in the spectral range of 1300–100 cm^{-1} and **b** laser-power dependence of CeZr collected at 0.017 and 4.0 mW. The CeZr spectrum at 0.017 mW is multiplied by the specified factor. The included figure is the SEM image of the nanocomposite

to the distinct spectral changes occurring in the lattice mode regions indicating the coexistence of the tetragonal ZrO_2 and MnO_x phases. Some of the Raman bands discussed before have broadened a little and a set of weak modes appears at 181, 267, 481, 539, 582, 625, 653 and 1329 cm^{-1} for the spectrum at 0.58 mW (Fig. 3c). These vibrational Raman modes match well with that of α-Mn_2O_3 structure [14]. Additionally, the vibrational bands at 481 and 625 cm^{-1} are assigned to out-of-plane bending modes and Mn–O–Mn symmetric stretching vibrational of α-Mn_2O_3, respectively [14]. The mode at 582 cm^{-1} is ascribed to Mn–O–Mn lattice vibrations in MnO_2 [14]. Also, the Raman band at 1329 cm^{-1} is related to superficial O–H vibrational bonds.

Figure 3a shows the Raman spectra of the ZrMn acquired to observe the spectral evolution of the solid upon increasing laser power from 0.017 to 4.0 mW. Even though the MnO_x species are not incorporated into the ZrO_2 lattice, the MnO_x are highly

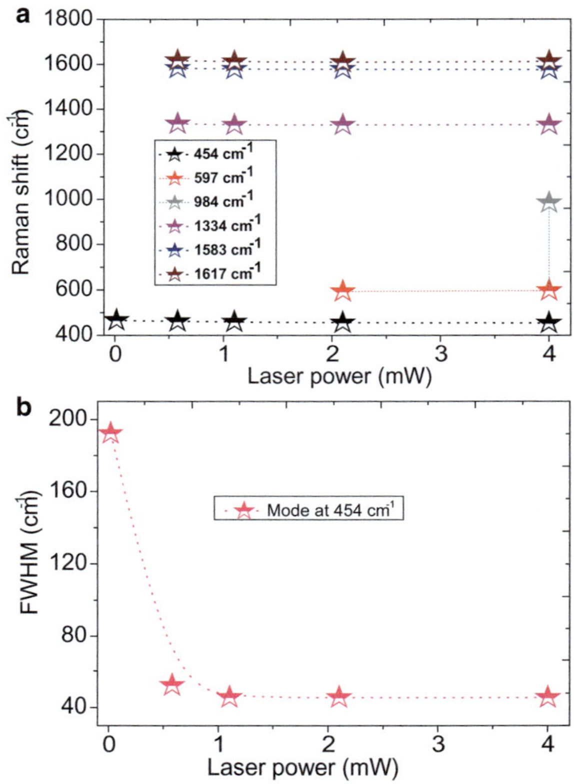

Fig. 2 **a** Variation of the Raman modes position with the laser power and **b** FWHM dependence of T_{2g} mode of CeO_2 present in CeZr with the laser power

dispersed on the solid surface, which is indicative of their low interaction with ZrO_2 mainly as the temperature increases with the laser power increments. Supportive evidence is found in the case of high temperatures inducing the formation of a highly dispersed and stabilized α-Mn_2O_3 amorphous phase [9]. It is obvious that a new spectral feature arises at 1.1 mW with the Raman bands broadened most due to the anharmonicity effects. From 1.1 to 2.1 mW (Fig. 3a), a broadening of the Raman bands is seen with some of these bands experiencing a redshift along with their increased intensities. The sintering and crystallization of the amorphous α-Mn_2O_3 phase with laser power irradiation seems to be the reason for these observations. It may exist due to the assumption of the occurrence of thermal-induced strain causing sintering instead of anharmonicity effects. Consistently, this is indeed the case with the Zr–O–Mn interactions to form the α-Mn_2O_3 phase and other MnO_2 species in ZrMn, as reported [15]. Thus, sintering at a certain temperature causes a decreased surface area of the solid, as found elsewhere [16]. At 4.0 mW, the disappearance of the Raman bands under intensive laser illumination complicates the identification of the mode's assignation. In particular, previous works carried out at temperatures above 450 °C demonstrate that elevated temperatures caused considerable sintering in ZrMn oxide with the use of this solid as a catalyst limited below this temperature

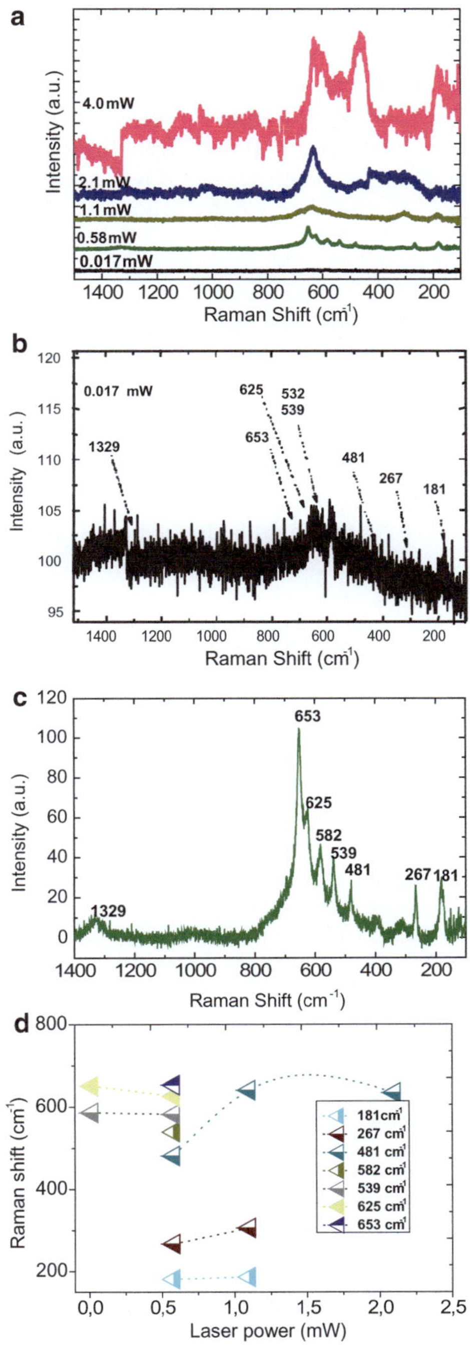

Fig. 3 **a** Raman spectra of ZrMn collected at various laser power values in the spectral range of 1300–100 cm^{-1} and **b** laser-power dependence of ZrMn collected at 0.017 mW and **c** laser-power dependence of ZrMn collected at 0.58 mW. **d** Variation of the position of the Raman modes with the laser power. The ZrMn spectrum at 0.017 mW is multiplied by a 3 × factor

[17]. Figure 3d reveals the laser power dependence of the position of various modes in the Raman spectra of ZrMn.

The Raman modes at 625, 653, 539, and 181 cm^{-1} do not change their positions. The bands at 267 and 481 cm^{-1} due to ZrO_2 and MnO_x raise their positions with increasing the laser power. This position-raising trend is consistent with the assumption that the α-Mn_2O_3 phase is formed at high temperatures upon increasing laser power.

4 Results and Discussion

Laser power influences on the structural features of the nanocomposite oxide catalysts were investigated by Raman spectroscopy. Laser-induced local heating upon varying laser power values from 0.017 to 4.0 mW. This resulted in the thermal-induced structural changes and particle growths of the ZrMn nanocomposite catalyst. The use of ZrMn should be limited to 1.1 mW in applications. The creation of oxygen vacancies defects was observed in CeZr upon raising the laser power.

Acknowledgements The authors thank Funcap financial support for this research project (FUNCAP BPI 03/2018 n° BP3-0139-00055.04.00/18 and the FUNCAP/Edital 07/2021 grant n° PS1-0186-00346.01.00/21.

References

1. Gouadec G., Colomban P., Raman spectroscopy of nanostructures and nanosized materials, J. Raman Spectrosc. 38 (2007) 598–603.
2. da Silva A.N., Neto A.B.S., Oliveira A.C., Junior M.C., Junior J.A.L., Freire P.T.C., Filho J. M., Oliveira A.C., Lang R., Raman studies of nanocomposites catalysts: temperature and pressure effects of CeAl, CeMn, and NiAl oxides, Spectrochim. Acta A Mol. Biomol. Spectrosc. 198 (2018) 160–167.
3. Sallum L.O., Vaz W.F., Borges N.M., de Campos C.E.M., Bortoluzzi A.J., Franco C.H.J., Ramos L.M., Napolitano H.B., Synthesis, conformational analysis and molecular docking studies on three novel dihydropyrimidine derivatives, J. Mol. Struct. 1192 (2019) 274–287.
4. Palneedi H., Park J.H., Maurya D., Peddigari M., Hwang G., Annapureddy V., Kim J., Choi J., Hahn B., Pryia S., Lee K.J., Ryu J., Laser Irradiation of Metal Oxide Films and Nanostructures: Applications and Advances, Adv. Mater. 30 (2018) 1705148.
5. Arora A.K., Rajalakshmi M., Ravindran T.R., Sivasubramanian V., Raman spectroscopy of optical phonon confinement in nanostructured materials, J. Raman Spectrosc. 38 (2007) 604–617.
6. Popović Z.V., Dohčević-Mitrović Z., Cros A., Cantarero A., Raman scattering study of the anharmonic effects in CeO_{2-y} nanocrystals, J. Phys.:Condens. Matter 19 (49) (2007).
7. Lima C.L., Vasconcelos S.J.S., Filho J.M., Neto B.C., Rocha M.G.C., Bargiela P., Oliveira A.C., Nanocasted oxides for gas phase glycerol conversion, Appl. Catal. A 399 (2011) 50–62.
8. Vasconcelos S.J.S., Lima C.L., Filho J.M., Oliveira A.C., Barros E.B., de Sousa F.F., Rocha M.G.C., Bargiela P., Oliveira A.C., Activity of nanocasted oxides for gas-phase dehydration of glycerol, Chem. Eng. J. 168 (2011) 656–664.

9. Saitzek S., Blach J.-F., Villain S., Gavarri J.-R., Nanostructured ceria: a comparative study from X-ray diffraction, Raman spectroscopy, and BET specific surface measurements, Phys. Status Solidi (a) 205 (2008) 1534–1539.
10. Mota G., do Carmo J.V.C., Paz C.B., Saraiva G.D., Campos A., Duarte G., da Silva Filho E.C., Oliveira A.C., Soares J.M., Rodríguez-Castellon E., Rodríguez- Aguado E., Influence of the Metal Incorporation into Hydroxyapatites on the Deactivation Behavior of the Solids in the Esterification of Glycerol, Catalysts 12 (2022) 10.
11. Schilling C., Hofmann A., Hess C., Ganduglia-Pirovano M.V., Raman Spectra of Polycrystalline CeO_2: A Density Functional Theory Study, J. Phys. Chem. C 121 (2017) 20834–20849.
12. de Sousa F.F., de Sousa H.S.A., Oliveira A.C., Junior M.C.C., Ayala A.P., Barros E.B., Viana B.C., Filho J.M., Oliveira A.C., Nanostructured Ni-containing spinel oxides for the dry reforming of methane: Effect of the presence of cobalt and nickel on the deactivation behavior of catalysts, Int. J. Hydrogen. Energy 37 (2012) 3201–3212.
13. Hirata T., Asari E., Kitajima M., Infrared and Raman Spectroscopic Studies of ZrO_2 Polymorphs Doped with Y_2O_3 or CeO_2, J. Solid State Chem. 110 (1994) 201–207.
14. da Silva A.N., Pinto R.C.F., Freire P.T.C., Junior J.A.L., Oliveira A.C., Filho J.M., Temperature and high-pressure effects on the structural features of catalytic nanocomposites oxides by Raman spectroscopy, Spectrochim. Acta A. 138 (2015) 763–773.
15. Ghule A.V., Ghule K., Punde T., Liu J., Tzing S., Chang J., Chang H., Ling Y., In situ monitoring of $NiO–Al_2O_3$ nanoparticles synthesis by thermo-Raman spectroscopy, Mater. Chem. Phys. 119 (2010) 86–92.
16. Gil-Calvo M., Jiménez-González C., de Rivas B., Gutiérrez-Ortiz J.I., López-Fonseca R., Effect of Ni/Al molar ratio on the performance of substoichiometric $NiAl_2O_4$ spinel-based catalysts for partial oxidation of methane, Appl. Catal. B-Environ. 209 (2017) 128–138.
17. Machida M., Uto M., Kijima T., Preparation of large surface area MnO_x-ZrO_2 for sorptive NO_x removal, Stud. Surf. Sci. Catal. 143 (2000) 855–862.
18. An M., Li H., Su M., Gao S., Wang M.C., Shen S.G., Gao Z.F., Dong J.X., Cu^{2+} enhanced fluorescent A_g nanoclusters with tunable emission from red to yellow and the application for Ag^+ sensing, Spectrochim. Acta A Mol. Biomol. Spectrosc. 252 (2021), 119484.

Intermediate Phase Glasses of the As-S-Ge Ternary: Self-Organization, Properties and Applications in Programmable Metallization Cells

Dumitru Tsiulyanu, Marina Ciobanu, and Andrei Afanasiev

Abstract A brief review is given of the elastic constraint theory applicable to chalcogenide glasses (ChG) and its experimental proving on the intermediate phase (IP) of As-S-Ge ternary. The last has been performed via Raman scattering experiments and analysis of the same at composition change along the IP tie-line $(GeS_4)_x(AsS_3)_{1-x}$. Such an approach allowed the revealing of the vibrational modes of the main structural blocks of the glass network and their scattering strength. It was pointed out that the scattering strength of the Raman modes strongly varies with composition change, at that the mode 336 cm^{-1} associated to symmetric stretch in flexible QT $S = As(S_{1/2})_3$ in charge of structural self-organization (SO) of the glass, reaches its maximal value in composition ($Ge_{7.7}As_{15.3}S_{77}$) but suddenly decreases in neighbor compositions. These results correlate with the compositional dependence of molar volume and elastic modulus of the glass because they reach the respective extrema, namely around that composition. The solid electrolytes with a mixed electronic and ionic conductivity based on glasses from the IP region of the As-S-Ge ternary have been fabricated via photo-dissolution of Ag in ChG that further were used as active parts of the programmable memory cells. It turned out that the SO of the glass structure leads to a significant increase in the electrical conductance of solid electrolytes based on them, which significantly reduces the erasing time of information stored in the memory cells. The given results reveal the positive effect of structural SO of the GhG on their properties for different applications.

Keywords Chalcogenide Glasses · Self-organization · Memory Cells

D. Tsiulyanu (✉) · M. Ciobanu · A. Afanasiev
Technical University of Moldova, Chisinau, Republic of Moldova
e-mail: tsiudima@gmail.com

A. Vaseashta et al. (eds.), *Proceedings of the Eighth International Symposium on Dielectric Materials and Applications (ISyDMA'8)*, Springer Proceedings in Physics 349, https://doi.org/10.1007/978-3-031-84809-4_14

1 Introduction

The concept of rigidity percolation in glassy random networks, including those of chalcogenides, has been largely studied and discussed in scientific literature for many years. The beginning of theoretical works in this field is based on the mechanical principles of the stability of frames [1] and free energy at the glass transition point (T_g) at which the random network is formed [2]. At first, it was shown [3] that dependent on the value of the mean coordination number of atoms $<r>$ the glassy network can be floppy which is soft or amorphous solid, i.e. rigid. The structure of the amorphous rigid regions can be compatible either with the continuous random network (CRN) model [4], which maintains the local chemical order in the absence of long-range order (ideal glasses), or with the model, in which inside of amorphous network there are non-random structural elements [5] that go beyond just simple chemical ordering (real glasses). The compatibility of the CRN model occurred because of the coincidence of results of X-ray diffraction experiments, from which the radial distribution function can be determined and the computing-built model. Nevertheless, these arguments still admit that at low temperatures near the melting point, minimizing the free energy can form local stressed blocks (inclusions) with non-random structure, which cannot be detected via XRD but may manifest themselves in other experiments such as elastic ones or at Raman scattering. The state of the art in this area began to become clearer at the end of the last century with the development of the Topological Constraint Theory (TCT), originally proposed by Phillips and Thorpe. In accordance with this theory, the rigidity transition in solids can be either the first order at $<r_1> = 2.4$ (Maxwell's counting with the associated mean-field transition) or the second order at $<r_2> = 2.375$ if the network is an ideal random [5]. The effect of stressed non-random blocks inhabited in the CRN was found to result in the appearance of two-phase transitions: the first order, related to floppy–rigid transition at the same $<r_2> = 2.375$, and the second-order one at $<r_1> = 2.392$, related to unstressed—stressed rigid transition, with an intermediate phase between them (Fig. 1), which is rigid but stress-free.

Experimentally, the intermediate phase (IP) in ChG was discovered by Boolchand and coworkers [6] via different experimental methods, including the differential scanning calorimetry (DSC), Raman and Mössbauer spectroscopy, viscosity measurements et al. It was pointed out that the IP is characterized by a vanishing non-reversing heat-flow (enthalpy), ΔH_{nr} (r) $\rightarrow 0$, suggesting that glass compositions in IP are close to their liquid counterparts, i.e. are self-organized (SO). The compositional width (and centroid) of the IP was found to be determined by glass structure and appears to be very different for different glasses. In particular, the IP of glassy As-S-Ge ternary is like an oblique cut rhombus with the small diagonal oriented approximately along $Ge_xAs_xS_{100-2x}$ tie-line and the long one oriented perpendicularly, so that connects the centers of IP regions of binary As-S and Ge-S systems [7]. The glasses from the intermediate phase were shown to exhibit several remarkable properties such as the absence of stress [8] or aging [9], low molar volume, and high elastic modulus correlated with the glass medium range ordering [10], so are

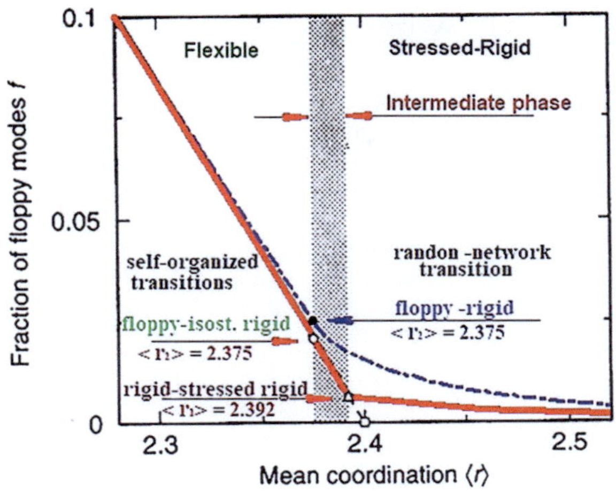

Fig. 1 The topological phases of a real glassy network (adapted from [6])

of great interests for application to fast ionic conductors [11]. The present work is devoted to studying the structural transformations in the IP region of ternary As-S-Ge system along the AsS_3–GeS_4 tie-line, addressed to seek the features process of self-organization at the gradual replacement of threefold coordinated As atoms by fourfold coordinated Ge ones. Besides, it was aimed to fabricate the solid electrolytes based on studied glasses from the intermediate phase and to trace how the structural self-organization of the chalcogenide glassy backbone influences their electrical superionic conductivity and finally, to check the potential application of these solid electrolytes for further advancing the switchers and the non-volatile memory cells based on the same.

2 Intermediate Phase Glasses of GeS_4–AsS_3 Pseudo-Binary System: The Structural Self-Organization and Properties

The global topological phase diagram in the As-S-Ge ternary (Fig. 2) has been constructed after a thorough investigation of the equimolar glassy $Ge_xAs_xS_{100-2x}$ tie-line along with relevant studies of binary As_xS_{100-x} and Ge_xS_{100-x} systems [7]. The glassy tie-line $(GeS_4)_x(AsS_3)_{1-x}$ that is the subject of the present work, lies near perpendicular on the equimolar $Ge_xAs_xS_{100-2x}$ line and intersects it at the point <r> = 2.33. The mean coordination number and chemical composition of the studied here glasses of the $(GeS_4)_x(AsS_3)_{1-x}$ system is listed in Table 1. It is seen that despite of small mean coordination number change of $2.25 \leq \langle r \rangle \leq 2.4$ it is entirely located inside the IP region of the As-S-Ge system. The alloys in question were prepared by the usual melt-quenching method, following the technology described earlier [12]. Determination of elemental composition and the structural investigations were

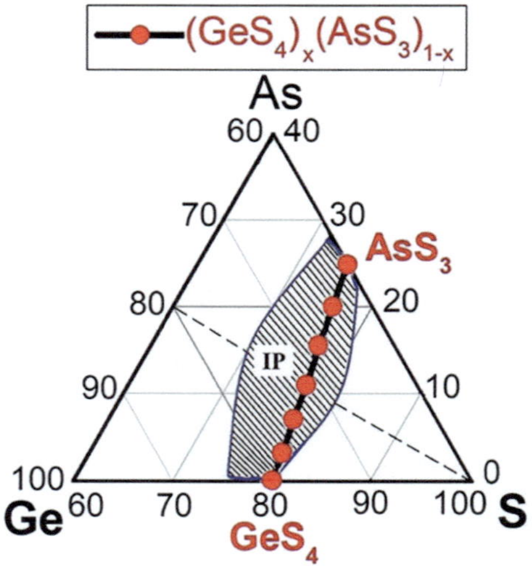

Fig. 2 The compositional triangle of the As-S-Ge system with intermediate phase region, shown as a hashed area that includes the GeS_4–AsS_3 glasses under investigation marked by red-filled circles. The interrupted line shows the equimolar $Ge_xAs_xS_{100-2x}$ tie-line

carried out by the EDX spectroscopy (INCA Energy 200 EDX, OXFORD Instruments), SEM (VEGA TESCAN TS 5130 MM, TESCAN), and XRD analyses (Bruker D8), but the results did not differ from those described in our previous work [13]. To study the structural transformations that occur in the $(GeS_4)_x(AsS_3)_{1-x}$ system in charge of an IP formation we have used the Raman spectroscopy, which provides such information through the analysis of fundamental vibrational modes of constituent structural blocks.

The measurements were performed on a Renishaw InVia Raman spectrometer attached to a Leica DM2700 microscope in the 75–550 cm^{-1} spectral range. The excitation beam (633 nm) was focused into a spot of 2.0 μm. The Raman spectra of studied glasses across the AsS_3-GeS_4 tie-line are shown in Fig. 3. Twelve (shown by arrows) modes have been identified but the most prominent can be classified into two groups, common for all compositions: the modes located around 151, 219, 473 cm^{-1} and those located inside of a broad band that covers the 280–400 cm^{-1} range. The first group of mentioned modes has been revealed in many works, which dealt with As-S-Ge glasses and were assigned to stretch modes of S–S in rings and chains. The second feature which is an intensive broadband, arrives because in the studied glasses the Raman bands corresponding to different As-S and Ge-S structural blocks arrive in the same frequency region. In this respect, truly useful results can be obtained only via performing the deconvolution of these broad spectral bands via fitting the Gaussian profiles followed by their thorough analysis. We made the Gaussian fitting, then the deconvoluted profiles were used both for the determination of wavenumbers of main vibrational modes—components of the mentioned broadband and their assignment to counterpart structural blocks.

The results for each studied glass are given in Table 1, alongside with normalized scattering strength of respective vibrational mode, calculated as ratio A_k/A_t, where A_k is the area of the individual (k) Raman mode and A_t is the total integrated area of

Table 1 Mean coordination number <r>, chemical composition, assignments, and strengths of some Raman modes revealed in $(GeS_4)_x(AsS_3)_{1-x}$ glasses

nr	<r>	[x]	Composition	Mode, (cm^{-1})	Assignment	A_k/A_t (%)	A_{QT}/A_t (%)
1	2.25	0	$As_{25}S_{75}$	328	mix. sym. str. $AsS_{3/2}$(PYR) & QT S = $As(S_{1/2})_3$	30	30
				353–355	sym. str. $AsS_{3/2}$	28	–
2	2.28	0.17	$Ge_4As_{20}S_{76}$	323	mix. sym. str. $AsS_{3/2}$(PYR) & QT S = $As(S_{1/2})_3$	14	18
				331–336	sym. stretch QT S = $As(S_{1/2})_3$	4	
				342	sym. str. GeS_4(CS)	6	–
				353–355	sym. str. $AsS_{3/2}$	47	–
3	2.31	0.33	$Ge_{7.7}As_{15.3}S_{77}$	331–336	sym. stretch QT S = $As(S_{1/2})_3$	23	41
				372–377	mix. sym. str GeS_4 (ES) asym str. QT S = $As(S_{1/2})_3$	12	
				353–355	sym. str. $AsS_{3/2}$	35	–
4	2.33	0.50	$Ge_{11.1}As_{11.1}S_{778}$	342	sym. str. GeS_4 (CS)	52	–
				372–377	mix. sym. str. GeS_4 (ES) asym. str. QT S = $As(S_{1/2})_3$	11	11
5	2.36	0.67	$Ge_{14.3}As_{7.1}S_{78.6}$	342	sym. str. GeS_4 (CS)	40	–
				372–377	mix. sym. str. GeS_4 (ES) asym. str. QT S = $As(S_{1/2})$	18	18
6	2.38	0.83	$Ge_{17.2}As_{3.5}S_{79.3}$	342	sym. str. GeS_4 (CS)	39	–
				363	asym. stretch QT S = $As(S_{1/2})_3$	7	15
				372–377	mix. sym. str GeS_4 (ES) asym str.QT S = $As(S_{1/2})$	8	

(continued)

Table 1 (continued)

nr	<r>	[x]	Composition	Mode, (cm^{-1})	Assignment	A_k/A_t (%)	A_{QT}/A_t (%)
7	2.4	1.00	$Ge_{20}S_{80}$	342	sym. str. GeS_4 (CS)	27	–
				372–377	mix. sym. str GeS_4 (ES) asym str. QT $S = As(S_{1/2})$	13	13

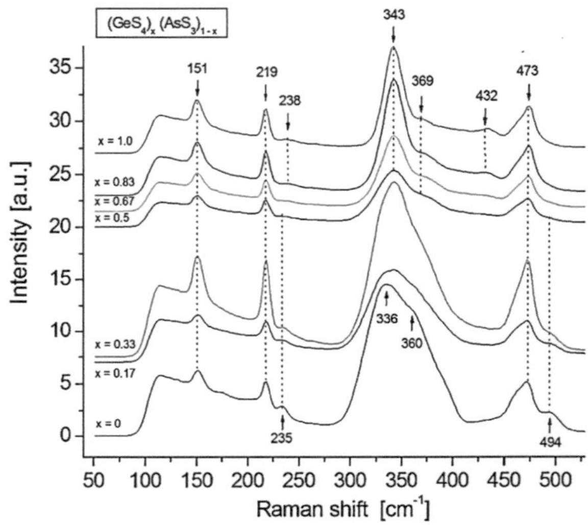

Fig. 3 Raman spectra of GeS_4–AsS_3 glasses

all modes revealed in the spectral range 75–550 cm^{-1}. There were revealed the modes caused by symmetric and asymmetric stretching vibrations assigned to three main structural units (s.u.) that are the pyramidal (PYR) structural building blocks $AsS_{3/2}$ (353–355 cm^{-1}), quasi-tetrahedral (QT) $S = As(S_{1/2})_3$ (331–336 cm^{-1}), which comprise fourfold coordinated As atoms [14], as well as the corner–shared (CS) tetrahedral GeS_4 blocks with characteristic vibrational mode at 342 cm^{-1} [15–17].

Besides, we have pointed out several vibrational modes 323–328 cm^{-1} assigned to mixed symmetric stretch of PYR $AsS_{3/2}$ with QT $S = As(S_{1/2})_3$ [14, 18]and the modes 372–377cm^{-1} ascribed to mixed symmetric stretch of the tetrahedral GeS_4 edge-sharing (ES) blocks and asymmetric stretch of $S = As(S_{1/2})_3$ quasi-tetrahedral units [19, 20]. Rely on TCT both the PYR ($AsS_{3/2}$) and tetrahedral GeS_4 structural blocks belong to isostatic rigid units with ideal connectivity as locally the number of constraints per atom equals the number of its degrees of freedom, i.e. 3 in the 3D network. Oppositely, the quasi-tetrahedral (QT) $S = As(S_{1/2})_3$ s.u. is ascribed to the most flexible species that can contribute to the self-organization of the glass network to avoid stress [21]. Noteworthy that the QT $S = As(S_{1/2})_3$ s.u. are inherent in one

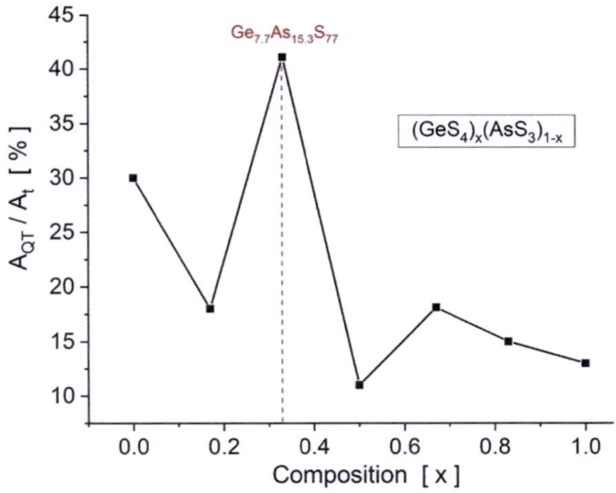

Fig. 4 Normalized scattering strengths of Raman modes A_{QT}/A_t versus glass composition of GeS_4–AsS_3 glasses

form or another in all studied compositions of $(GeS_4)_x(AsS_3)_{1-x}$ ternary system. As one can follow from Table 1, dependent on the glass composition, the Raman scattering modes inherent to QT S = $As(S_{1/2})_3$ units can either originate directly from symmetrical stretch vibrations of the same units or their mixing with modes belonging to vibrations of symmetrical stretch of the $AsS_{3/2}$ pyramids. Likewise, the strong Raman modes belonging to stretch asymmetric vibrations of QT S = $As(S_{1/2})_3$ blocks mixed with modes of a symmetric stretch of the edge-sharing (ES) tetrahedral GeS_4 blocks, are displayed in the same table. Raman scattering strength A_k/A_t of modes inherent in one form or another to flexible QT S = $As(S_{1/2})_3$ blocks points to glass structural self-organization and can be used as a measure of the same. To assess the degree of self-organization of the glass, we calculated the normalized scattering force A_{QT}/A_t (Table 1), which is the ratio of the total area of modes, one way or another associated with vibrations in the structural blocks QT S = As $(S_{1/2})_3$, divided by the integral area of all modes in a given spectral range, and plot it versus glass composition (Fig. 4).

One can observe two distinct maxima of scattering strength, the highest being located at composition nr. 3 that comprises 7.7 at.% Ge (<r> = 2.31), and the other, lower one appears around composition nr. 5 with 14.3 at.% Ge (<r> = 2.36). We assume that the revealed maxima show nothing else but the enhanced degree of SO of these glasses (especially around the composition $Ge_{7.7}As_{15.3}S_{77}$) that have to influence or even enhance their physical properties. Our previous investigations [13] have shown that namely around these two compositions the molar volume of the pseudo-binary glasses of $(GeS_4)_x(AsS_3)_{1-x}$ system exhibits a global (double-humped) minimum (Fig. 5a), observed also in homogenized binary As_xS_{100-x} melts/glasses [22]. Alongside, as shown in Fig. 5a the elastic modulus exhibits two maxima as well, the higher being located around composition $Ge_{7.7}As_{15.3}S_{77}$ (<r> = 2.31) and the lower one, arises in the vicinity of composition $Ge_{14.3}As_{7.1}S_{78.6}$ (<r> = 2.36) what entire concur with data related to compositional dependence of degree of the structural self-organization of the glass, graphically represented in Fig. 4. In such

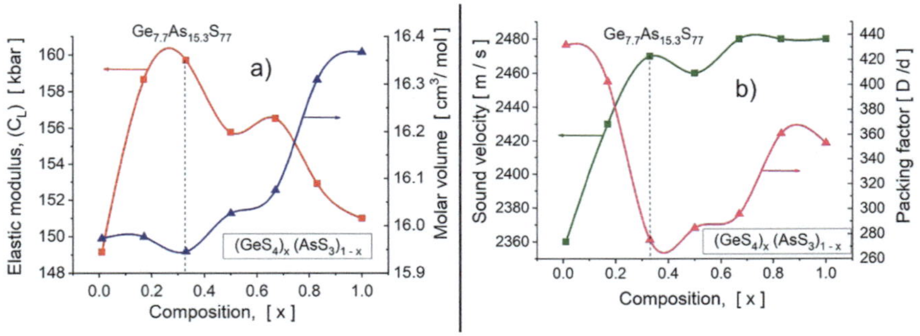

Fig. 5 Variation of molar volume and Young's modulus (**a**), as well as sound velocity and packing factor with composition change of glassy $(GeS_4)_x(AsS_3)_{1-x}$ ternary. All dependencies highlight the composition $Ge_{7.7}$ $As_{15.3}$ S_{77}

circumstances the structural self-organization of the glass backbone has to influence also their acoustic properties such as the velocity of the sound propagation and acoustic loss. Figure 5b illustrates the compositional dependence of the velocity of ultrasound (15 MHz) propagation along the compositional tie-line $(GeS_4)_x(AsS_3)_{1-x}$.

Two weak maxima of sound velocity, one can clearly distinguish around compositions that comprise 7.7 at.% and 14.3 at.% Ge, and which were ascribed a high degree of structural SO compared to neighboring compositions (Fig. 4). However, after the second maximum, the sound velocity saturates and there are no tendencies to diminish at further Ge concentration increase. Such behavior is due to the gradual decrease of the density of the glasses from the $(GeS_4)_x(AsS_3)_{1-x}$ line by Ge concentration increase [13]. The explanation of the last seems to be that the germanium-based chalcogenides comprise free volumes (voids) in their structure [23], so the increasing of Ge concentration leads to indecisive consequences: packing of the local polymeric texture of disordered floppy network and formation of voids. Obtained results show also an evident correlation between the compositional dependence of the Raman scattering normalized strengths and the glass middle range ordering (MRO), applied to sizes of MRO molecular domains (D) and correlated inter-domain distances (d). Both these parameters exhibit the minima around compositions nr. 3 ($Ge_{7,7}As_{15,3}S_{77}$) and nr. 5 ($As_{7,1}S_{78,6}Ge_{14,3}$) indicating their high degree of the "packing" of the middle order structure [13]. Figure 5b illustrates the compositional dependence of the so-called "packing factor", defined as the ratio $p = D/d$ [24]. One can observe that the packing factor non-linearly varies with composition change reaching the minimum around composition nr. 3; $Ge_{7,7}As_{15,3}S_{77}$, as for composition nr. 5; $As_{7,1}S_{78,6}Ge_{14,3}$ the minimum transfers into the shoulder, which concurs with data on the degree of glass self-organization derived from Raman spectra (Fig. 4).

3 Programmable Metallization Cells Based on Self-organized GeS_4–AsS_3 Glasses

One of the most remarkable properties of chalcogenide glasses is the possibility of realization in them of an ionic charge transport and what is more—its transition to a metallic one. The introduction of some metals (e.g. Ag) into chalcogenide glasses leads to a dramatic change in their physical and chemical properties. Already at around 7% of Ag, the Ag: GCh mixture becomes a solid electrolyte with high ionic conductivity that exceeds the electronic one by 5–10 orders of magnitude [25] which is a reason for them to be called superionic solid electrolytes. Fabrication of ChG-based superionic solid electrolytes can be synthesized via the usual melt-quenching method or by both thermo-diffusion or photo-dissolution (PD) of Ag into chosen chalcogenide glass. The Photo-dissolution (or photodoping) phenomenon was discovered at the early stage of the ChG study but remains the most applicable due to the simplicity of both technology of fabrication and monitoring of solid electrolyte formation [26]. In the present work, to fabricate the superionic solid electrolytes based on pseudo-binary glasses $(GeS_4)_x(AsS_3)_{1-x}$ from the intermediate phase of the As-S-Ge ternary, we have used the PD phenomenon as well. Figure 6 schematic illustrates the technological process of the solid electrolyte fabrication addressed to their further utilization as active parts of the programmable memory cells (PMC). The last (Fig. 6c) consists of a silicon substrate on which an insulating film of SiO_2 with a resistivity greater than 10^8 Ω·cm is beforehand grown, the passive metal electrode (electrochemically neutral) made of Pt, the solid electrolyte based on glassy chalcogenide with a self-organized structure and the active silver electrode, which serves also as a source of Ag ions. The solid electrolyte layer has been formed in two stages: In the first stage, its growth is achieved by thermal evaporation in a vacuum of the pre-synthesized ChG material with SO structure from the As-S-Ge system, and in the second stage, the transformation of this material into a solid electrolyte through the partial photo-dissolution in it of the Ag active electrode, deposited on it beforehand (Fig. 6a, b). The first that was noticeable is the strong influence of the glass composition of ChG on the electrical conductivity of fabricated solid electrolytes. Figure 7a shows the electrical conductance of fabricated solid electrolytes as a function of material composition. One can observe that this dependence well correlates with the degree of structural self-organization of the ChG backbone (Fig. 4): the composition nr. 3; $(Ge_{7.7}As_{15.3}S_{77})$ with maximal degree of SO, exhibits a maximal conductivity. A Schematic of the fabricated PMC is given in Fig. 6c, but the review of basic electrochemistry that enables its memory operation is the same as described earlier [27]. Initially, the "off" electrical resistance of the cell is high $R_o \sim 10^7\,\Omega$. When the direct polarization is applied between the cathode (Pt) and anode (Ag), in the solid electrolyte film the Ag+ ions are reduced at the cathode, turning into metallic Ag and the Ag atoms on the surface of the anode it oxidizes, i.e. turn into positive ions and migrates towards the cathode. The result of this electrochemical reaction is the rapid formation of a stable conductive metal electrodeposit (Ag) extending from the cathode to the anode that makes the electrical resistance of the cell sharply drop, going from the "off" state value (R_{off}~10^7 Ω) to the "on" state value, which is less than 10 Ω (Fig. 6d). The "on" state is maintained even after the applied field is reduced to zero, i.e. the power supply is disconnected, that is the memorization process is non-volatile.

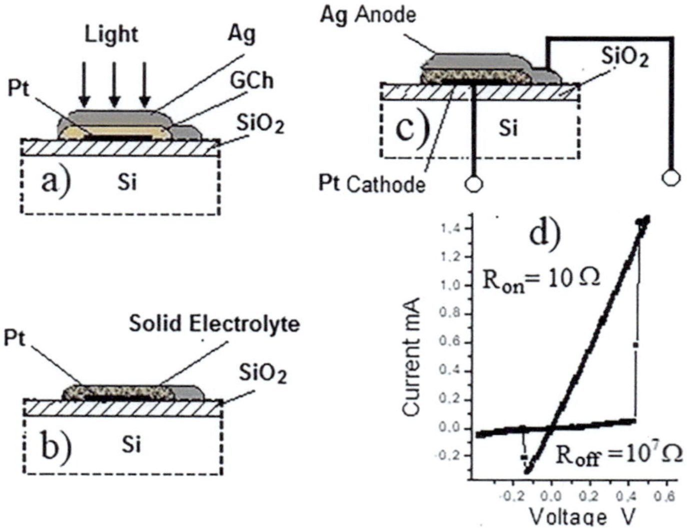

Fig. 6 **a–c** Schematic of PMC fabrication process based on intermediate phase glasses of the pseudo-binary $(GeS_4)_x(AsS_3)_{1-x}$ system; **d** The processes of storage—erasing information in PMC (details are in the text)

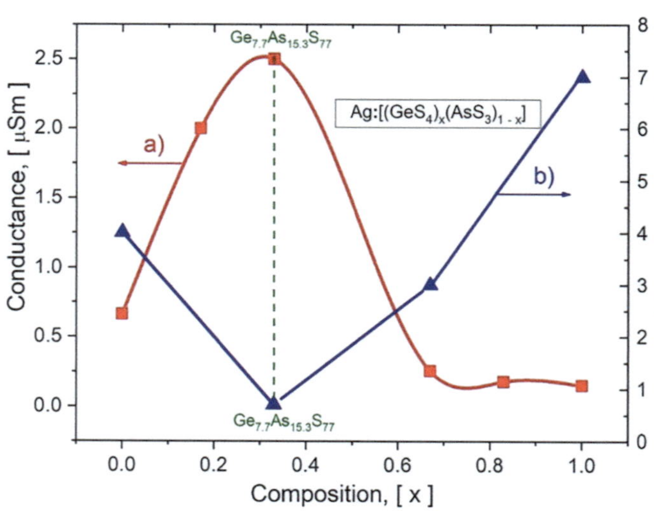

Fig. 7 Conductance of solid electrolytes (**a**) and the time of information erasing in PMC based on the same (**b**) versus composition of $(GeS_4)_x(AsS_3)_{1-x}$ glasses from the intermediate phase

To return the memory cell to the initial state with high resistance a reverse polarization is applied, which will cause an opposite process, disconnecting the metallic connection between the electrodes. The time of the information writing appeared to be very short <<1 μs, as the solid electrolyte consists of large local superionic conducting areas [25] which simplifies the process of the formation of the metallic filaments connecting electrodes on the drift of ions only between close superionic areas. The time of erasing is much longer and depends on many factors, including the geometry of PMC, glass composition, and others. Figure 7b shows the erasing time of the PMC (b) versus the composition of the studied $(GeS_4)_x(AsS_3)_{1-x}$ glasses, from an intermediate phase.

Noteworthy this parameter also depends on the (and correlates with) degree of structural self-organization of the glass, as the information erasing time also exhibits a global minimum around composition #3; $Ge_{7.7}As_{15.3}S_{77}$, which has the highest degree of structural SO among the glasses under investigation.

4 Conclusions

The following conclusions were drawn. Inside the intermediate phase region of chalcogenide glasses, including the As-S-Ge ternary, can form very narrow compositional areas with a high level of structural self-organization, mainly due to a sudden increase in the concentration of highly flexible species. Along the AsS_3–GeS_4 tie-line, a strongly self-organized glass builds around composition $As_{15.3}S_{77}Ge_{7.7}$ due to a high concentration of flexible quasi-tetrahedral $S = As(S_{1/2})_3$ atomic clusters. In consequence, this glass exhibits a minimal molar volume, the highest values of elastic modulus, and the velocity of sound propagation. The ionic conductivity of solid electrolytes fabricated from glasses of the intermediate phase exhibit global maxima at compositions with a high degree of self-organization that allows the improvement of operating parameters of programmable metallization cells, based on the same.

Acknowledgements This work was supported by the Ministry of Education and Research of the Republic of Moldova; Institutional subprogram **021201**. The authors express their gratitude to Dr. Ion Stratan for help in the PMC study and fruitful discussions.

References

1. Maxwell, J.: On the Calculation of the Equilibrium and Stiffness of Frames. Philos. Mag. 27, 294–299 (1864).
2. Phillips, J.: Topology of covalent non-crystalline solids I: Short-range order in chalcogenide alloys. J. Non-Cryst. Solids 34 (2), 153–181(1979).
3. Thorpe, M.: Continuous deformations in random networks. J. Non-Cryst. Solids 57(3), 355–370 (1983).

4. Zachariasen, W.: The atomic arrangement in glass. *J. Am. Chem. Soc.* 54(10), 3841–3851 (1932).
5. Thorpe, M., Jacobs, D., Chubynsky, M., Phillips, J.: Self-organization in network glasses. J. Non-Cryst. Solids 266–269, 859–866 (2000).
6. Boolchand, P., Georgiev, D., Goodman B.: Discovery of the intermediate phase in chalcogenide glasses. J. Optoelectronics and Advanced Materials 3(3), 703–720 (2001).
7. Chakravarty, S., Almutairi, B., Chbeir, R., Chakraborty, S., Bauchy, M., Micoulaut, M., Boolchand, P.: Progress, challenges, and rewards in probing melt dynamics, configurational entropy change, and topological phases of Group V- and Group IV-based multicomponent sulfide glasses. Phys. Status Solidi B 257 (2000116), 2–17 (2020). https://doi.org/10.1002/pssb.202000116.
8. Wang, F., Mamedov, S., Boolchand, P., Goodman, B., Chandrasekhar, M.: Pressure Raman effects and internal stress in network glasses. Phys. Rev. B 71 (174201), 1–8 (2005). https://doi.org/10.1103/PhysRevB.71.174201.
9. Chakravarty, S., Georgiev, D., Boolchand, P., Micoulaut, M.: Aging, fragility and the reversibility window in bulk alloy glasses. J. Phys.: Condens. Matter. 17, L1–L7 (2005). https://doi.org/10.1088/0953-8984/17/1/L01.
10. Tsiulyanu, D., Kozyukhin, S., Ciobanu, M.: Middle range order and elastic properties of non-stoichiometric chalcogenide glasses in the AsS3 - GeS4 system. J. Non-Cryst. Solids 575 (1), 121207 (1–7) (2022). https://doi.org/10.1016/j.jnoncrysol.2021.121207
11. Micoulaut, M., Phillips, J.: Rings and rigidity transitions in network glasses. Phys. Rev. B 67, (104204), 1–9 (2003). https://doi.org/10.1103/Phys.RevB.67.104204.
12. Tsiulyanu, D., Dragich, A., Gumeniuc, N.: Elastic properties of micro-nonhomogeneous As-S-Ge alloys. J. Non-Cryst. Solids 155(2), 180–184 (1993).
13. Tsiulyanu, D., Ciobanu, M. Afanasiev, A.: Physical-chemical properties of sulfur enriched As–S–Ge glasses related to middle-range order structure. In: 6th International Symposium on Dielectric Materials and Applications (ISyDMA'6) Proceedings, pp. 3–17. Springer, Cham. (2022).
14. Chen, P., Holbrook, C., Boolchand, P., Georgiev, D., Jackson, K., Micoulaut, M.: Intermediate phase, network de-mixing, boson, and floppy modes, and compositional trends in glass transition temperatures of binary AsxS1−x system. Phys. Rev. B 78, 224208 (2008). https://doi.org/10.1103/PhysRevB.78.224208.
15. Feng, X., Bresser, W., Boolchand, P.: Direct evidence for stiffness threshold in chalcogenide glasses. Phys. Rev. Lett. 78, 4422–4425 (1997).
16. Tanaka, K., Yamaguchi, M.: Resonant Raman scattering in GeS2. J. Non-Cryst. Solids 227-230, 757–760 (1998). https://doi.org/10.1016/S0022-3093(98)00184-7.
17. Sakaguchi,Y., Hanashima, T., Ohara, K., Simon, A., Mitkova, M.: Structural transformation in GexS100 − x ($10 \leq x \leq 40$) network glasses: Structural varieties in short-range, medium-range, and nanoscopic scale. Phys. Rev. Materials 3, (035601), 1–18 (2019). https://doi.org/10.1103/PhysRevMaterials.3.035601.
18. Golovchak, R., Shpotyuk, O., Mccloy, J., Riley, B., Windisch, C., Sundaram, S., Kovalskiy A., Jain, H.: Structural model of homogeneous As–S glasses derived from Raman spectroscopy and high-resolution XPS. Phil. Mag., 90, 4489–4501 (2010).
19. Jackson, K., Briley, A., Grossman, S., Porezag, D., Pederson, M.: Raman-active modes of a-GeSe2 and a-GeS2: A first-principles study. Phys. Rev. B. 60, R14985–R14989 (1999). https://doi.org/10.1103/PhysRevB.60.R14985.
20. Stronski, A., Kavetskyy, T., Revutska, L., Shportko K., Popovych, M., Kaban, I.: Structural order in (As2S3)x(GeS2)1-x glasses. J. Non-Cryst. Solids 572, 121075 (2021). https://doi.org/10.1016/j.jnoncrysol.2021.121075.
21. Micoulaut, M.: Rigidity and intermediate phases in glasses driven by speciation. Phys. Rev. B 74 (184208), 1–5 (2006). https://doi.org/10.1103/PhysRevB.74.184208.
22. Chakravarty, S., Chbeir, R., Chen, P., Micoulaut, M., Boolchand, P.: Correlating melt dynamics and configurational entropy change with topological phases of As_xS_{100-x} glasses and the crucial

role of melt/glass homogenization. In: Boolchand, P., Micoulaut, M. (eds). Topology of Disordered Networks and their Applications. Frontiers Media SA. Lausanne. https://doi.org/10.3389/fmats.2019.00166.
23. Phillips, J.: Topology of covalent non-crystalline solids II: Medium–range order in chalcogenide alloys and a-Si (Ge). J. Non-Cryst. Solids 43 (1), 37–77(1981). https://doi.org/10.1016/0022-3093(81)90172-1.
24. Kavetskky, T., Shpotyuk, O., Popescu, M., Lorinczi, A., Sava, F.: FSDP-related correlations in chalcogenide glasses, J. Optoelectronics and Advanced Materials 9 (10), 3079–3081 (2007).
25. Bychkov E., Tveryanovich Yu., Vlasov Yu.: Ion conductivity and sensors. In: Kwillardson, R., Weber, E. (eds). Semiconductors and Semimetals 80, pp. 103–168, Elsevier Inc. (2004). https://doi.org/10.1016/S0080-8784(04)80027-9.
26. Tsiulyanu, D., Stratan, I.: On the photo-dissolution kinetics of silver in glassy As_2S_3. J. Non-Cryst. Solids 356 (3), 147–152 (2010). https://doi.org/10.1016/j.jnoncrysol.2009.10.003.
27. Mitkova, M., Kozicki, M.: Fourfold coordinated silver-containing chalcogenide glasses – basic science and applications in optical programmable metallization cell (PMC) technologies. In: Lucovsky, G., Popescu, M. (eds). Non-Crystalline Materials for Optoelectronics, INOE Publishing House. Bucharest (2004).

Eco-Friendly Chalcogenides Semiconductors Cu_2XSnS_4 (X: Zn, Ni, Fe, Co): Elaboration, Characterization and Applications in Photovoltaics

A. El kissani, D. Ait Elhaj, S. Drissi, A. Abali, H. Badr, H. Dahabi, S. Elmassi, H. Amiri, M. Bousseta, L. Nkhaili, Z. Aqachmar, K. El Assali, and A. Outzourhit

Abstract The Eco-Friendly quaternary chalcogenides semiconductors with chemical formula Cu_2XSnS_4 where X is the zinc or nickel or cobalt (X = Ni, Fe, Co) were successfully synthesized by facile sol–gel route, using spin coating technique and without sulfurization step. X-ray diffraction, Raman spectroscopy, scanning electron microscopy, energy dispersive x-ray spectroscopy, and UV–visible-NIR double beam spectrophotometer determined their structures, surface morphologies, composition, and optical properties. Also, to test the possibility of using the elaborated materials for photovoltaic applications, the solar cells with the Cu_2XSnS_4 (X = Ni, Fe, Co) as absorbing materials were prepared. Current density–voltage (J–V), measurements were used to characterize the cell devices under dark and standard AM 1.5 sunlight conditions.

Keywords Chalcogenides Cu_2XSnS_4 · Sol–gel · Spectroscopy · Characterization · Solar cells

1 Introduction

The quaternary semiconductors Cu_2XSnS_4 (CXTS) where the X is Ni, Fe, or Co noted with formula I_2–II–IV–VI_4, have attracted great interest in recent years due to the abundance, and non-toxicity of their elementary constituents, and their excellent optical, electrical, thermoelectric properties [1, 2]. These materials have a p-type conductivity, and an optical band gap in the range of 1.1–1.7 eV; optimal for the

A. El kissani (✉) · D. A. Elhaj · S. Drissi · A. Abali · H. Badr · H. Dahabi · S. Elmassi · H. Amiri · M. Bousseta · L. Nkhaili · Z. Aqachmar · K. El Assali · A. Outzourhit
Materials, Energy and Environment Laboratory (LaMEE), Faculty of Sciences Semlalia, Cadi Ayyad University, Marrakech, Morocco
e-mail: ab.elkissani@uca.ac.ma

A. Vaseashta et al. (eds.), *Proceedings of the Eighth International Symposium on Dielectric Materials and Applications (ISyDMA'8)*, Springer Proceedings in Physics 349, https://doi.org/10.1007/978-3-031-84809-4_15

absorption of the majority of the solar spectrum [3]. Also, they are promising materials that can be used in different applications such as solar-cell absorbers, batteries, photodetectors [4], photocatalysts [5], thermoelectric [6], and non-linear optics [7]. Several methods have been used for the synthesis of Cu_2XSnS_4 (X: Ni, Fe, Co) materials such as thermal evaporation [8–10], RF sputtering [11], electrodeposition [12–14], the sol–gel method [15–17] and the spray technique [18, 19]. All of these methods use the process of annealing with sulfurization. In this work, we have developed a new and facile route to synthesize the Cu_2XSnS_4, semiconductor materials using the sol–gel spin-coating method without the sulfurization step. In addition, the CXTS-based solar cells with structure FTO/i-ZnO/n-CdS/p-CXTS/Au have been fabricated and characterized.

2 Experimental Details

2.1 CXTS Thin Films Synthesis

The Cu_2XSnS_4 (CXTS) sol–gel solution where X is Nickel Ni, iron Fe or cobalt Co were prepared from aqueous solution containing copper chloride CuCl (2 M), tin chloride dehydrate $SnCl_2$, $2H_2O$ (1.3 M), thiourea $SC(NH_2)_2$ (8 M), and XCl_2 (1.4 M), which are dissolved in the mixture of 2-methoxy ethanol (70%) and deionized water (30%) and stirred on hot plate at 50 °C for 20 min. Figure 1 shows the images of obtained sol–gel solution precursors.

After that, the CXTS thin films were deposited by spin coating technique, using the sol–gel solution precursors dropped onto soda lime glass and fluorine tin oxide glass (FTO) substrates, which were cleaned in an ultrasonic bath using acetone, ethanol, and distilled water for 5 min, and rotated at 3500 rpm for 30 s. Then the deposited thin films were dried in an air atmosphere at 110 °C for 5 min to remove residual organic materials. Subsequently, the formed CXTS thin films were annealed

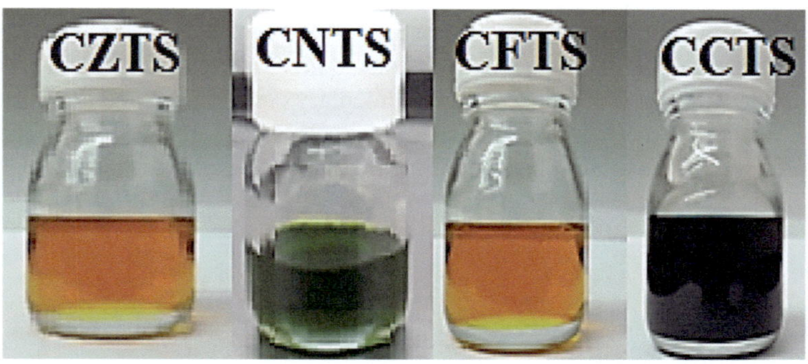

Fig. 1 Cu_2XSnS_4 (CXTS) sol–gel solution where X is Zinc Zn, Nickel Ni, iron or cobalt Co

Table 1 Presents optimal conditions of annealing time and temperature for the formation of the Cu_2XSnS_4 thin films where X is Nickel Ni, iron Fe, or Cobalt Co

	CZTS	CNTS	CFTS	CCTS
Temperature, °C	470	300	400	300
Time, min	60	60	40	60

in a nitrogen atmosphere using a tubular furnace at a ramp-up rate of 15 °C. The optimal annealing conditions are presented in Table 1.

2.2 *Characterization Technique*

X-ray diffraction (Rigaku, Smart Lab SE diffractometer, Japan) equipped with a Cukα (λ = 1.54059 Å), operating at 40 kV and 40 mA was used to examine the crystalline structure of CXTS thin films. Raman Microscopy (Sol Instrument LTD) was applied to identify the existence of crystalline phases in CXTS thin films. Scanning electron microscopy (Tescan Vega 3 Orsay, France) equipped with an energy dispersive analysis (EDX, USA) was used to observe the surface morphology and elemental composition of the prepared thin films. The optical transmittance was taken using a UV–Vis-NIR spectrophotometer (Shimadzu UV1800, Tokyo, Japan).

3 Results and Discussion

3.1 *X-ray Diffraction and Raman Analysis*

Figure 2 shows the X-ray diffraction (XRD) patterns of the prepared CXTS thin films under optimal annealing conditions of temperature and time. It can be seen that the synthesized thin films are polycrystalline, with peaks corresponding to the CXTS phases clearly observed. Additionally, the XRD peaks of the spin-coated thin films are quite sharp and narrow, indicating the high crystalline quality of the elaborated CXTS thin films. The average crystallite sizes (D), micro-strain (ε) and the dislocation density (δ) were calculated for the different peaks (hkl) planes of CXTS thin films using the following formulas:

$$D = K\lambda/\beta cos\theta \tag{1}$$

$$\varepsilon = \beta\cos(\theta)/4 \tag{2}$$

$$\delta = {}^{1}/_{D^2} \tag{3}$$

where λ is the wavelength of X-ray diffraction, K a constant with a value of 0.9, β is the full width at half maximum, and θ is the Bragg's angle of the related peak. The obtained values are summarized in Table 2.

To confirm the purity of prepared CXTS (X: Ni, Fe, and Co) phases, the Raman spectroscopy analysis was carried out. Figure 3 shows the Raman spectra of the prepared CXTS thin films. It showed that only the peak characteristics of CXTS phases have been detected, which indicated that the absence of CuS, SnS, NiS, and CoS secondary phases or their alloys.

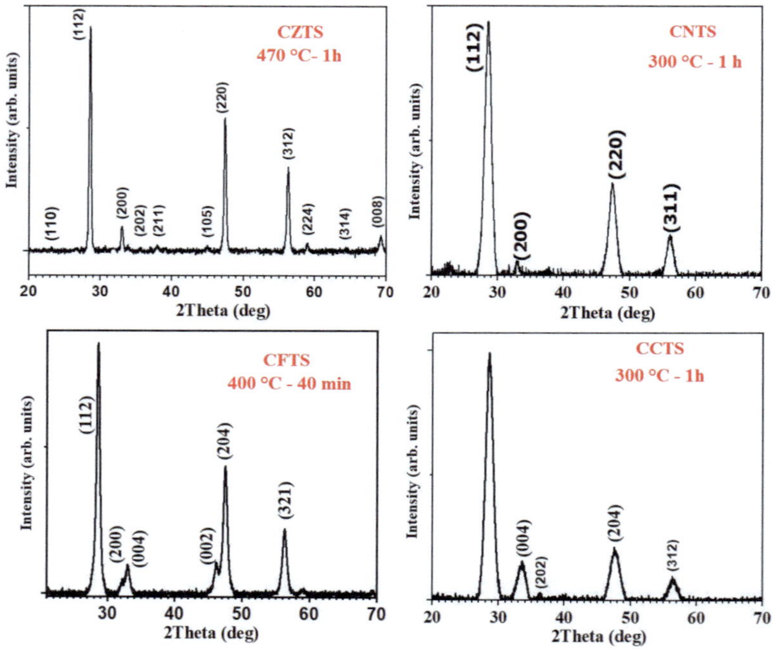

Fig. 2 X-ray diffraction patterns of **a** Cu_2NiSnS_4 (CNTS), **b** Cu_2FeSnS_4 (CFTS) and Cu_2CoSnS_4 (CCTS) thin films synthesized by sol–gel spin coating technique and without sulfurization

Table 2 Crystallite size, microstrain, and dislocation density of the prepared CXTS thin films at optimal annealing temperature and time conditions

	Crystallite sizes D (nm)	Micro strain $\varepsilon(10^{-4})$	Dislocation density $\delta(10^{14}m^{-2})$
CZTS	63	5.5	2.5
CNTS	8	0.7	156.25
CFTS	60	5.2	2.77
CCTS	13	1.1	59.17

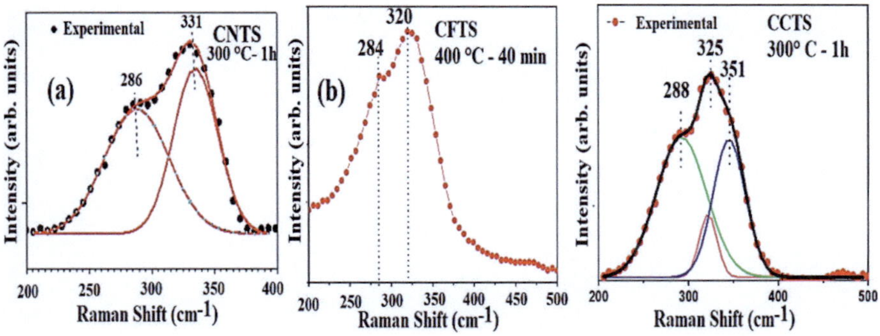

Fig. 3 Raman patterns of **a** Cu_2NiSnS_4 (CNTS), **b** Cu_2FeSnS_4 (CFTS) and Cu_2CoSnS_4 (CCTS) thin films synthesized by sol–gel spin coating technique and without sulfurization at optimal conditions of temperature and time

3.2 Morphological and Compositional Analysis

Figure 4 shows the surface morphology and the cross-section images of CXTS (X:Zn, Ni, Fe or Co) thin films observed by Scanning electronic microscopy. It is shown that the surfaces of synthesized thin films are granular, homogenous, smooth, dense, and relatively void-free surfaces. In addition, the grains of CNTS thin films are spherical and well-defined. Based on the cross-section the thickness values of the synthesized CNTS, CFTS, and CCTS thin films are 1.8, 1.5, and 2 nm respectively

To understand the chemical composition of the synthesized CXTS (X: Zn, Ni, Fe, or Co) thin films, the energy dispersive spectroscopy (EDS) analysis was performed. Figure 5 Shows the EDS spectra of the synthesized CXTS thin films. It's observed that only the chemical elements that are involved in the composition of CXTS thin films are present. Table 3 Reports the average chemical composition of synthesized CXTS thin films. It's revealed that the compositional ratios of CXTS thin film are near 2:1:1:4, which corresponds to the stoichiometric approximately Cu_2XSnS_4.

3.3 Optical Characterization

To evaluate the optical band gap, the absorption coefficients (α) have been determined from the measured optical transmissions using the following equation:

$$\alpha = \frac{\ln(1/\mathrm{T})}{\mathrm{e}}$$

where T is the transmittance and e is the thickness of CXTS (X: Zn, Ni, Sn or Co) thin films. For a direct band gap semiconductor, the absorption coefficient and optical band gap are related by the following relationship:

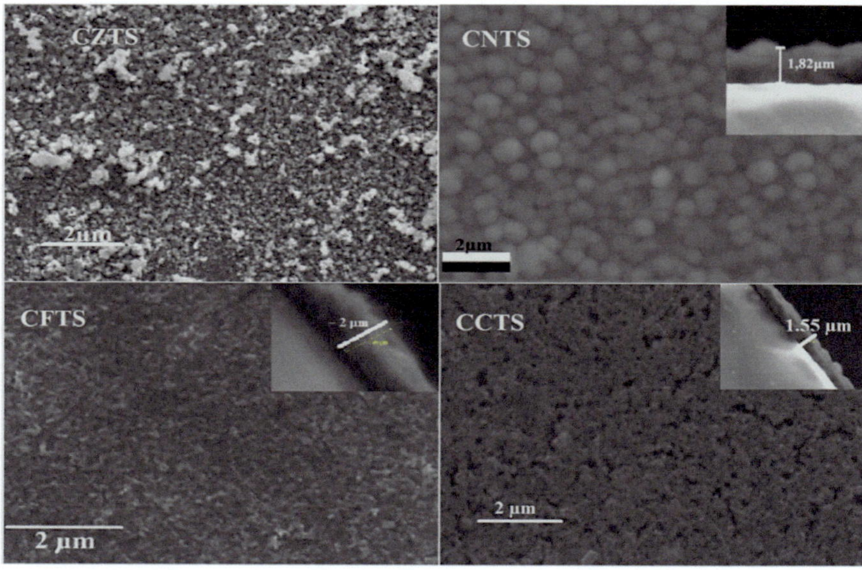

Fig. 4 SEM images of CXTS (X: Zn, Ni, Fe, Co) thin films deposited by spin-coating and without sulfurization

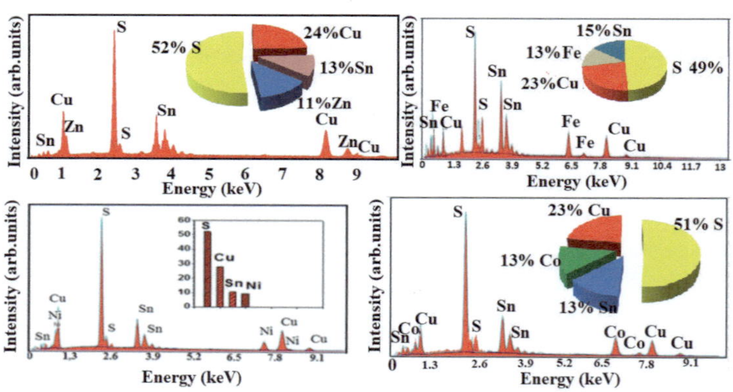

Fig. 5 SEM images of CXTS (X: Ni, Fe, Co) thin films deposited by spin-coating without sulfurization at optimal annealing conditions of temperature and time

Table 3 Elemental analysis of synthesized CXTS (X: Ni, Fe or Co) thin films at the optimal annealing condition of temperature and time

	%Atomic				Ratio		
	%Cu	%X (Ni, Fe, Co)	%Sn	%S	X/Sn	Cu/(Sn + X)	S/(Cu + X + Sn)
CNTS	28	9	11	52	0.8	1.4	1
CFTS	23	13	14	49	0.9	0.9	1
CCTS	23	13	14	51	0.9	0.9	1

$$(\alpha h\vartheta)^2 = A\left(h\vartheta - E_g\right),$$

where α is absorption coefficient, $h\vartheta$ photon energy, E_g band gap, and A is a constant. Figure 6 shows the variation of absorption coefficients (α) versus the wavelength (λ). It's observed that the CXTS thin films exhibit high coefficient absorption in the visible region. The Inset of Fig. 6 shows the plot of $(\alpha h\vartheta)^2$ as a function of photon energy ($h\nu$). The optical band gap can be obtained by extrapolating the linear part of curves to intercept on the horizontal axis corresponding to photon energy $h\vartheta$. The extracted band gap values of CXTS (X: Zn, Ni, Sn or Co) thin film are presented in Table 4. The obtained value of the direct optical band gap for synthesized CXTS thin films is suitable for solar cell applications.

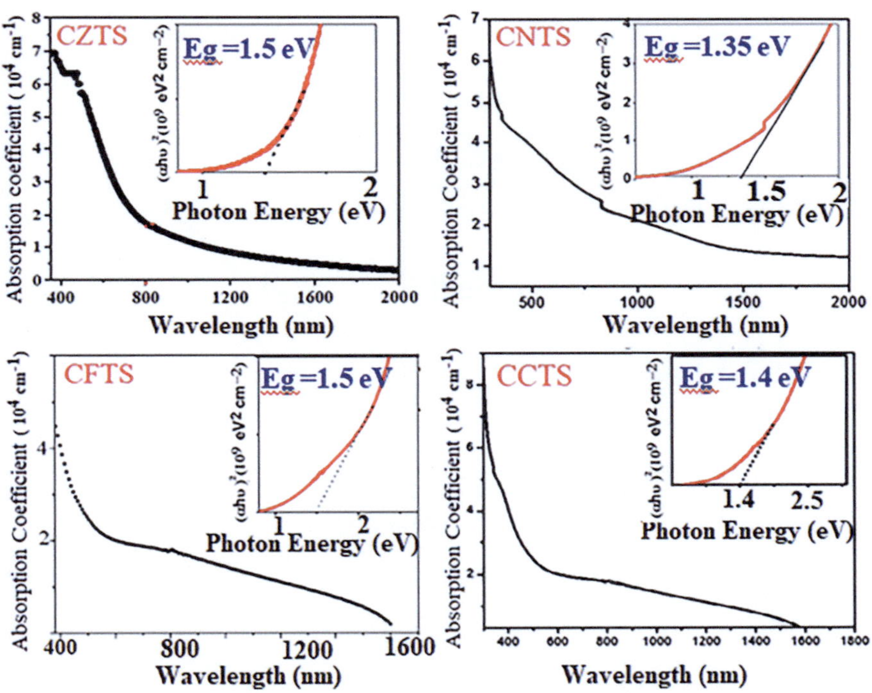

Fig. 6 Shows the variation of absorption coefficients versus the wavelength of CXTS (X: Zn, Ni, Fe, or Co) thin films

Table 4 Optical band gaps of CXTS (X: Zn, Ni, Sn or Co) thin films prepared at optimal conditions of annealing temperature and time

CXTS thin films	CZTS	CNTS	CFTS	CCTS
Eg (eV)	1.5	1.35	1.5	1.4

3.4 CXTS-Based Solar Cells

The CXTS-based solar cells with structure FTO/i-ZnO/n-CdS/p-CXTS/Au (inset of Fig. 7), and with an active area of 0.5 cm^2 were prepared and characterized. The current density–voltage (J–V) characteristics of the best solar cells are presented in Fig. 7. The performance parameters are given in Table 5. It showed that a higher open-circuit voltage V_{oc} of 0.8 V was obtained under simulated AM 1.5 illumination conditions, compared with literary works [20, 21]. The Shunt and series resistances, R_{sh} and R_s, are restively higher. In addition, the low value of Fill Factor FF caused the quite low fill factor.

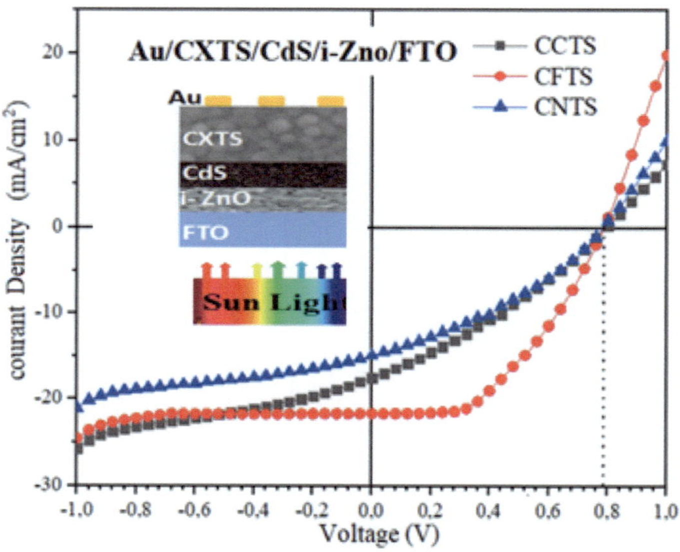

Fig. 7 Current density- voltage characteristics of CXTS (X: Ni, Fe or Co) based solar cells with structure FTO/i-ZnO/n-CdS/p-CXTS/Au

Table 5 Performance parameters of prepared CXTS-based solar cells

CXTS	Voc (V)	Jsc (mA cm^{-2})	FF (%)	η (%)	Rs (Ω cm^2)	Rsh (Ω cm^2)
CNTS	0.78	14.9	35	4.06	22.85	110
CFTS	0.79	21.8	46	7.92	12.5	3850
CCTS	0.78	17.5	33	4.83	28	74

4 Conclusion

In summary, the earth-abundant, environment-friendly Cu2XSnS4 where X is Zn, Ni, Fe or Co (CXTS) thin films have been successfully deposited via a sol–gel method, spin-coating technique and without sulfurization step in a nitrogen atmosphere. Highly pure, stoichiometric, uniform and compact CXTS thin films were synthesized. The optimal annealing temperature of CZTS, CNTS, CFTS, and CCTS thin films are 470 for 1 h, 300 for 1 h, 400 for 40 min, and 300 °C for 1 h respectively. These films possess a high optical absorption coefficient and optical band gap near the ideal band gap of the absorber for single junction solar cells. The performance parameters of CXTS-based solar cells with structure FTO/i-ZnO/n-CdS/ p-CXTS/ Au were measured and investigated. The optimization experiments are in progress and are expected to lead to a significant improvement.

References

1. I.M. El Radaf, H.I. Elsaeedy, H.A. Yakout, M.T. El Sayed, Junction Parameters and Electrical Characterization of the Al/n-Si/Cu_2CoSnS_4/Au Heterojunction, Journal of Electronic Materials. 48, 6480–6486 (2019). https://doi.org/10.1007/s11664-019-07445-7.
2. M.C. Benachour, R. Bensaha, R. Moreno, Optik Annealing duration influence on dip-coated CZTS thin films properties obtained by sol-gel method, Optik - International Journal for Light and Electron Optics. 187, 1–8 (2019). https://doi.org/10.1016/j.ijleo.2019.05.015.
3. D.B. Khadka, J. Kim, Structural, optical and electrical properties of Cu_2FeSnX_4 (X = S, Se) thin films prepared by chemical spray pyrolysis, Journal of Alloys and Compounds. 638, 103–108 (2015). https://doi.org/10.1016/j.jallcom.2015.03.053.
4. R.R. Prabhakar, N. Huu Loc, M.H. Kumar, P.P. Boix, S. Juan, R.A. John, S.K. Batabyal, L.H. Wong, Facile water-based spray pyrolysis of earth-abundant Cu_2FeSnS_4 thin films as an efficient counter electrode in dye-sensitized solar cells, ACS Applied Materials and Interfaces. 6, 17661–17667 (2014). https://doi.org/10.1021/am503888v.
5. H. Guan, X. Wang, Y. Huang, Optical, Photocatalytic And Thermoelectric Properties Of Cu_2MeSnS_4 (Me=Mn^{2+}, Fe^{2+}, Co^{2+}) Nanocrystals Via Microwave-Assisted Solvothermal Method, Chalcogenide Letters, Vol. 15, 435–440, (2018).
6. H. Şükür Kılıç, Y. Gündoğdu, S. Kılıç, S. Y. Gezgin, Nonlinear optical properties of Cu_2ZnSnS_4 nanocrystal thin films and its constituents thin films, Optical and Quantum Electronics 53,19 1–11, (2021). https://doi.org/10.1016/j.ijleo.2020.166054.
7. H. Hammami, M. Marzougui, H. Oueslati, M. Ben Rabeh, M. Kanzari, Synthesis, Growth and Characterization of Cu_2CoSnS_4 Thin Films via Thermal Evaporation Method, Optik. 227 166054 (2021). https://doi.org/10.1016/j.ijleo.2020.166054.
8. E. Garcia-Llamas, J.M. Merino, R. Gunder, K. Neldner, D. Greiner, A. Steigert, S. Giraldo, V. Izquierdo-Roca, E. Saucedo, M. León, S. Schorr, R. Caballero, Cu_2ZnSnS_4 thin film solar cells grown by fast thermal evaporation and thermal treatment, Solar Energy. 141, 236–241 (2017). https://doi.org/10.1016/j.solener.2016.11.035.
9. C. Shi, G. Shi, Z. Chen, P. Yang, M. Yao, Deposition of Cu 2ZnSnS 4 thin films by vacuum thermal evaporation from single quaternary compound source, Materials Letters. 73 89–91 (2012). https://doi.org/10.1016/j.matlet.2012.01.018.
10. M.Y. Zaki, F. Sava, A.T. Buruiana, I.D. Simandan, N. Becherescu, A.C. Galca, C. Mihai, A. Velea, Synthesis and characterization of Cu_2ZnSnS_4 thin films obtained by combined

magnetron sputtering and pulsed laser deposition, Nanomaterials. 11 (2021). https://doi.org/10.3390/nano11092403.

11. C.L. Yang, Y.H. Chen, M. Lin, S.L. Wu, L. Li, W.C. Liu, X.S. Wu, F.M. Zhang, Structural, optical and magnetic properties of Cu_2NiSnS_4 thin films deposited by facile one-step electrodeposition, Materials Letters. 166, 101–104 (2016). https://doi.org/10.1016/j.matlet.2015.12.054.
12. M. Beraich, M. Taibi, A. Guenbour, A. Zarrouk, A. Bellaouchou, M. Fahoume, Synthesis of Tetragonal Cu_2NiSnS_4 Thin Film via Low-Cost Electrodeposition Method: Effect of Ni^{2+} Molarity, Journal of Electronic Materials. 49 728–735 (2020). https://doi.org/10.1007/s11664-019-07707-4.
13. E.M. Mkawi, Kesterite Cu_2ZnSnS_4 thin films synthesized utilizing electrodeposition: Influence of metal doping on the properties, International Journal of Energy Research. 45, 1908–1917 (2021). https://doi.org/10.1002/er.5873.
14. K. Tanaka, N. Moritake, H. Uchiki, Preparation of Cu_2ZnSnS_4 thin films by sulfurizing sol-gel deposited precursors, Solar Energy Materials and Solar Cells. 91, 1199–1201 (2007). https://doi.org/10.1016/j.solmat.2007.04.012.
15. B. Murali, M. Madhuri, S.B. Krupanidhi, Solution-processed Cu2CoSnS4 thin films for photovoltaic applications, Crystal Growth and Design. 14 3685–3691 (2014). https://doi.org/10.1021/cg500622f.
16. M. İlhan, M.M. Koç, B. Coşkun, F. Yakuphanoğlu, Optical, Electrical and Photoresponsive Properties of Cu_2NiSnS_4 Solar Detectors, Journal of Electronic Materials. 49, 4457–4465 (2020). https://doi.org/10.1007/s11664-020-08197-5.
17. S. Dridi, N. Bitri, M. Abaab, Synthesis of quaternary Cu_2NiSnS_4 thin films as a solar energy material prepared through «spray» technique, Materials Letters. 204, 61–64 (2017). https://doi.org/10.1016/j.matlet.2017.06.028.
18. M. Adelifard, Preparation and characterization of Cu2FeSnS4 quaternary semiconductor thin films via the spray pyrolysis technique for photovoltaic applications, Journal of Analytical and Applied Pyrolysis. 122, 209–215 (2016). https://doi.org/10.1016/j.jaap.2016.09.022.
19. S.A. Vanalakar, P.S. Patil, J.H. Kim, Recent advances in the synthesis of Cu_2FeSnS_4 materials for solar cell applications: A review, Solar Energy Materials, and Solar Cells. 182, 204–219 (2018). https://doi.org/10.1016/j.solmat.2018.03.021.
20. P. Baskaran, K.D. Nisha, S. Harish, R. Ramesh, H. Ikeda, J. Archana, M. Navaneethan, Improved electrochemical performance of Cu2NiSnS4 hierarchical nanostructures as counter electrode in dye-sensitized solar cells, Materials Letters. 307, 130946 (2022). https://doi.org/10.1016/j.matlet.2021.130946.
21. P.S. Maldar, A.A. Mane, S.S. Nikam, S.D. Dhas, A. V. Moholkar, Spray deposited Cu_2CoSnS_4 thin films for photovoltaic application: Effect of film thickness, Thin Solid Films. 709 (2020). https://doi.org/10.1016/j.tsf.2020.138236.

Raman Spectroscopy of Pristine and Filled Single-Walled Carbon Nanotubes

Arbaoui Zakariya, Chenouf Jamal, and Rahmani Abdelhai

Abstract Raman spectroscopy has supplied an extremely strong tool to reveal the unique structure and outstanding electronic and vibrational properties of single-walled carbon nanotubes (SWCNTs). We report here the computed Raman spectra of pristine and filled SWCNTs through a combination of molecular mechanics and density functional theory. The calculated Raman spectra of pristine SWCNTs with different diameters and chiralities revealed the diameter dependence of RBM where the RBM wavenumbers are shown to be inversely proportional to the SWCNTs diameters, as well as the dependence of tangential modes on the chirality which are in agreement with experimental observations. The filling effects on Raman active modes of (8,6) nanotube encasing oligothiophenes are analyzed. The disappearance of libration modes of oligothiophenes after filling afforded the first hints on the stability of the filled (8,6) nanotube characterized by its optimal diameter. The downshift of RBM gives additional evidence on the stability of the filled nanotube. Moreover, changes in the peak positions of the G-band after encapsulation reveal the appearance of charge transfer between the nanotube and oligothiophenes.

Keywords Raman spectroscopy · SWCNTs · Oligothiophenes

1 Introduction

Owing to their unusual properties, carbon nanotubes (CNTs) have received a lot of attention in the field of nanotechnology. Since Iijima's discovery [1], researchers throughout the world have been attracted by their various chemical and physical properties. These distinguishing properties have opened the way for a multitude of possible applications in various fields, including mechanical reinforcement and electronics, energy storage materials, and biological sensors, with a significant impact

A. Zakariya · C. Jamal · R. Abdelhai (✉)
Laboratory for the Study of Advanced Materials and Applications (LEM2A), Moulay Ismail University FSM-ESTM, Meknes, Morocco
e-mail: ab.rahmani@umi.ac.ma

A. Vaseashta et al. (eds.), *Proceedings of the Eighth International Symposium on Dielectric Materials and Applications (ISyDMA'8)*, Springer Proceedings in Physics 349, https://doi.org/10.1007/978-3-031-84809-4_16

on photovoltaic devices [2]. CNTs can be multi-walled (MWCNTs), double-walled (DWCNTs), or single-walled (SWCNTs) [3]. Among these, MWCNTs with diameters of 10 nm, have exceptional electrical, mechanical, and chemical capabilities similar to SWCNTs, but at a cheaper cost and with less carbonaceous and metallic impurities. SWCNTs versatility is demonstrated by their functions as model systems for researching low-dimensional features and as unique objects with excellent mechanical and electronic properties, both of which have contributed to advances in nanoscale science and industry [4].

Several synthesis methods have been developed for producing SWCNTs including chemical vapor deposition, laser ablation, electrochemical synthesis, electric arc discharge, and ball milling–annealing processes [5]. Over the years, researchers have introduced novel synthesis approaches aimed at creating a diverse array of nanotubular materials. These include filled SWCNTs with fullerene molecules, and simple substances like halogenides ($ZrCl_4$), metals (K, Bi, Eu), nonmetals (Se), and oxides (Re_xO_y) [5]. Such advancements have expanded the scope of nanotube applications and opened avenues for innovative materials with tailored properties.

SWCNTs have emerged as an excellent system for studying Raman spectra in one-dimensional structures. Simultaneously, Raman spectroscopy has become a useful technique for investigating the vibrational characteristics and electronic structures of SWCNTs [6–8]. This approach has excellent sensitivity, allowing for a detailed evaluation of SWCNTs samples in terms of diameter and overall quality. Raman spectroscopy allows researchers to acquire insights into the distinct vibrational modes and electronic transitions seen in various types of SWCNTs, offering light on their fundamental features and prospective applications. In this paper, we propose learning about pristine and filled SWCNTs from their Raman features using a combination of molecular mechanics and density functional theory. We calculate the Raman spectra of empty SWCNTs with different diameters and chiralities and filled SWCNTs with a series of oligothiophenes nT (n = 2, 4, and 6). The diameter and chirality dependence of Raman spectra of pristine SWCNTs, as well as the filling effects on Raman active modes are investigated.

2 Model and Computational Method

The differential Raman scattering cross section describing the time average of the energy flux of the Raman-scattered light with a wavenumber between ω_k and $\omega_k + d\omega_k$ in a given direction and in a solid angle $d\Omega$, is written as.

$$\frac{d^2\sigma}{d\Omega d\omega_k} = \frac{1}{8\pi^2 c^2}\omega_k^3\omega_i[B(\omega)+1]\hbar \times \sum_{\alpha\beta\gamma\lambda} v_\alpha v_\beta H_{\alpha\gamma\beta\lambda}(\omega) w_\gamma w_\lambda \quad (1)$$

where,

$$H_{\alpha,\gamma,\beta,\lambda}(\omega) = \sum_j a^*_{\alpha,\gamma}(j) a_{\beta,\lambda}(j) \frac{1}{2\omega_j}\left[\delta(\omega - \omega_j) - \delta(\omega + \omega_j)\right] \tag{2}$$

With,

$$a_{\alpha,\gamma}(j) = \sum_{n\delta} \frac{\pi^n_{\alpha,\gamma;\delta}}{\sqrt{M_n}} \langle \delta n | j \rangle \tag{3}$$

ω_i is the incident light wavenumber, $\overrightarrow{v}$ is the polarization unit vector of scattered light and $\overrightarrow{w}$ is that of the incident light, $B(\omega)$ and M_n are the Bose factor,and the mass of the nth atom respectively. ω_j is the wavenumber of the jth mode, and its (δn) component is $\langle \delta n | j \rangle$. The coefficients connecting the polarization fluctuations $(\pi^n_{\alpha,\gamma;\delta})$ to the motions of the atoms are acquired by extending the polarizability tensor $\overset{\sim n}{\pi}$ in terms of the displacements of the atoms u^n_δ, with

$$\pi^n_{\alpha,\gamma;\delta} = \sum_m \left(\frac{\partial \pi^m_{\alpha,\gamma}}{\partial u^n_\delta} \right)_0 \tag{4}$$

Usually, calculation of Raman spectrum requires the injection of ω_j and $\langle \delta n | j \rangle$ values in the previous expressions obtained from solving the equation,

$$\tilde{D}|j\rangle = \omega_j^2 |j\rangle \tag{5}$$

where $\tilde{D}$ represent the dynamical matrix of the system, with the elements $D_{\alpha\beta}(nm) = \langle \alpha n | \tilde{D} | \beta m \rangle$.

The polarizability tensor $\left(\overset{\sim n}{\pi}\right)$ and the dynamical matrix $(\tilde{D})$ with the elements $D_{\alpha\beta}(nm) = \langle \alpha n | \tilde{D} | \beta m \rangle$ are the two required quantities to compute Raman spectra of a molecular system. The molecular mechanics constitutes an adequate method to construct the dynamical matrix $\tilde{D}$ of such systems. The carbon–carbon intra-nanotube interactions are described using the model of the force constants introduced by Saito et al. [9] where the accounted interactions up to 4th nearest neighbor for deriving the $\tilde{D}$, and the intra-molecular interactions between atoms of oligothiophenes are described using density functional theory (DFT) calculations carried out with SIESTA code [10–13]. Whereas the van der Waals interactions between atoms of the encapsulated oligothiophenes and the nanotube are described using Lennard–Jones potential.

$$U(r_{ij}) = 4\varepsilon_{ij}\left[\left(\frac{\sigma_{ij}}{r_{ij}} \right)^{12} - \left(\frac{\sigma_{ij}}{r_{ij}} \right)^6 \right], \tag{6}$$

With,

Table 1 Lennard–Jones parameters

	ε_i (meV)	σ_i (Å)
C	2.964	3.407
S	10.7	3.560
H	0.67	1.069

$$\varepsilon_{ij} = \sqrt{\varepsilon_i \varepsilon_j}\, and\ \sigma_{ij} = \frac{\sigma_i + \sigma_j}{2}, \tag{7}$$

where r_{ij} is the distance between the ith and jth interaction sites. ε_{ij} and σ_{ij} represent the parameters of Lennard–Jones. The Lennard–Jones parameters used for the different atoms are listed in Table 1.

Block by block, the $\tilde{D}$ is constructed from the $\tilde{D}$ of each subsystem, where the bidiagonal block is formed by the $\tilde{D}$ of the nanotubes and oligothiophenes, while the off-diagonal block is formed by the $\tilde{D}$ related to the nanotube/oligothiophenes interactions.

On the other hand, the computation of the polarizability tensor $\overset{\sim n}{\pi}$ is performed pursuant to the bond polarizability model [14], where the modulation of the polarizability is solely applied by the nearest neighbors' bond, whose parameters are functions only of the lengths of the bonds. For a particular bond, the polarizability π is written as,

$$\pi_{\alpha,\beta}(r) = \frac{1}{3}(\alpha_l + 2\alpha_p)\delta_{\alpha,\beta} + (\alpha_l - \alpha_p)\left(\hat{r}_\alpha \hat{r}_\beta - \frac{1}{3}\delta_{\alpha,\beta}\right), \tag{8}$$

r is the vector that connects the n and m atoms, with the unit vector $\hat{r}$.

The Cartesian components are indicated by α and β. Whereas α_p and α_l represent the perpendicular and longitudinal bond polarizability parameters, respectively. The derivatives $\pi^n_{\alpha,\beta;\gamma}$ are given by,

$$\begin{aligned}\pi^n_{\alpha,\beta;\gamma} = &\sum_m \frac{1}{3}\left(2\alpha'_p + \alpha'_l\right)\delta_{\alpha,\beta}\hat{r}_\gamma + \left(\alpha'_l - \alpha'_p\right)\left(\hat{r}_\alpha \hat{r}_\beta - \frac{1}{3}\delta_{\alpha,\beta}\right)\hat{r}_\gamma \\ &+ \frac{\alpha_l - \alpha_p}{r}\left(\delta_{\alpha,\gamma}\hat{r}_\beta + \delta_{\beta\gamma}\hat{r}_\alpha - 2\hat{r}_\alpha \hat{r}_\beta \hat{r}_\gamma\right),\end{aligned} \tag{9}$$

where $\alpha_{p'} = (\partial\alpha_p/\partial r)|_{r=r_0}$, and $\alpha_{l'} = (\partial\alpha_l/\partial r)|_{r=r_0}$ with r_0 is the equilibrium bond distance. For each type of bond, the bond polarizability model of the system is thus defined entirely by the three parameters $\overline{\alpha} = 2\alpha_{p'} + \alpha_{l'}$, $\overline{\beta} = \alpha_{l'} - \alpha_{p'}$, and $\overline{\gamma} = (\alpha_l - \alpha_p)/r$. The values of empirical bond polarizability for free oligothiophenes are taken from Ref. [15] and combined with those of pristine SWCNTs [9]. Finally, Raman spectra are carried out using spectral moments method which was shown to be a powerful tool to compute Raman scattering spectra of very large harmonic systems [16–23].

3 Results and Discussion

3.1 Raman Spectra of Pristine SWCNTs

We present here the computed Raman spectra of pristine zigzag, armchair, and chiral SWCNTs of different chiralities and diameters. From the peak position in the computed Raman spectra, the wavenumber of each mode is directly obtained, and for each peak, the linewidth of the line shape supposed to be Lorentzian is fixed at 1.7 cm^{-1}, and the intensities for each SWCNTs are normalized. The computed Raman spectra of pristine [(12,0), (12,3), (12,6), (12,9), and (12,12)] nanotubes in the wavenumbers range (0–1800 cm^{-1}) are displayed in Fig. 1. In general, the computed Raman spectrum of pristine SWCNTs consists of two lines: (i) Radial breathing modes (RBM) at wavenumbers range between 120 and 350 cm^{-1}, assigned to synchronous vibrations of carbon atoms in the radial direction. (ii) G-band at higher wavenumbers range between 1500 and 1600 cm^{-1}, corresponding to tangential carbon–carbon bond vibrations. Learning about SWCNTs from their Raman features is discussed below.

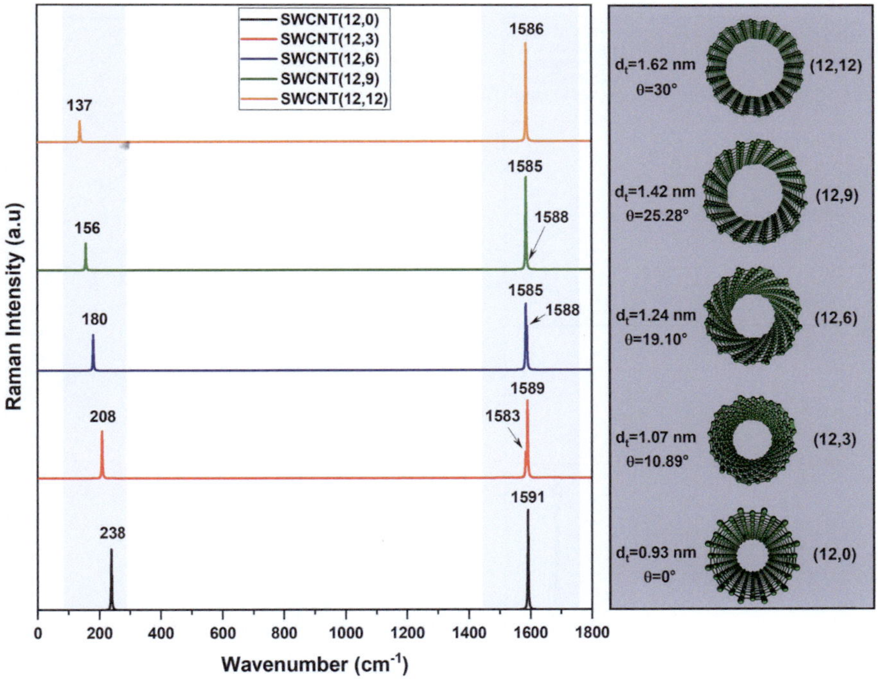

Fig. 1 Raman spectra of pristine [(12,0), (12,3), (12,6), (12,9), and (12,12)] SWCNTs in the wavenumber range of (0–1800 cm^{-1})

Let us focus first on the RBM of pristine SWCNTs located in the lower wavenumbers range. Actually, from an experimental point of view, the RBM is used to investigate the SWCNTs diameter through their wavenumber positions, to scout the electronic behavior by their intensities, and to carry out the assignments (n, m) from analysis of both the intensity of the RBM and the diameter [6–8]. The RBM feature corresponding to radial vibrations of carbon atoms is unique to CNTs and arises at wavenumbers ranging from 120 to 350 cm^{-1} for SWCNTs with diameters between 0.7 and 2 nm. The presence of the RBM is useful to identify whether certain carbon-based materials include SWCNTs, and to characterize the SWCNTs diameters in the sample by their wavenumber positions where the positions in Raman spectra of RBM peaks are inversely proportional to the SWCNTs diameters. As depicted in Fig. 2, an illustration of the RBM behavior with the diameter is given, where the RBM dependence with the diameter inverse is established by our approach. Starting with the zigzag nanotube (12,0) characterized by its diameter $d_t = 0.93$ nm, the RBM featured by one line centered at 238 cm^{-1}, and as can be seen, the RBM peaks are shifted toward lower wavenumbers to 208, 180, 156, and 137 cm^{-1} for the chiral (12,3), (12,6), (12,9), and armchair (12,12) nanotubes characterized by their diameters of 1.07 nm, 1.24 nm, 1.42 nm, and 1.62 nm, respectively.

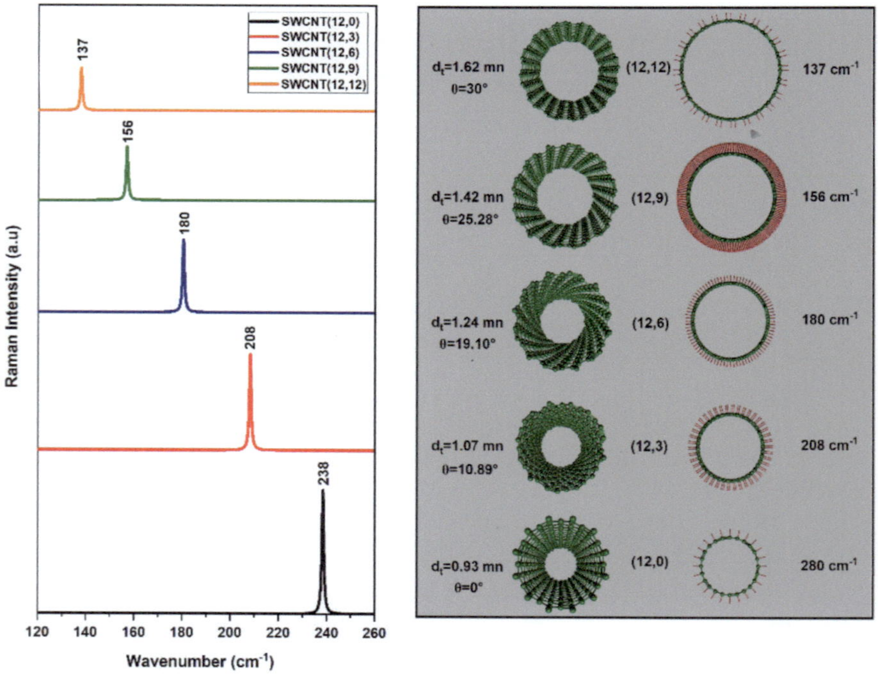

Fig. 2 Calculated Raman spectra of pristine [(12,0), (12,3), (12,6), (12,9), and (12,12)] SWCNTs in the RBM range

Concerning the tangential modes in CNTs which occur in the higher **wavenumbers** range (1500–1600 cm^{-1}), basically, the G-band in CNTs is derived from the Raman-allowed optical mode between the two inequivalent carbon atoms in the unit cell of 2D graphite [6–8]. The G-band in graphite associated with the tangential modes shows one Lorentzian peak centerd at 1581 cm^{-1}, in contrast, the G-band in SWCNTs comprise various lines by dint of the confinement of the phonon wave vector in the circumferential direction of the nanotube, and the symmetry breaking effect related to nanotube curvature. From an experimental point of view, the wavenumbers of G-band are used to differentiate between semi-conducting SWCNTs and metallic, through strong differences in their Raman line shapes [24, 25], and to probe the charge transfer that arises from SWCNTs doping. In general, the G-band features in empty SWCNTs are fitted with two components: (i): The first one peaked at the lower wavenumber (G^-) belonging to the transverse optical phonon mode (TO), and (ii): the higher wavenumber peak (G^+) assigned to the longitudinal optical phonon mode (LO). The G-band of [(12,0), (12,3), (12,6), (12,9), and (12,12)] nanotubes in the wavenumbers range (1578–1600 cm^{-1}) are displayed in Fig. 3. Thus, an illustration of the G-band dependence with the chirality is given. As expected from symmetry considerations [9, 26], the tangential modes bunch in zig-zag (12,0) and armchair (12,12) SWCNTs spectra show one peak as compared to two peaks in the chiral [(12,3), (12,6), (12,9)] SWCNTs.

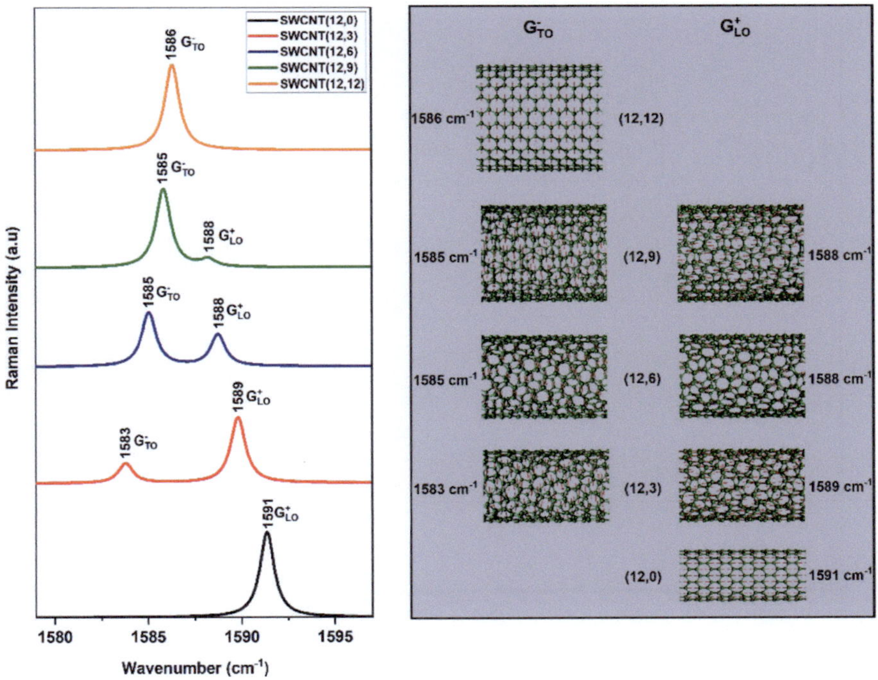

Fig. 3 The G-band Raman spectra of pristine [(12,0), (12,3), (12,6), (12,9), and (12,12)] SWCNTs

3.2 Raman Spectra of Filled SWCNTs with Oligothiophenes

Raman spectroscopy plays an important role in investigations on SWCNTs filled with various compounds where the filling effect on their Raman spectra is found to be pertinent to probe the appearance of the charge transfer and the stability of filled SWCNTs [5]. In Fig. 4, the computed Raman spectra of empty and filled (8,6) nanotube with oligothiophenes nT (n = 2,4, and 6) are exposed. The (8,6) nanotube is chosen as the host since its diameter of 0.95 nm is close to the optimal diameter allowing the accommodation of one molecular chain [11–13]. At first glance, the Raman spectra of the filled nanotube [nT@(8,6)] appear as the superposition of the Raman spectra of the (8,6) nanotube and oligothiophenes. However, careful analysis exposes different changes characterizing the intermolecular interactions subsisting between the encapsulated oligothiophenes and the (8,6) nanotube.

Starting with the filling effects on the Raman features of the oligothiophenes reported in our previous works [11–13] which agree with the experimental data. An upshift of 2–14 cm^{-1} is spotted of all Raman modes in the filled nT@(8,6) nanotubes compared to corresponding ones in the free oligothiophenes, which shows the first hints on the significant interactions that subsist in the filled systems between the encapsulated molecules and (8,6) nanotube. Furthermore, the libration modes located

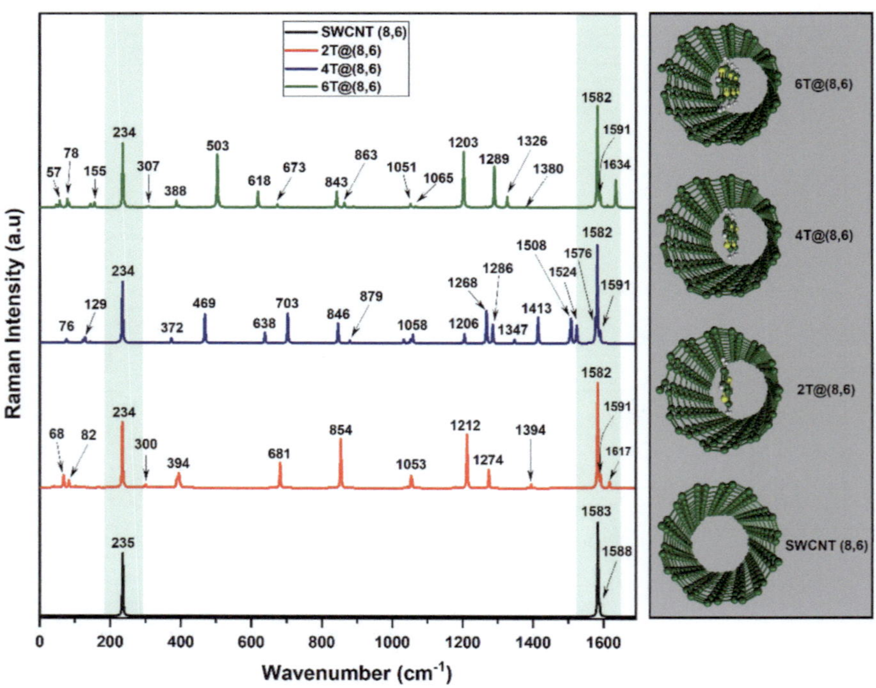

Fig. 4 Raman spectra of pristine and filled (8,6) nanotube with a series of oligothiophene nT (n = 2,4, and 6)

in the lower wavenumbers range at 7 cm^{-1}, 20 cm^{-1}, and 8 and 25 cm^{-1} for 2 T, 4 T, and 6 T respectively, which are very sensitive to the filling effects, disappear in the filled systems. The deactivation of these modes describes the stability of the oligothiophenes in the (8,6) nanotube host specified by its optimal diameter.

Concerning the Raman features of the nanotube, the RBM and G-band of the (8,6) nanotube are also evidenced in the filled nT@(8,6) nanotubes. First, the RBM wavenumber position in the pristine (8,6) nanotube centered at 235 cm^{-1} is downshifted in filled nT@(8,6) nanotubes by about 1 cm^{-1} as displayed in Fig. 5 (top). The shift of the RBM induced by the filling of nanotubes can originate from a competition between van der Waals interactions and charge transfer effects [27]. Nevertheless, as reported in our previous works [11, 20, 28], the appearance of the charge transfer phenomenon from RBM shifts is excluded, where van der Waals interactions were found to be the only ones accountable for the shifts of the RBM. The downshifts of the RBM of a few cm^{-1} affirm the stability of the filled nT@(8,6) nanotubes.

The G–band on the other hand of pristine (8,6) nanotube observed in the higher wavenumbers range (1570–1600 cm^{-1}) consisting of two components, one peaked at lower wavenumber (1583 cm^{-1}) belongs to carbon atoms vibrations in the circumferential direction of the (8,6) nanotube (G^-_{TO}), and the higher wavenumber peak at about (1588 cm^{-1}) assigned to carbon atoms vibrations in the (8,6) nanotube axis

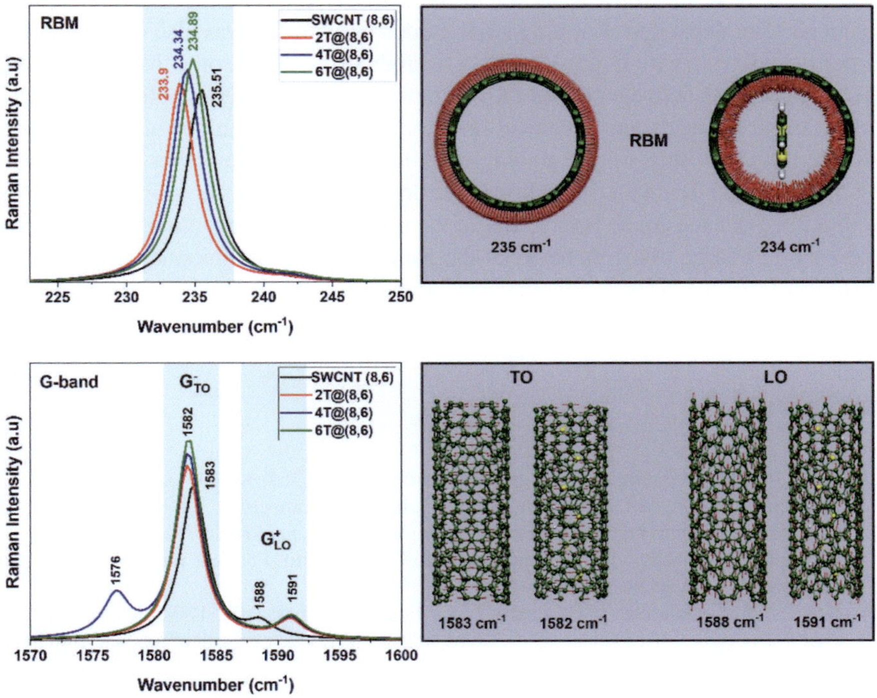

Fig. 5 Raman spectra of pristine (8,6) nanotube and filled nT@(8,6) nanotubes in the RBM range (top), and G-band (bottom)

direction (G^+_{LO}) (Fig. 5 (bottom)). Actually, the G–band in filled SWCNTs is very sensitive to charge transfer [5–8, 29]. Indeed, an electron or hole doping prompts a modification of the force constants in the carbon atoms, which leads to downshifts or upshifts of G-band. As can be seen in Fig. 5 (bottom), the G^-_{TO} feature of the empty (8,6) nanotube is downshifted by about 1 cm^{-1} and the G^+_{LO} feature is upshifted by about 3 cm^{-1} in the filled nT@(8,6) nanotubes under the filling effects. However, conclusions about the direction of the charge transfer in the filled SWCNTs cannot be drawn exclusively from Raman spectroscopy as reported by Kalbac et al. [30]. Thus, we can conclude only that the G-band shifts testify to a charge transfer occurring in the filled nT@(8,6) nanotubes between the encapsulated oligothiophenes and the (8,6) nanotube.

4 Conclusion

A combination of molecular mechanics and density functional theory was used to compute the Raman spectra of pristine SWCNTs with different diameters and chiralities, as well as Raman spectra of filled (8,6) nanotubes with oligothiophenes. The diameter dependence of RBM was clearly observed where the wavenumber of RBM peaks in Raman spectra was shown to be inversely proportional to the SWCNTs diameter. The dependence of tangential modes on chirality was also observed, and experimental observations agree with these predictions. On the other hand, the filling effects on Raman active modes of (8,6) nanotube encasing oligothiophenes are carefully analyzed. The disappearance of libration modes of oligothiophenes after filling afforded the first hints on the stability of filled nT@(8,6) nanotubes characterized by their optimal diameter. In addition, the downshift of RBM of a few cm^{-1} gives additional evidence about the stability of the filled nT@(8,6) nanotubes. Further, the presence of charge transfers between the nanotube and oligothiophenes is evinced by the shifts of the G-band after encapsulation.

References

1. Iijima, S.: Helical microtubules of graphitic carbon. Nature 354, 56 (1991).
2. Shoukat, R., Khan, M.I.: Carbon nanotubes: a review on properties, synthesis methods and applications in micro and nanotechnology. Microsystem Technologies, 1–10 (2021).
3. Dresselhaus, G., Mildred S. Dresselhaus, and Riichiro Saito. Physical properties of carbon nanotubes. World Scientific, (1998).
4. Hermes, P.H., Gabriela, M.P., Ileana, V.R., Fusaro, C., Fernando, L.-V., Mariana, M.A., Padilla-Rodriguez, C., Fabian, F.L.: Carbon nanotubes as plant growth regulators. In: Green Nanoparticles, pp. 77–115. Springer (2020).
5. Kharlamova, M.V.: Advances in tailoring the electronic properties of single-walled carbon nanotubes. Progress in Materials Science 77, 125–211 (2016).
6. Saito, R., Hofmann, M., Dresselhaus, G., Jorio, A., Dresselhaus, M.: Raman spectroscopy of graphene and carbon nanotubes. Advances in Physics 60(3), 413–550 (2011).

7. Dresselhaus, M.S., Dresselhaus, G., Saito, R., Jorio, A.: Raman spectroscopy of carbon nanotubes. Physics reports 409(2), 47–99 (2005).
8. Jorio, A., Saito, R.: Raman spectroscopy for carbon nanotube applications. Journal of Applied Physics 129(2) (2021).
9. Saito, R., Takeya, T., Kimura, T., Dresselhaus, G., Dresselhaus, M.: Raman intensity of single-wall carbon nanotubes. Physical Review B 57(7), 4145 (1998).
10. Soler, J.M., Artacho, E., Gale, J.D., Garcia, A., Junquera, J., Ordej´on, P., Sanchez-Portal, D.: The siesta method for ab initio order-n materials simulation. Journal of Physics: Condensed Matter 14(11), 2745 (2002).
11. Chenouf, J., Boutahir, M., Rahmani, A., Chadli, H., Hermet, P., Mejía-López, J., Rahmani, A.: Charge transfer evidence in donor-acceptor single-walled carbon nanotubes filled with sexithiophene oligomers: Nanotube diameter dependence. J. of Raman Spec.52(8), 1381–1394 (2021).
12. Chenouf, J., Boutahir, M., Fakrach, B., Rahmani, A., Chadli, H., Hermet, P., Mejía-López, J., Rahmani, A.: Encapsulation effect of π-conjugated quaterthiophene on the radial breathing and tangential modes of semiconducting and metallic single-walled carbon nanotubes. Journal of Computational Chemistry 41(28), 2420–2428 (2020).
13. Chenouf, J., Boutahir, M., Fakrach, B., AH, R., Chadli, H., Rahmani, A.: Raman analysis of conjugated bithiophene (2t) encapsulated in semiconducting and metallic single-walled carbon nanotubes. Journal of Raman Spectroscopy 51(8), 1315–1323 (2020).
14. Guha, S., Menendez, J., Page, J., Adams, G.: Empirical bond polarizability model for fullerenes. Physical Review B 53(19), 13106 (1996).
15. Hermet, P., Izard, N., Rahmani, A., Ghosez, P.: Raman scattering in crystalline oligothiophenes: a comparison between density functional theory and bond polarizability model. The Journal of Physical Chemistry B 110(49), 24869–24875 (2006).
16. Benoit, C., Royer, E., Poussigue, G.: The spectral moments method. Journal of Physics: Condensed Matter 4(12), 3125 (1992).
17. Fakrach, B., Rahmani, A., Chadli, H., Sbai, K., Sauvajol, J.-L.: Raman spectrum of single-walled boron nitride nanotube. Physica E: Low-dimensional Systems and Nanostructures 41(10), 1800–1805 (2009).
18. Chadli, H., Rahmani, A., Sbai, K., Hermet, P., Rols, S., Sauvajol, J.-L.: Calculation of Raman-active modes in linear and zigzag phases of fullerene peapods. Phy. Rev. B 74(20), 205412 (2006).
19. Sbai, K., Rahmani, A., Chadli, H., Sauvajol, J.-L.: Finite-size effect on the Raman-active modes of double-walled carbon nanotubes. J. of Phy.: Cond. Matter 20(1), 015204 (2007).
20. Chenouf, J., Boutahir, M., Mejía-López, J., Rahmani, A., Fakrach, B., Chadli, H., Rahmani, A.: Predicting the structure configuration and Raman analysis of caffeine molecules encapsulated into single-walled carbon nanotubes: Evidence for charge transfer. Solar Energy 232, 204–211 (2022).
21. Elhadfi, S., Chenouf, J., Abdelkader, S. A. A., Fakrach, B., Rahmani, A. H., Chadli, H., Rahmani, A.: Single-wall boron nitride nanotubes encapsulating conjugated bithiophene molecule: Raman analysis. In: E3S Web of Conf., vol. 469, p.24. EDP Sciences, (2023).
22. Chenouf, J., Boutahir, M., Fakrach, B., Rahmani, A., Chadli, H., Rahmani, A.: Noncovalent functionalization of carbon nanotubes by polyaniline molecules: Raman analysis. In: IOP Conference Series: Materials Science and Engineering, vol. 783, p. 012022. IOP Publishing, (2020).
23. Jamal, C., Mourad, B., Brahim, F., Abdelhai, R., Hassane, C., Abdelali, R.: Vibrational study of bithiophene encapsulated in single wall carbon nanotube using the infrared calculations. In: 2020 1st International Conference on Innovative Research in Applied Science, Engineering and Technology (IRASET), pp. 1–6. IEEE, (2020).
24. Jorio, A., Pimenta, M., Souza Filho, A., Saito, R., Dresselhaus, G., Dresselhaus, M.: Characterizing carbon nanotube samples with resonance Raman scattering. New Journal of Physics 5(1), 139 (2003).

25. Dresselhaus, M., Dresselhaus, G., Jorio, A., Souza Filho, A., Pimenta, M., Saito, R.: Single nanotube Raman spectroscopy. Accounts of Chem. Research 35(12), 1070–1078 (2002).
26. Rahmani, A., Sauvajol, J.-L., Rols, S., Benoit, C.: Nonresonant Raman spectrum in infinite and finite single-wall carbon nanotubes. Physical Review B 66(12), 125404 (2002).
27. Almadori, Y., Alvarez, L., Le Parc, R., Aznar, R., Fossard, F., Loiseau, A., Jousselme, B., Campidelli, S., Hermet, P., Belhboub, A., et al.: Chromophore ordering by confinement into carbon nanotubes. The Journal of Physical Chem. C 118(33), 19462–19468 (2014).
28. Kensi, Y., Chenouf, J., Fakrach, B., Abdelkader, S.A.A., Rahmani, A.H., Chadli, H. Rahmani, A.: 1d van der Waals heterostructures based on single-walled carbon nanotubes encapsulating linear carbon chains: Stability and charge transfer evidence from Raman spectroscopy. Applied Surface Science 611, 155692 (2023).
29. Alvarez, L., Almadori, Y., Arenal, R., Babaa, R., Michel, T., Le Parc, R., Bantignies, J.-L., Jousselme, B., Palacin, S., Hermet, P., et al.: Charge transfer evidence between carbon nanotubes and encapsulated conjugated oligomers. The J. Phys. Chem. C 115(24), 11898–11905 (2011).
30. Kalbac, M., Kavan, L., Dunsch, L., Dresselhaus, M.: Development of the tangential mode in the Raman spectra of SWCNT bundles during electrochemical charging. Nano letters 8(4), 1257–1264(2008).

Next-Generation Sensors Using Artificial Intelligence as Enabling Technologies for Biomedical and Healthcare Applications

Samina Noorali, James C. Baldwin, and Ashok Vaseashta

Abstract Infectious diseases, especially the ones that are antibiotic-resistant, pose a serious threat to global health. Currently, bacterial contamination is detected using conventional culturing techniques, which are limited by the processing time and on-site availability. Thus, real-time and continuous monitoring of pathogen levels is necessary to analyze information that could assist prevention and containment of outbreaks. Nanotechnology-based smart sensors present new avenues for rapid detection of such pathogens at the patient's point of care. Hence, nanomaterial-based composites play a vital role, especially with high K dielectrics, in bacterial sensing due to their unique characteristics due to reduced dimensions. Such characteristics in conjunction with the fourth industrial revolution, viz. artificial intelligence (AI), have led to the development and application of health monitoring sensors by digitalization and intelligence, which will help identify the sites and sources of disease outbreaks and the next variant of novel coronavirus. The addition of piezoelectric materials to composites provides energy harvesting capability for stand-alone and monitoring at remote locations. These sensors have wide-ranging applications in medical care, personal health management, elderly care, sports, and other fields, providing convenient and real-time healthcare services. However, these sensors have limitations such as noise, drift, difficulty in data extraction, and lack of feedback or control signals. AI provides powerful tools and algorithms for data processing and analysis, enabling intelligent health monitoring, and achieving high-precision predictions and decisions.

S. Noorali
Department of Biology, Claflin University, Orangeburg, SC, USA

J. C. Baldwin
Applied Technology and Genomics Center, USAFSAM/FHT, Wright-Patterson AFB, Dayton, OH, USA

A. Vaseashta (✉)
Office of Applied Research, International Clean Water Institute, Manassas, VA, USA
e-mail: prof.vaseashta@ieee.org

IEEN, Academy of Sciences of Moldova, Chisinau, Moldova

Institutul de Cercetare, Universitatea Din Bucureşti, Bucureşti, Romania

A. Vaseashta et al. (eds.), *Proceedings of the Eighth International Symposium on Dielectric Materials and Applications (ISyDMA'8)*, Springer Proceedings in Physics 349, https://doi.org/10.1007/978-3-031-84809-4_17

Integrating the Internet of Things (IoT), AI, and health monitoring sensors, will establish a closed-loop system with real-time monitoring functionality, data collection, online analysis, diagnosis, and treatment recommendations. We provide an overview of the development of sensor platforms using exponential technologies for healthcare. We further introduce the advances in wearable sensors for cardiovascular care, nerve signal acquisition, neurotransmitter monitoring; and precise therapy. A fusion of AI, IoTs, and next-generation sensors will provide more intelligent, convenient, and secure services for healthcare and biomedical applications.

Keywords Healthcare · Artificial · Wearable sensors · Implantable sensors · Human–machine interfaces · Artificial intelligence · Internet of things

1 Introduction

Resistance to antibiotics is a growing global health crisis since bacteria are evolving and developing resistance to the antibiotics used previously to treat infections. Multidrug-resistant bacteria is a global crisis, increasing morbidity and mortality of infected patients and negatively impacting the outcomes of a wide range of groups, including those in intensive care units, undergoing surgery, transplantation, or cancer treatment. Furthermore, drug resistance complicates treatment options and leads to longer hospitalization, higher medical costs as the physicians do not know which antibiotics will be effective in treating the patients, and increased mortality. Conventional mitigation strategies include improved infection control and hygiene, surveillance, and monitoring, public awareness, and promoting vaccination. All of these methods are inherently slow. One way to combat this problem is by employing novel sensors and AI technologies for more effective detection and diagnosis. The next-generation sensors, augmented by AI, offer unparalleled capabilities in monitoring, diagnosing, and treating various medical conditions. The synergy between AI and sensors enables the collection, analysis, and interpretation of complex biological data in real time, driving personalized medicine, early disease detection, and more effective healthcare interventions. Some of the novel sensors are:

Biosensors: These devices use biological elements, such as enzymes and antibodies, to detect specific pathogens or resistance markers [1]. Recently, researchers have developed biosensors that can quickly identify antibiotic-resistant bacterial strains by identifying specific genetic markers or protein expressions associated with resistance [2].

Optical Sensors: Optical sensors, including surface plasmon resonance (SPR) and fluorescence-based sensors, can provide real-time detection of bacterial infections and resistance mechanisms. They often offer high sensitivity and can be used for point-of-care testing [3–5].

Electrochemical Sensors: These sensors can detect changes in the electrical properties when bacteria or their byproducts interact with the sensor's surface. Their operation is based on the interaction between the sensor's surface and a chemical species, which leads to a measurable electrical change such as voltage, current, or impedance. These sensors are highly sensitive, selective, and often cost-effective, and hence are often used for rapid and inexpensive diagnostic purposes [6, 7].

Microfluidic Devices: These devices can manipulate the physical and chemical properties of tiny amounts of fluids and perform complex assays on small samples. They can be used for rapid screening of bacterial cultures to determine antibiotic susceptibility, offering quicker results to physicians than the traditional methods [8].

Surface-Enhanced Raman Spectroscopy (SERS): SERS is a promising spectroscopic analysis method that can provide detailed information about the chemical composition of bacterial cells and be used to detect resistant markers and identify bacterial strains rapidly [9–11].

Genomic and Molecular Sensors: These sensors use molecular techniques such as PCR (Polymerase Chain Reaction) or isothermal amplification to detect specific resistance genes at the DNA or RNA level. These sensors are used for Real-time PCR or Loop-mediated Isothermal Amplification (LAMP) assays for detecting genes like mecA (methicillin resistance) [12] or bla (beta-lactam resistance) [13]. The strategy is useful for identifying specific mutations associated with resistance to drugs (e.g., mutations in the rpoB gene for rifampicin resistance) [14].

AI-Based Wearable Sensors: These are being developed to monitor physiological changes associated with infections. These sensors could potentially alert patients and healthcare providers about infections before they become severe, allowing timely interventions [15–17].

1.1 Use of AI in Detection and Diagnosis of Antibiotic Resistance

Leveraging machine learning algorithms, big data analytics, and predictive modeling, AI can significantly improve our ability to track and respond to the spread of resistant pathogens. AI is contributing to the fight against antibiotic resistance by using AI in diagnostics. AI is transforming the way antibiotic resistance is diagnosed by analyzing complex datasets from patient samples, microbial genomes, and clinical records. (a) Rapid Detection of Resistant Pathogens by Machine Learning (ML) for Diagnostics: AI models, particularly machine learning algorithms, can be trained to recognize antibiotic-resistant pathogens from data like microbial genome sequences, patient records, and even images of bacterial colonies. These models can predict whether a bacterial strain is resistant based on its genetic makeup or phenotypic behavior [18]. Furthermore, Deep Learning for Image-Based Diagnostics using AI models, especially convolutional neural networks (CNNs), can analyze images of bacterial growth to detect resistance [19]. They can recognize specific growth patterns that correlate with antibiotic resistance, making diagnosis faster and more accurate. (b) Genomic Prediction of Resistance: Genomic Prediction of Resistance using Whole Genome Sequencing (WGS) + AI –AI can analyze the entire genome of a pathogen to identify known antibiotic resistance genes and mutations [20]. It can predict whether a pathogen will be resistant to specific antibiotics by analyzing large datasets of genomic information. This approach enables clinicians to prescribe more effective antibiotics based on the precise genetic profile of the infection. Furthermore, AI-powered CRISPR systems are being developed to detect specific antibiotic-resistance genes in bacterial populations [21], which can help in real-time diagnostics and treatment decisions. (c) AI Predicting Antibiotic Susceptibility: The use of AI models can help predict the susceptibility of bacteria to different antibiotics, reducing the need for traditional susceptibility testing (like culture-based methods that can take days) by Predictive Modeling using ML algorithms that can be trained on large datasets that contain information about bacterial species, genetic markers, and patient outcomes. These models can predict which antibiotics are likely to be effective against a specific bacterial strain, even if lab testing is not immediately available. Also, AI-based decision support systems can integrate patient data (e.g., symptoms, history of infections) and suggest the best antibiotic treatment by predicting which bacterial strains are likely resistant, streamlining the clinical decision-making process. AI-powered systems can assist in optimizing antibiotic use, helping to reduce the development of resistance by analyzing patient data ensuring antibiotics are only prescribed when necessary and appropriate, and reducing unnecessary prescriptions. They can provide personalized treatment recommendations based on the likelihood of resistance to a given pathogen. Also, AI can analyze population-level data to predict future resistance trends in hospitals or communities. This helps in planning more effective antibiotic use and infection control policies [22].

AI plays a vital role in public health by monitoring and predicting antibiotic resistance patterns on a larger scale. AI algorithms can process data from hospitals, labs, and environmental sources to predict the spread of resistant strains geographically and temporally. For example, AI models have been used to track the spread of *Methicillin-Resistant Staphylococcus aureus* (MRSA) across regions, helping to implement timely interventions. Additionally, AI-driven tools can process and analyze vast amounts of data, including patient outcomes, environmental samples, and hospital records. This allows for the real-time detection of emerging resistance patterns and hotspots, enabling faster public health responses. AI can enhance biosensors used in environmental monitoring to detect antibiotic-resistant bacteria in water sources, soil, and food supplies. By using machine learning models, these sensors can identify patterns or markers of resistance and provide real-time alerts for public health action. AI is advancing the discovery of new antibiotics and treatments to mitigate resistant bacteria. Machine learning models are being used to screen large compound libraries and identify new antibiotics that could work against resistant bacteria. AI can predict which molecular structures are most likely to disrupt bacterial function, helping to accelerate the discovery of effective new drugs. AI can also help design new antibiotics by predicting how bacterial resistance mechanisms can be overcome. This leads to the development of antibiotics that are less susceptible to existing resistance mechanisms. Yet another use is for repurposing Existing Drugs, as AI models can analyze the molecular structures of existing drugs to find new uses for them, such as repurposing drugs for combating resistant bacteria. By identifying off-label uses of drugs, AI helps in finding alternative treatments without needing to develop entirely new antibiotics. Natural Language Processing (NLP) algorithms can mine vast amounts of medical literature, clinical records, and research papers to find patterns related to antibiotic resistance by using Automated Literature Review and by integration with Electronic Health Records (EHRs). AI systems integrated with EHRs can help detect early signs of antibiotic resistance by analyzing patient records, identifying resistance patterns from clinical data, and providing real-time treatment recommendations [23]. These AI models can predict the likelihood of antibiotic resistance, helping physicians choose the most effective treatment [24]. By leveraging novel sensors and AI, healthcare providers, and public health management can enhance the detection and management of antibiotic-resistant diseases, addressing one of the pressing challenges that pose a threat to global public health in modern medicine.

2 Antibiotic Resistant Diseases

Antibiotic resistance has become a significant global health concern. New strains of bacteria are constantly evolving due to overuse of antibiotics resulting in bacteria developing resistance to antibiotics, making treatment of infections problematic. Some of the bacterial infections associated with antibiotic resistance are listed below:

Methicillin-Resistant Staphylococcus Aureus (MRSA): MRSA infection is caused by *Staphylococcus aureus* (*S. aureus*), a gram-positive coccus, which has become resistant to methicillin antibiotic used to treat ordinary Staphylococcus infections. MRSA infection in hospitals and healthcare facilities acquired. MRSA is a global threat to public health [25].

Vancomycin-Resistant Staphylococcus Aureus (VRSA): Vancomycin is one of the oldest antibiotics and has been in clinical use for over 60 years. Vancomycin is used for infections caused by gram-positive bacteria, such as Staphylococci, Enterococci, Streptococci, Pneumococci, Listeria, Corynebacterium, and Clostridia. Vancomycin is the last resort to treat MRSA infections and to treat patients allergic to semisynthetic penicillin or cephalosporins [26]. VRSA infections can be treated by daptomycin, linezolid, telavancin, ceftaroline, and quinupristin-dalfopristin [27].

Carbapenem-Resistant Enterobacteriaceae (CRE): This group of gram-negative bacilli-resistant bacteria includes *Escherichia coli* and *Klebsiella pneumonia*, which are human gastrointestinal normal flora [28, 29]. Both these bacteria are common human pathogens causing a wide spectrum of hospital and community-acquired infections, such as urinary tract infection (UTI) [30], septicemia, pneumonia, peritonitis, meningitis, etc. [31, 32]. CRE has emerged because carbapenem is often used as a last resort for multi-drug-resistant (MDR) bacterial infections [33]. Antibiotics used to treat CRE infections are polymyxin or colistin, tigecycline, fosfomycin, aminoglycosides (gentamicin or tobramycin), rifampin, and carbapenem paired with an inhibitor (avibactam, vaborbactam, and relebactam) [34–42].

Multidrug-Resistant Tuberculosis (MDR-TB): Tuberculosis (TB) is caused by gram-positive bacillus *Mycobacterium tuberculosis* and is an infectious disease of the lungs, but other parts of the body (kidney, spine, and brain) can also be affected. TB is resistant to the two most powerful first-line anti-TB drugs: isoniazid and rifampicin. MDR-TB regimen includes fluoroquinolone (later generation: levofloxacin or moxifloxacin), bedaquiline and linezolid. These three drugs should be included in the initial therapy unless patients show drug resistance or there is a risk of toxicity [43].

Gonorrhea: Gonorrhea is a sexually transmitted disease (STD) and is caused by gram-negative diplococci *Neisseria gonorrhoeae. N. gonorrhoeae* infects the mucus membrane of the reproductive tract and over the years has developed drug resistance to penicillin, tetracycline, and fluoroquinolones, as a result, has contributed to the spread of the disease. The third-generation cephalosporin, ceftriaxone or cefixime, and cefuroxime are the recommended treatments for gonorrhea that have developed resistance to prescribed antibiotics [44].

Acinetobacter baumannii Infections: *Acinetobacter baumannii* is a gram-negative coccobacillus bacterium. It is an opportunistic pathogen in humans, affecting immunocompromised individuals, and is particularly increasingly important as hospital-derived infections (ventilator-associated pneumonia and bloodstream infection). MDR *A. baumannii* is associated with wound infection in severely injured soldiers [45]. *A. baumannii* has developed resistance to multiple antibiotics including carbapenems and polymyxins [45], hence it is difficult to treat the infection. Currently, broad-spectrum cephalosporins (ceftazidime or cefepime), fluroquinolones (ciprofloxacin), or a combination of antibiotics are given to treat the infection, such as piperacillin-tazobactam, ampicillin-sulbactam, carbapenems (meropenem or imipenem-cilastatin), polymyxins (polymyxin B-colistin) and tetracycline derivatives (minocycline, doxycycline, and tigecycline) [46].

Pseudomonas aeruginosa Infections: *Pseudomonas aeruginosa* (*P. aeruginosa*) is a gram-negative bacillus bacterium and common cause of hospital-acquired infections. *P. aeruginosa* shows resistance to multiple antibiotics, including aminoglycosides, quinolones and β-lactams [47]. *P. aeruginosa* should be classified into carbapenem-resistant, fluoroquinolone-resistant, aminoglycoside-resistant, cephalosporin-resistant, and piperacollin/tazobactam-resistant (EUCAST 2023). The first choice of antibiotics used to treat drug-resistant Pseudomonas urinary tract infections are newer beta-lactams (ceftazidime-avibactam, ceftolozane-tazobactam, imipenem-cilastatin-relebactam, and cefiderocol), followed by a single dose of aminoglycosides as a second choice. But in complicated and non-complicated infections, the first and second choices are the same, with complicated infections requiring the addition of colistin as an alternative therapy. In infections are outside of the urinary tract, the first choices are ceftazidime-avibactam, ceftolozane-tazobactam, imipenem-cilastatin-relebactam, and as an alternative therapy, cefiderocol could also be used [48].

Clostridium difficile Infections: *Clostridium difficile* (*C. difficile*) is a gram-positive spore-forming bacillus bacterium that causes infection of the colon. *C. difficile* produces toxins and causes severe diarrhea which is often associated with antibiotic use [49]. Overuse of antibiotics (aminoglycosides, lincomycin, tetracycline, erythromycin, penicillin, cephalosporins, clindamycin, and fluoroquinolones) can lead to resistance to treatment [50, 51]. A therapy known as bezlotoxumab (Zinplava), a human antibody or fecal microbiota transplant (FMT) is recommended [52].

Enterococcal Infections: Enterococcus are gram-positive facultative anaerobic cocci bacteria, commonly live in our small and large intestines, are usually harmless, constitute 1% of human fecal microbiota, with *Enterococcus faecalis* (*E. faecalis*) and *Enterococcus faecium* (*E. faecium*) being the most common inhabitants [53–57]. However, it can cause serious infections, such as *E. faecium* and *E. faecalis*, which have developed resistance to multiple antibiotics, such as penicillin, ampicillin, cephalosporins, and semi-synthetic penicillin's [58]. Ampicillin is the first antibiotic of choice. If Enterococcus is vancomycin-resistant, then the two antibiotics used are linezolid and daptomycin [59].

The list is not exhaustive; however, it provides a spectrum of evolving strains that are drug-resistant and hence it highlights a need for new generational sensor platforms involving AI to detect bacteria and its potential spread.

3 Recent Progress in AI and Its Use in Scientific Research

Artificial Intelligence has a significant and continually growing impact on practically all aspects of humanity, including scientific research by creating new directions and opportunities by enhancing data analysis, discovery, and experimentation across various fields. While an exhaustive discussion is beyond the scope of this chapter, a few examples include data analysis and pattern recognition, where AI, particularly ML and deep learning (DL), can process massive datasets much faster than traditional methods. This is important in genomics since AI can analyze genetic data to identify mutations linked to diseases like cancer or to predict the effects of genetic variations. AI is used to enhance the rate of drug discovery by predicting how compounds will interact with biological targets, which drastically reduces the time and cost involved in developing new drugs. AI can also analyze medical images to assist in diagnostics, rendering healthcare more accurate and accessible. AI can predict the 3D structures of proteins, accelerating research in molecular biology and drug design and tailoring treatments based on individual genetic information, lifestyle, and environmental factors, leading to more effective healthcare strategies. ML algorithms explore vast knowledge of design and discovery of new materials with specific properties, essential in developing drugs, advanced materials, or efficient catalysts. AI suggests more efficient synthesis pathways for materials synthesis and device fabrication, thereby saving time and resources in laboratories. Furthermore, AI enhances the accuracy and efficiency of complex simulations, such as those involving chemical reactions or molecular dynamics, facilitating a deep understanding of mechanisms at the atomic and molecular levels. Thus, AI helps solve complex problems in quantum physics and chemistry, such as predicting molecular binding energies or the electronic behavior of chemical systems. There are many more examples to be cited in automation and robotics, data analysis, and interpretation to identify patterns, predict chemical behaviors, and interpret complex results. AI is used to design more efficient catalysts that reduce the energy required for chemical reactions and increase the efficiency of industrial processes. Hence, AI is profoundly transforming chemistry, accelerating discoveries, improving process precision and efficiency, and opening new avenues for innovation in this essential field. The specific examples of AI for detecting and mitigating antibiotic resistant diseases are provided below.

3.1 Antibiotic Resistant Diseases and Artificial Intelligence

Antibiotic resistance is a complex problem and requires multifaceted solutions. The role of AI is steadily increasing in addressing this issue. AI algorithms are used to analyze large datasets to identify potential drugs, predict their pharmacological properties, and assist in optimizing drug design. Hence, AI algorithms can predict which molecules are effective against antibiotic-resistant bacteria, accelerate the drug discovery process, and reduce costs. Additionally, AI algorithms can analyze patient data, antibiotic resistance microbiological information, and antibiotic prescription patterns, thereby providing clinicians with real-time guidance on antibiotic selection and dosage. Hence, the use of AI algorithms can help prevent the dependence, overuse, and misuse of antibiotics, thereby reducing the development of antibiotic resistance. As diagnostics, AI can identify bacterial pathogens and determine antibiotic susceptibility profiles. Thereby enabling quicker and more accurate diagnosis, helping clinicians to prescribe the most appropriate antibiotics, and reducing the likelihood of the emergence of bacterial-resistant strains. AI models can help in predicting the likelihood of antibiotic resistance based on bacterial genomic data. Thereby, identifying high-risk patients and regions, allowing healthcare providers to implement measures to prevent the development and transmission of antibiotic-resistant bacterial infections. AI-led drug repurposing approaches can help identify existing drugs that have antibacterial properties or synergistic effects with antibiotics. This can help researchers speed up the development of new treatment options for bacterial-resistant infections. Additionally, AI algorithms can analyze large-scale epidemiological data to track the spread of antibiotic-resistant bacteria and assist in identifying emerging antibiotic-resistant bacterial trends. Monitoring antibiotic resistance patterns can help public health authorities implement targeted interventions to prevent outbreaks and contain the spread of antibiotic-resistant bacterial strains.

4 Sensors Platforms and AI for Biomedical Applications

Biosensors are analytical devices that combine a biological specimen with a physicochemical sensing element to measure the presence or concentration of chemicals, materials, or other biological markers. Typical elements of such devices are biorecognition surface/interface as detector, and transducer, followed by a signal amplifier and/or analyzer with a readout display to read absolute or normalized quantity of biological specimen. A basic schematic of a biosensor is shown in Fig. 1. The figure is for illustration only, as biosensors can be quite complex depending upon the analyte or sample matrix.

An exhaustive review of biosensors is beyond the scope of this article, however novel, nanotechnology-based sensors, with high-K dielectric for low noise are briefly described here. The addition of a machine learning core (MLC) adds the capability of data storage and processing, adding a capability of edge computing for faster processing and artificial intelligence. Materials in reduced dimensions provide unique characteristics in terms of sensitivity, more specifically the materials display higher surface activation energy due to a high surface area/diameter ratio [4, 11, 17, 60]. Using the process of functionalization, nanomaterials can be functionalized with specific receptors, providing high sensitivity and specificity. Nanomaterials. With recent advances in material synthesis and device fabrication techniques,

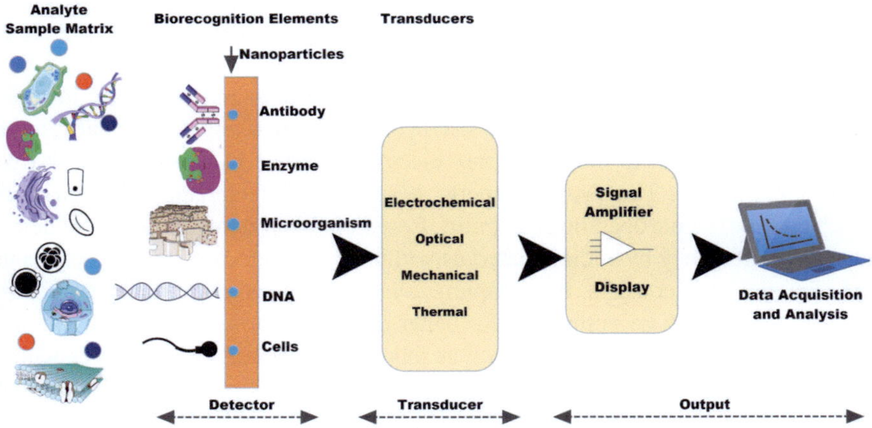

Fig. 1 A generalized diagram of a biosensor, showing biorecognition element, transducer, and signal displayed on a data acquisition and display system

nanomaterials-based sensors leverage the unique characteristics of nanomaterials—such as their small size, high surface area, and enhanced sensitivity—to improve detection accuracy, lower detection limits, and enable faster responses compared to traditional sensors [5, 61–63]. In addition to sensing elements, nanomaterials are also used to enhance the transducer characteristics. Typical nanomaterials that are used as the sensing element or to enhance the properties of the transducer are carbon nanotubes (CNT), both single wall (SWCNT) and multiwall (MWCNT), quantum dots (QD), nanoparticles (typically gold, and silver), nanowires, and graphene [64]. Various transduction mechanisms convert the interaction between the analyte and the nanomaterial into a measurable signal, that includes electrochemical, optical, mechanical, or thermal. Nanomaterials-based sensors now serve as transducers. Finally, the detected signal is processed, conditioned for noise, and analyzed as readable data.

4.1 Novel Sensor Platforms

Figure 2 below shows a spectrum of sensor platforms. The schematic shows nano biosensors in various configurations.

The three main categories are mass-based, electro-chemical, electro-mechanical, and optical-based sensors. Mass-based sensors typically detect proteins, DNA, and bacteria using quartz crystal microbalance (QCM) and surface acoustic wave (SAW) devices, as variation in frequency. Quantum dot (QD) based sensors normally detect the presence of biological targets primarily due to their fluorescence. Nanoscale Optical Sensors detect changes in light properties, such as Surface-Enhanced Raman Scattering (SERS), and surface plasmon resonance (SPR) due to changes in refractive index when the nanomaterials attract biological targets. SPR and SERS utilize metal nanoparticles to enhance the Raman scattering signal of molecules, allowing ultra-sensitive chemical detection [5, 11].

These biological sensors have been used to sense biomarkers and are increasingly used to sense/detect antibiotic-resistant disease biomarkers. Additionally, there were recent attempts to introduce piezoelectric materials in nanomaterials-based biosensors [65–67]. This provides energy harvesting characteristics to the sensors which provide remote operation capability. Research is in progress to enhance energy harvesting efficiency by adding materials with high piezoelectric charge coefficient and high dielectric constant. By optimum variation of material constants, the figure of merit (FOM) can be optimized, for better device performance. For introducing artificial intelligence, capabilities such as finite state machine (FSM) and machine learning core (MLC) are also conjoined with biosensor configuration. FSM is an in-sensor behavioral model composed of a finite number of states and transitions between states and is able to generate the interrupt signals activated by user-defined motion patterns. MLC is an in-sensor classification engine based on a decision tree logic and can increase accuracy with better context detectability, offloading the main processor while the built-in sensors identify motion data. Using these capabilities

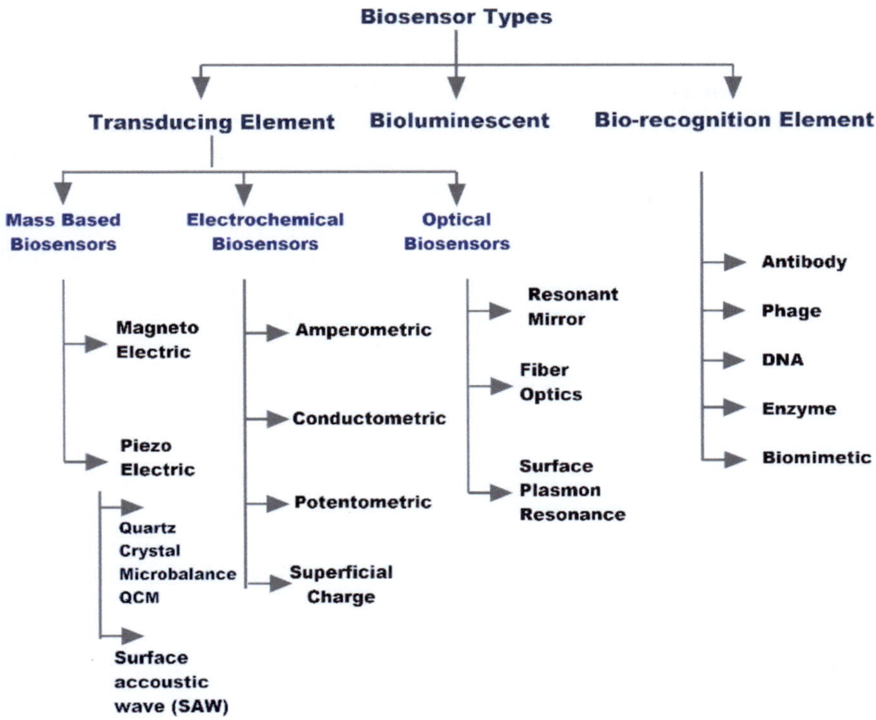

Fig. 2 Spectrum of biosensors

along with charge variability (Qvar), as electrostatic sensors, intelligent biosensors will become more autonomous and adaptable. These sensors would be able to adjust their sensing and operating parameters and behaviors based on changing conditions. This adaptability will improve their performance in dynamic environments and enable them to meet specific application needs more effectively. Intelligent biosensors will incorporate more computational power and onboard AI capabilities for disease identification at faster speed. This trend is driven by the need for faster processing and decision-making at the edge of networks, reducing latency and reliance on cloud-based systems.

4.2 AI-Based Disease Surveillance Program

The authors have proposed a wide-spectrum disease surveillance prototype based on bioinformatics for defense and security applications [68]. Integration of AI in the bio-threats identification and classification prototype will transform healthcare and disease surveillance programs with wide-ranging applications and may prevent yet another COVID-19 [69] scenario. Additionally, AI is increasingly being used to enhance the accuracy of sensor technologies. The cutting-edge developments in AI have enabled rapid and accurate detection of abnormalities, early signs of the disease, diagnostics, and personalized medicine.

AI is shaping the development of sensors for biomedical applications in terms of enhanced sensitivity and accuracy in identifying patterns and anomalies normally overlooked by the human eye by processing vast data sets in real time. This increased accuracy can reduce misdiagnoses, identify risk factors, early detection of disease, and ensure that patients receive appropriate treatment. AI also enables the integration of data by fusing data from multiple sources. AI-based sensors can monitor patients' vital signs and detect health issues, allowing timely intervention and preventative care for healthcare management and predictive analytics. AI also offers the capability to analyze large genomic datasets to identify patterns, mutations, and associations with diseases for both patient and infectious agents. Hence, AI-powered sensors can tailor healthcare interventions by aiding in personalized treatment plans, predicting risk factors for a particular treatment, and developing targeted therapies based on the individual patient's genetic profile using sequencing of DNA. Piezoelectric-based sensors provide continuous monitoring of patients, providing instant feedback to healthcare providers and patients themselves, enabling proactive management of health conditions and early detection of disease and complications. AI-based sensors having energy harvesting capability facilitate remote monitoring of patients, enabling healthcare providers to monitor the health status of patients without frequent in-person visits. This telehealth is valuable for managing chronic conditions and providing health care to underserved populations and patients living in remote areas far away from healthcare facilities.

5 Conclusion and Path Forward

With an increase in antibiotic drug-resistant disease propagation, an advanced sensor platform must be used. With recent advances in the synthesis of nanomaterials and associated device fabrication, Nanomaterial-based therapies are promising tools to combat difficult bacterial infections, featuring the capacity to evade existing mechanisms. This study highlights the recent progress in nanotechnology-based biosensor platforms. In addition, using AI, the disease propagation can be easily detected and its consequences mitigated. We further examine human–machine interfaces, AI-enhanced multimode sensors, and AI-enhanced self-sustainable systems for disease surveillance. A review of several known diseases is provided and by appropriate functionalization, such sensors can be developed. AI-driven sensors can offer more reliable and holistic measurements, but ethical considerations associated with their implantation in the healthcare domain should be considered.

References

1. Castillo-Henríquez, Luis, Mariana Brenes-Acuña, Arianna Castro-Rojas, Rolando Cordero-Salmerón, Mary Lopretti-Correa, and José Roberto Vega-Baudrit. 2020. Biosensors for the Detection of Bacterial and Viral Clinical Pathogens. *Sensors* 20: 6926. https://doi.org/10.3390/s20236926.
2. Yamin, Dina, Vuk Uskoković, Abubakar Wakil, Mohammed Goni, Shazana Shamsuddin, Fatin Mustafa, Wadha Alfouzan, et al. 2023. Current and Future Technologies for the Detection of Antibiotic-Resistant Bacteria. *Diagnostics* 13: 3246. https://doi.org/10.3390/diagnostics13203246.
3. Guliy, Olga I., Olga A. Karavaeva, Andrey V. Smirnov, Sergei A. Eremin, and Viktor D. Bunin. 2023. Optical Sensors for Bacterial Detection. *Sensors* 23: 9391. https://doi.org/10.3390/s23239391.
4. Vaseashta, A. 2006. Carbon Nanotubes Based Devices and Sensors. *Materials and Manufacturing Processes* 21: 710–716. https://doi.org/10.1080/10426910600613595.
5. Stamatin, Ioan, C. Berlic, and A. Vaseashta. 2006. On the computer-aided modeling of analyte–receptor interactions for an efficient sensor design. *Thin Solid Films* 495: 312–315. https://doi.org/10.1016/j.tsf.2005.08.299.
6. Baranwal, Jaya, Brajesh Barse, Gianluca Gatto, Gabriela Broncova, and Amit Kumar. 2022. Electrochemical Sensors and Their Applications: A Review. *Chemosensors* 10: 363. https://doi.org/10.3390/chemosensors10090363.
7. Vaseashta, A., A. Erdem, and J. Irudayaraj. 2005. Nanoparticles, Nanoporous and Carbon Nanotube Based Devices for Bio-Molecular Detection. In *Nanostructured and Advanced Materials for Applications in Sensor, Optoelectronic and Photovoltaic Technology*, ed. A. Vaseashta, D. Dimova-Malinovska, and J. M. Marshall, 204:391–394. NATO Science Series II: Mathematics, Physics and Chemistry. Dordrecht: Springer Netherlands. https://doi.org/10.1007/1-4020-3562-4_42.
8. Zhang, Kaixiang, Shangshang Qin, Sixuan Wu, Yan Liang, and Jinghong Li. 2020. Microfluidic systems for rapid antibiotic susceptibility tests (ASTs) at the single-cell level. *Chemical Science* 11: 6352–6361. https://doi.org/10.1039/D0SC01353F.
9. Bickerstaff-Westbrook, Elinor, Anastasiia Tukova, Nana Lyu, Chao Shen, Alison Rodger, and Yuling Wang. 2024. Advancing SERS label-free detection of bacteria: Sensing in liquid vs drop-cast. *Materials Today Sustainability* 27: 100912. https://doi.org/10.1016/j.mtsust.2024.100912.
10. Wang, Kaidi, Shenmiao Li, Marlen Petersen, Shuo Wang, and Xiaonan Lu. 2018. Detection and Characterization of Antibiotic-Resistant Bacteria Using Surface-Enhanced Raman Spectroscopy. *Nanomaterials* 8: 762. https://doi.org/10.3390/nano8100762.
11. Vaseashta, Ashok, and Joseph Irudayaraj. 2005. Nanostructured and nanoscale devices and sensors. *Journal of Optoelectronics and Advanced Materials* 7: 35–42.
12. Khosravi, A.D., S. Khoshnood, E. Abbasi Montazeri, N. Jomehzadeh, M. Moradi, and F. Shahi. 2022. The application of the loop-mediated isothermal amplification method for rapid detection of methicillin-resistant Staphylococcus aureus. *New Microbes and New Infections* 45: 100960. https://doi.org/10.1016/j.nmni.2022.100960.
13. Kim, Eun Jin, Jiwon Lee, Youngbae Yoon, Donghyun Lee, Yeongjun Baek, Chika Takano, Jun Sakai, et al. 2023. Development of a novel loop-mediated isothermal amplification assay for ß-lactamase gene identification using clinical isolates of Gram-negative bacteria. *Frontiers in Cellular and Infection Microbiology* 12: 1000445. https://doi.org/10.3389/fcimb.2022.1000445.
14. Grandjean Lapierre, Simon, and Michel Drancourt. 2018. rpoB Targeted Loop-Mediated Isothermal Amplification (LAMP) Assay for Consensus Detection of Mycobacteria Associated With Pulmonary Infections. *Frontiers in Medicine* 5: 332. https://doi.org/10.3389/fmed.2018.00332.

15. Shajari, Shaghayegh, Kirankumar Kuruvinashetti, Amin Komeili, and Uttandaraman Sundararaj. 2023. The Emergence of AI-Based Wearable Sensors for Digital Health Technology: A Review. *Sensors* 23: 9498. https://doi.org/10.3390/s23239498.
16. Vaseashta, Ashok, and Nimet Karagülle-Bölgen. 2018. Loaded Nanofibers: Force Protection, Filtration, Decontamination. In *Advanced Nanotechnologies for Detection and Defence against CBRN Agents*, ed. Plamen Petkov, Dumitru Tsiulyanu, Cyril Popov, and Wilhelm Kulisch, 241–252. NATO Science for Peace and Security Series B: Physics and Biophysics. Dordrecht: Springer Netherlands. https://doi.org/10.1007/978-94-024-1298-7_23.
17. Vaseashta, Ashok, and Surik Khudaverdyan, ed. 2013. *Advanced Sensors for Safety and Security*. NATO Science for Peace and Security Series B: Physics and Biophysics. Dordrecht: Springer Netherlands. https://doi.org/10.1007/978-94-007-7003-4.
18. De La Lastra, José M. Pérez, Samuel J. T. Wardell, Tarun Pal, Cesar De La Fuente-Nunez, and Daniel Pletzer. 2024. From Data to Decisions: Leveraging Artificial Intelligence and Machine Learning in Combating Antimicrobial Resistance – a Comprehensive Review. *Journal of Medical Systems* 48: 71. https://doi.org/10.1007/s10916-024-02089-5.
19. Hayashi-Nishino, Mitsuko, Kota Aoki, Akihiro Kishimoto, Yuna Takeuchi, Aiko Fukushima, Kazushi Uchida, Tomio Echigo, et al. 2022. Identification of Bacterial Drug-Resistant Cells by the Convolutional Neural Network in Transmission Electron Microscope Images. *Frontiers in Microbiology* 13: 839718. https://doi.org/10.3389/fmicb.2022.839718.
20. Ren, Yunxiao, Trinad Chakraborty, Swapnil jad, Linda Falgenhauer, Jane Falgenhauer, Alexander Goesmann, Anne-Christin Hauschild, Oliver Schwengers, and Dominik Heider. 2022. Prediction of antimicrobial resistance based on whole-genome sequencing and machine learning. Edited by Inanc Birol. *Bioinformatics* 38: 325–334. https://doi.org/10.1093/bioinformatics/btab681.
21. Tao, Shuan, Huimin Chen, Na Li, and Wei Liang. 2022. The Application of the CRISPR-Cas System in Antibiotic Resistance. *Infection and Drug Resistance* Volume 15: 4155–4168. https://doi.org/10.2147/IDR.S370869.
22. Branda, Francesco, and Fabio Scarpa. 2024. Implications of Artificial Intelligence in Addressing Antimicrobial Resistance: Innovations, Global Challenges, and Healthcare's Future. *Antibiotics* 13: 502. https://doi.org/10.3390/antibiotics13060502.
23. Sheikhalishahi, Seyedmostafa, Riccardo Miotto, Joel T Dudley, Alberto Lavelli, Fabio Rinaldi, and Venet Osmani. 2019. Natural Language Processing of Clinical Notes on Chronic Diseases: Systematic Review. *JMIR Medical Informatics* 7: e12239. https://doi.org/10.2196/12239.
24. Sakagianni, Aikaterini, Christina Koufopoulou, Georgios Feretzakis, Dimitris Kalles, Vassilios S. Verykios, Pavlos Myrianthefs, and Georgios Fildisis. 2023. Using Machine Learning to Predict Antimicrobial Resistance—A Literature Review. *Antibiotics* 12: 452. https://doi.org/10.3390/antibiotics12030452.
25. Rasigade, Jean-Philippe, and François Vandenesch. 2014. Staphylococcus aureus: A pathogen with still unresolved issues. *Infection, Genetics and Evolution* 21: 510–514. https://doi.org/10.1016/j.meegid.2013.08.018.
26. Rubinstein, Ethan, and Yoav Keynan. 2014. Vancomycin revisited - 60 years later. *Frontiers in Public Health* 2: 217. https://doi.org/10.3389/fpubh.2014.00217.
27. Cha, Raymond, William J. Brown, and Michael J. Rybak. 2003. Bactericidal activities of daptomycin, quinupristin-dalfopristin, and linezolid against vancomycin-resistant Staphylococcus aureus in an in vitro pharmacodynamic model with simulated endocardial vegetations. *Antimicrobial Agents and Chemotherapy* 47: 3960–3963. https://doi.org/10.1128/AAC.47.12.3960-3963.2003.
28. Ayatollahi, Jamshid, Alireza Vahidi, Seyed Hossein Shahcheraghi, Alireza Bagheripour, Marzieh Lotfi, Malihe Lotfi, and Saeid Reza Lotfi. 2013. Investigating the Resistance of Escherichia coli Against Some Selected Antimicrobials in Bam. *Jundishapur Journal of Microbiology* 6. https://doi.org/10.5812/jjm.7407.
29. Rodríguez-Baño, Jesús, Juan C. Alcalá, Jose M. Cisneros, Fabio Grill, Antonio Oliver, Juan P. Horcajada, Teresa Tórtola, et al. 2008. Community Infections Caused by Extended-Spectrum β-Lactamase–Producing Escherichia coli. *Archives of Internal Medicine* 168: 1897. https://doi.org/10.1001/archinte.168.17.1897.

30. Dormanesh, Banafshe, Farhad Safarpoor Dehkordi, Hassan Momtaz, Reza Mirnejad, Mohammad Javad Hoseini, Emad Yahaghi, Vahideh Tarhriz, and Ebrahim Khodaverdi Darian. 2014. Virulence Factors and O-Serogroups Profiles of Uropathogenic Escherichia Coli Isolated from Iranian Pediatric Patients. *Iranian Red Crescent Medical Journal* 16. https://doi.org/10.5812/ircmj.14627.
31. Centers for Disease Control and Prevention (CDC). 2013. Vital signs: carbapenem-resistant Enterobacteriaceae. *MMWR. Morbidity and mortality weekly report* 62: 165–170.
32. Allocati, Nerino, Michele Masulli, Mikhail Alexeyev, and Carmine Di Ilio. 2013. Escherichia coli in Europe: An Overview. *International Journal of Environmental Research and Public Health* 10: 6235–6254. https://doi.org/10.3390/ijerph10126235.
33. Meletis, Georgios. 2016. Carbapenem resistance: an overview of the problem and future perspectives. *Therapeutic Advances in Infectious Disease* 3: 15–21. https://doi.org/10.1177/2049936115621709.
34. Capone, A., M. Giannella, D. Fortini, A. Giordano, M. Meledandri, M. Ballardini, M. Venditti, et al. 2013. High rate of colistin resistance among patients with carbapenem-resistant Klebsiella pneumoniae infection accounts for an excess of mortality. *Clinical Microbiology and Infection* 19: E23–E30. https://doi.org/10.1111/1469-0691.12070.
35. Giacobbe, D.R., V. Del Bono, E.M. Trecarichi, F.G. De Rosa, M. Giannella, M. Bassetti, A. Bartoloni, et al. 2015. Risk factors for bloodstream infections due to colistin-resistant KPC-producing Klebsiella pneumoniae: results from a multicenter case–control–control study. *Clinical Microbiology and Infection* 21: 1106.e1–1106.e8. https://doi.org/10.1016/j.cmi.2015.08.001.
36. Yang, Haifei, Guosheng Chen, Lifen Hu, Yanyan Liu, Jun Cheng, Ying Ye, and Jiabin Li. 2018. Enhanced efficacy of imipenem-colistin combination therapy against multiple-drug-resistant Enterobacter cloacae : in vitro activity and a Galleria mellonella model. *Journal of Microbiology, Immunology and Infection* 51: 70–75. https://doi.org/10.1016/j.jmii.2016.01.003.
37. Allesina, Stefano, and Si Tang. 2015. The stability–complexity relationship at age 40: a random matrix perspective. *Population Ecology* 57: 63–75. https://doi.org/10.1007/s10144-014-0471-0.
38. Tseng, Sung-Pin, Sheng-Fan Wang, Ling Ma, Ting-Yin Wang, Tsung-Ying Yang, L. Kristopher Siu, Yin-Ching Chuang, et al. 2017. The plasmid-mediated fosfomycin resistance determinants and synergy of fosfomycin and meropenem in carbapenem-resistant Klebsiella pneumoniae isolates in Taiwan. *Journal of Microbiology, Immunology and Infection* 50: 653–661. https://doi.org/10.1016/j.jmii.2017.03.003.
39. Rodríguez-Baño, Jesús, Belén Gutiérrez-Gutiérrez, Isabel Machuca, and Alvaro Pascual. 2018. Treatment of Infections Caused by Extended-Spectrum-Beta-Lactamase-, AmpC-, and Carbapenemase-Producing Enterobacteriaceae. *Clinical Microbiology Reviews* 31: e00079-17. https://doi.org/10.1128/CMR.00079-17.
40. Zasowski, Evan J., Jeffrey M. Rybak, and Michael J. Rybak. 2015. The β-Lactams Strike Back: Ceftazidime-Avibactam. *Pharmacotherapy: The Journal of Human Pharmacology and Drug Therapy* 35: 755–770. https://doi.org/10.1002/phar.1622.
41. Petty, Lindsay, Oryan Henig, Twisha Patel, Jason Pogue, and Keith Kaye. 2018. Overview of meropenem-vaborbactam and newer antimicrobial agents for the treatment of carbapenem-resistant Enterobacteriaceae. *Infection and Drug Resistance* Volume 11: 1461–1472. https://doi.org/10.2147/IDR.S150447.
42. Zhanel, George G., Courtney K. Lawrence, Heather Adam, Frank Schweizer, Sheryl Zelenitsky, Michael Zhanel, Philippe R. S. Lagacé-Wiens, et al. 2018. Imipenem–Relebactam and Meropenem–Vaborbactam: Two Novel Carbapenem-β-Lactamase Inhibitor Combinations. *Drugs* 78: 65–98. https://doi.org/10.1007/s40265-017-0851-9.
43. Jang, Jong Geol, and Jin Hong Chung. 2020. Diagnosis and treatment of multidrug-resistant tuberculosis. *Yeungnam University Journal of Medicine* 37: 277–285. https://doi.org/10.12701/yujm.2020.00626.

44. Ison, C A. 2004. Which cephalosporin for gonorrhoea? *Sexually Transmitted Infections* 80: 386–388. https://doi.org/10.1136/sti.2004.012757.
45. Dijkshoorn, Lenie, Alexandr Nemec, and Harald Seifert. 2007. An increasing threat in hospitals: multidrug-resistant Acinetobacter baumannii. *Nature Reviews Microbiology* 5: 939–951. https://doi.org/10.1038/nrmicro1789.
46. Viehman, J. Alexander, M. Hong Nguyen, and Yohei Doi. 2014. Treatment options for carbapenem-resistant and extensively drug-resistant Acinetobacter baumannii infections. *Drugs* 74: 1315–1333. https://doi.org/10.1007/s40265-014-0267-8.
47. Hancock, Robert E. W., and David P. Speert. 2000. Antibiotic resistance in Pseudomonas aeruginosa: mechanisms and impact on treatment. *Drug Resistance Updates: Reviews and Commentaries in Antimicrobial and Anticancer Chemotherapy* 3: 247–255. https://doi.org/10.1054/drup.2000.0152.
48. Tamma, Pranita D., Samuel L. Aitken, Robert A. Bonomo, Amy J. Mathers, David van Duin, and Cornelius J. Clancy. 2022. Infectious Diseases Society of America 2022 Guidance on the Treatment of Extended-Spectrum β-lactamase Producing Enterobacterales (ESBL-E), Carbapenem-Resistant Enterobacterales (CRE), and Pseudomonas aeruginosa with Difficult-to-Treat Resistance (DTR-P. aeruginosa). *Clinical Infectious Diseases: An Official Publication of the Infectious Diseases Society of America* 75: 187–212. https://doi.org/10.1093/cid/ciac268.
49. Mada, Pradeep Kumar, and Mohammed U. Alam. 2024. Clostridioides difficile infection. In *StatPearls*. Treasure Island (FL): StatPearls Publishing.
50. Johanesen, Priscilla A., Kate E. Mackin, Melanie L. Hutton, Milena M. Awad, Sarah Larcombe, Jacob M. Amy, and Dena Lyras. 2015. Disruption of the Gut Microbiome: Clostridium difficile Infection and the Threat of Antibiotic Resistance. *Genes* 6: 1347–1360. https://doi.org/10.3390/genes6041347.
51. Spigaglia, Patrizia. 2016. Recent advances in the understanding of antibiotic resistance in *Clostridium difficile* infection. *Therapeutic Advances in Infectious Disease* 3: 23–42. https://doi.org/10.1177/2049936115622891.
52. Alhifany, Abdullah A, Abdulaali R Almutairi, Thamer A Almangour, Alaa N Shahbar, Ivo Abraham, Mohammed Alessa, Faris S Alnezary, and Ejaz Cheema. 2019. Comparing the efficacy and safety of faecal microbiota transplantation with bezlotoxumab in reducing the risk of recurrent *Clostridium difficile* infections: a systematic review and Bayesian network meta-analysis of randomized controlled trials. *BMJ Open* 9: e031145. https://doi.org/10.1136/bmjopen-2019-031145.
53. Vollaard, E J, and H A Clasener. 1994. Colonization resistance. *Antimicrobial Agents and Chemotherapy* 38: 409–414. https://doi.org/10.1128/AAC.38.3.409.
54. Sghir, Abdelghani, Genevieve Gramet, Antonia Suau, Violaine Rochet, Philippe Pochart, and Joel Dore. 2000. Quantification of Bacterial Groups within Human Fecal Flora by Oligonucleotide Probe Hybridization. *Applied and Environmental Microbiology* 66: 2263–2266. https://doi.org/10.1128/AEM.66.5.2263-2266.2000.
55. Eckburg, Paul B., Elisabeth M. Bik, Charles N. Bernstein, Elizabeth Purdom, Les Dethlefsen, Michael Sargent, Steven R. Gill, Karen E. Nelson, and David A. Relman. 2005. Diversity of the Human Intestinal Microbial Flora. *Science* 308: 1635–1638. https://doi.org/10.1126/science.1110591.
56. Baghdayan, A. S., N. Shankar, and P. M. Tendolkar. 2003. Pathogenic enterococci: new developments in the 21st century. *Cellular and Molecular Life Sciences (CMLS)* 60: 2622–2636. https://doi.org/10.1007/s00018-003-3138-0.
57. Lebreton, Francois, Rob J. L. Willems, and Michael S. Gilmore. 2014. Enterococcus Diversity, Origins in Nature, and Gut Colonization. In *Enterococci: From Commensals to Leading Causes of Drug Resistant Infection*, ed. Michael S. Gilmore, Don B. Clewell, Yasuyoshi Ike, and Nathan Shankar. Boston: Massachusetts Eye and Ear Infirmary.
58. Kristich, Christopher J., Louis B. Rice, and Cesar A. Arias. 2014. Enterococcal Infection—Treatment and Antibiotic Resistance. In *Enterococci: From Commensals to Leading Causes of Drug Resistant Infection*, ed. Michael S. Gilmore, Don B. Clewell, Yasuyoshi Ike, and Nathan Shankar. Boston: Massachusetts Eye and Ear Infirmary.

59. Narayanan, Navaneeth, Rena Rai, Parth Vaidya, Avani Desai, Tanaya Bhowmick, and Melvin P. Weinstein. 2019. Comparison of linezolid and daptomycin for the treatment of vancomycin-resistant enterococcal bacteremia. *Therapeutic Advances in Infectious Disease* 6: 2049936119828964. https://doi.org/10.1177/2049936119828964.
60. Vaseashta, Ashok, Surik Khudaverdyan, and Ashok Vaseashta. 2013. *Advanced Sensors for Safety and Security*. NATO Science for Peace and Security Series B: Physics and Biophysics. Dordrecht: Springer Netherlands Springer e-books Imprint: Springer. https://doi.org/10.1007/978-94-007-7003-4
61. Vaseashta, Ashok, and Ioan Stamatin. 2007. Electrospun polymers for controlled release of drugs, vaccine delivery, and system-on-fibers. *Journal of Optoelectronics and Advanced Materials* 9: 1606–1613.
62. Demir, Didem, Nimet Bolgen, and Ashok Vaseashta. 2023. Electrospun Nanofibers for Biomedical, Sensing, and Energy Harvesting Functions. *Polymers* 15: 4253. https://doi.org/10.3390/polym15214253.
63. Vaseashta, Ashok, and Nimet Bölgen, ed. 2022. *Electrospun Nanofibers: Principles, Technology and Novel Applications*. Cham: Springer International Publishing. https://doi.org/10.1007/978-3-030-99958-2.
64. Vaseashta, A. 2005. Nanostructured Materials Based Next Generation Devices and Sensors. In *Nanostructured and Advanced Materials for Applications in Sensor, Optoelectronic and Photovoltaic Technology*, ed. A. Vaseashta, D. Dimova-Malinovska, and J. M. Marshall, 204:1–30. NATO Science Series II: Mathematics, Physics and Chemistry. Dordrecht: Springer Netherlands. https://doi.org/10.1007/1-4020-3562-4_1.
65. Batra, Ashok, James Sampson, Angela Davis, James Currie, and Ashok Vaseashta. 2023. Electrospun nanofibers doped with PVDF and PLZT nanoparticles for potential biomedical and energy harvesting applications. *Journal of Materials Science: Materials in Electronics* 34: 1654. https://doi.org/10.1007/s10854-023-11066-6.
66. Sharvare Palwai, Ashok Batra, Sushma Kotru, and Ashok Vaseashta. 2022. Electrospun Polyvinylidene Fluoride Nanofiber Membrane-Based Flexible Capacitive Tactile Sensors for Biomedical Applications. *Surface Engineering and Applied Electrochemistry* 58: 194–201. https://doi.org/10.3103/S1068375522020089.
67. Vaseashta, Ashok, and Ashok Batra. 2022. Electrospun Nanofibers of High-Performance Electret Polymers for Tactile Sensing and Wearable Electronics. In *Proceedings of the Sixth International Symposium on Dielectric Materials and Applications (ISyDMA'6)*, ed. Ashok Vaseashta, Mohammed Essaid Achour, Mustapha Mabrouki, Didier Fasquelle, and Amina Tachafine, 31–52. Cham: Springer International Publishing. https://doi.org/10.1007/978-3-031-11397-0_3.
68. Baldwin, James, Samina Noorali, and Ashok Vaseashta. 2019. Wide Spectrum Bio-threats Identification and Classification. In *Modernizing the Future Fight: Accelerate & Adapt*. Wilmington, DE USA.
69. Baldwin, James, Samina Noorali, and Ashok Vaseashta. 2023. Biology and Behavior of Severe Acute Respiratory Syndrome Coronavirus Contagion with Emphasis on Treatment Strategies, Risk Assessment, and Resilience. *COVID* 3: 1259–1303. https://doi.org/10.3390/covid3090089.

Contributed Articles

Smart Nanocomposites for Energy Harvesting and Biomedical Sensing Applications

Angela Davis, Ashok Batra, Padmaja Guggilla, Clyde Varner, and Ashok Vaseashta

Abstract Demand for thin films of various functional materials is increasing due to the miniaturization of electronic devices to nanometer scales. Nano-sized thin films can be defined as material in reduced dimensions, having at least one dimension ~100 nm or less. Due to reduced dimensions, the materials exhibit unique characteristics, and the characteristics are completely different from bulk dimensions, thus categorized as SMART materials. Conjoining these characteristics, the present study involved the use of partially fluorinated polymers, such as PVDF, for their piezoelectric characteristics. The study reported in this presentation is to assess the optimum concentration of PVDF that is required to exhibit a high degree of polarity and consequently improved piezoelectric properties. Under the current investigation, PVDF thin films are fabricated with various concentrations of PVDF and characterized for their electrical, mechanical, and optical properties for their potential use as smart materials for energy harvesting and biomedical applications.

Keywords PVDF · Smart materials · Sensors · Energy harvesting · Biomedical

1 Introduction

Smart nanocomposites are an emerging class of materials that combine nanoscale components with traditional materials to create materials and systems with enhanced or novel properties [1, 2]. These materials have garnered significant attention due

A. Davis (✉) · A. Batra · P. Guggilla · C. Varner
Department of Physics, Chemistry, and Mathematics, Materials Science Group, Alabama A&M University, Normal, Huntsville, AL, USA
e-mail: angela.davis@bulldogs.aamu.edu

A. Vaseashta
Applied Research Division, International Clean Water Institute, Manassas, VA, USA

IEEN, Academy of Sciences of Moldova, Chisinau, Moldova

Institutul de Cercetare, Universitatea Din Bucureşti, Bucureşti, Romania

A. Vaseashta et al. (eds.), *Proceedings of the Eighth International Symposium on Dielectric Materials and Applications (ISyDMA'8)*, Springer Proceedings in Physics 349, https://doi.org/10.1007/978-3-031-84809-4_18

to their potential applications in various fields. The term "smart" refers to the material's ability to respond dynamically to external stimuli, such as vibrations, temperature, pressure, or electromagnetic fields. Energy harvesting involves capturing and converting energy from various sources—mechanical vibrations, thermal gradients, or light—into usable electrical energy. Introducing certain materials in a matrix (such as a polymer, metal, or ceramic) allows for the manipulation of physical, chemical, and biological properties, leading to improved functionalities for specific applications [3]. One such aspect is energy harvesting. Energy harvesting involves capturing and converting energy from various sources—such as mechanical vibrations, thermal gradients, or light—into usable electrical energy. Smart nanocomposites are particularly well-suited for this due to their enhanced mechanical, electrical, and thermal properties, viz. piezo electrics, thermos electrics, and photovoltaics.

For this investigation, we focus on piezoelectric nanocomposites which are materials that generate an electric charge in response to mechanical stress. Incorporating piezoelectric nanoparticles into polymers can create flexible and efficient energy harvesters for powering small electronic devices, especially as sensors in Internet of Things (IoT) devices. Polyvinylidene fluoride (PVDF) is a semi-crystalline and highly versatile polymer that has gained significant attention in the field of energy harvesting due to its unique properties, particularly its piezoelectric and pyroelectric characteristics [4, 5]. Research into PVDF-based energy harvesting is focused on improving the material's performance and expanding its applications, such as by doping PVDF with nanoparticles or blending it with other polymers, to enhance its piezoelectric response. This investigation involves finding an optimum concentration of PVDF that is required to exhibit a high degree of polarization and consequently improved piezoelectric properties. In our earlier investigations, PVDF was also doped with various nanoparticles to observe the degree of centrosymmetric β phase, which is primarily responsible for piezoelectric characteristics [6, 7]. The characterization of such composites yields a figure of merit that is an indication of its performance for a specific application. In this study, the figure of merit was assessed for several composites, as a means to optimize the piezoelectric response.

2 Experimental Methods

2.1 Material Preparation

PVDF is generally synthesized by the free radical polymerization of 1,1-difluoroethylene ($CH_2 = CF2$). The polymerization takes place in the suspension or emulsion at 10–150 °C or Pressure of 10–300 atm (Fig. 1).

Polyvinylidene Fluoride (PVDF) (–($C_2H_2F_2$)n–) was a polymer procured from Sigma Aldrich and used to fabricate thin films. Dimethylformamide (DMF) (C_3H_7NO) was selected as the solvent. The PVDF-DMF solutions were prepared with PVDF concentrations of 0.025, 0.15 and 0.25 g. The amount of solvent used

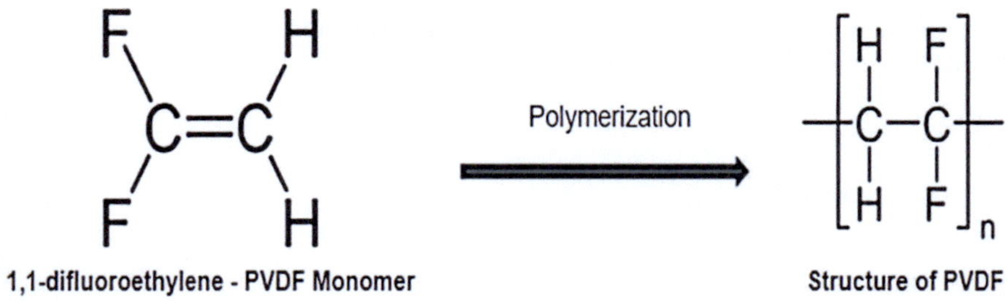

Fig. 1 PVDF Structure

was 5 mL for each. The thin films were fabricated utilizing the solution cast method, which can be seen in Fig. 2. The PVDF powder was dissolved in Dimethylformamide (DMF) and stirred on a Stuart HS 100–060 Hot plate and stirred at room temperature (25–30 °C) for 50 min at 300 rpm utilizing a magnetic stirrer. After the solution was stirred for 50 min the temperature was increased in 10-degree increments and stirred for 10 min until the temperature reached 75 °C. The PVDF/DMF solution was then removed from the stirrer upon reaching 75 °C & with a consistent pour transferred to a Pyrex petri dish. The PVDF was annealed in an oven at 76 °C from 6 to 12 h, then removed and cooled to ambient temperature.

Three different PVDF films with concentrations of 0.025, 0.15, and 0.25 g were fabricated. These PVDF films fabricated in DMF were developed to determine the effect PVDF concentrations have on its Figures of Merit (FOM), optical properties, and electrical properties. Several advantages appear from utilizing the solution casting method. Research has shown a polymer dissolved in a solvent on a temporary

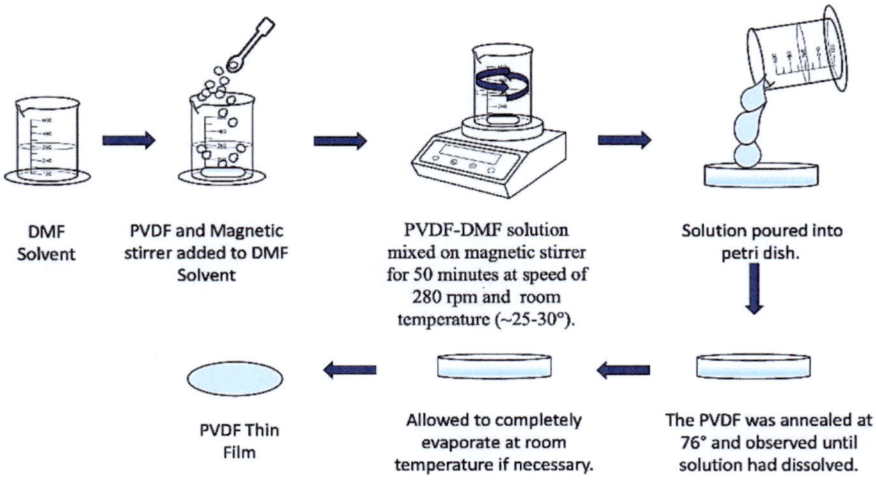

Fig. 2 Systematic representation of solution cast synthesis of PVDF thin films

substrate demonstrates anticipated advantages which include more uniform thickness distribution and higher dimension stability in addition to optical purity.

2.2 *Characterization*

2.2.1 Electrical Characterization

In 1969 Kawai [5] was the first to find high piezoelectricity in stretched and poled poly(vinylidene fluoride) (PVDF) films. The pyroelectric property of PVDF was then found by Berman et al. [8] and two years later by Nakamura and Wada [9]. Due to its pyro-, piezo-, and ferroelectric effects, PVDF has been investigated widely within the last decades [6, 7, 10, 11]. The most interesting classes of polymers are electroactive polymers functioning as smart materials in various and numerous applications, such as actuators, sensors, biomaterials in the biomedical field, and energy harvesting among others [10, 12]. The concentrations of PVDF were altered to determine whether or not varying PVDF concentrations inhibit or assist in PVDF's ability to function as biomedical sensors.

Dielectric, pyroelectric, and impedance studies on the PVDF films were carried out by a Lab View interfaced system, where LCR Bridge and an electrometer are connected as shown below in Fig. 3. The optical properties of the material give information regarding the composition nature of the material. Because molecules absorb different wavelengths of light depending on their structures, monitoring the percentage of absorbance or reflectance can identify the molecular structure of a compound. The testing of samples may provide explicit results for resistivity to be higher or lowered.

Dielectric surface currents, as a function of temperature and low, applied voltages, were measured using the combined Keithley Model 6517 Electrometer and Keithley Model 8009 resistivity test fixture with the Amprobe wireless thermometer coupled to

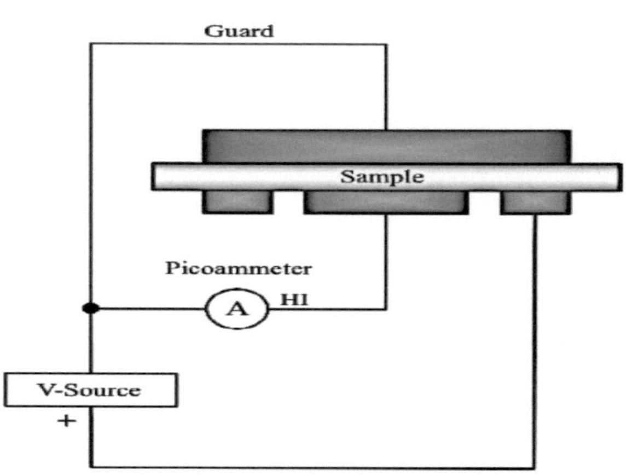

Fig. 3 Illustration of current path with a sample and electrodes

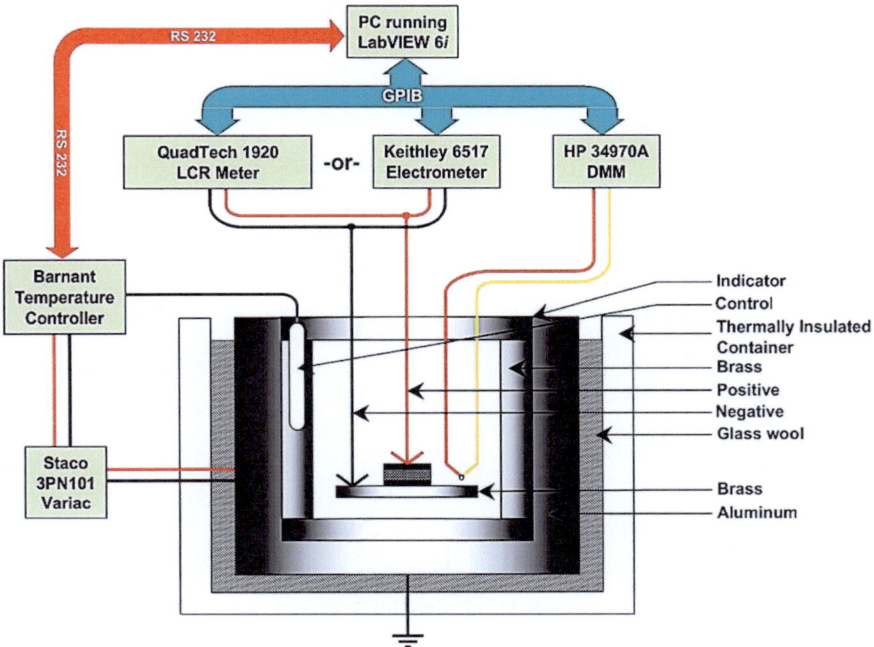

Fig. 4 Capacitance measurement system (LCR meter)

a J-type thermocouple. Figure 4, below, shows the electrode configuration of the test fixture, depicting the surface current measurement method, with a guarded electrode at the top, and two circular electrodes at the bottom with the sample (PVDF film) sandwiched between.

2.2.2 Optical Characterization

Optical properties (optical band gap, refractive indices, and extinction coefficients) of the polymer composite films were studied by UV–VIS spectroscopy. Fourier Transformation Infrared (FTIR) spectroscopy was also used to study the form and nature of bonding in the PVDF composite films. Fourier transform infrared spectra (FTIR) were collected using a Thermo Fisher Nicolet in ATR mode with a diamond crystal in attenuated total reflection (ATR) mode. For each measurement, 16 scans were performed with a measurement resolution of 2 cm^{-1}. The UV–Vis were taken on a Hitachi U-2900 spectrometer with 10 nm resolution between 200–600 nm.

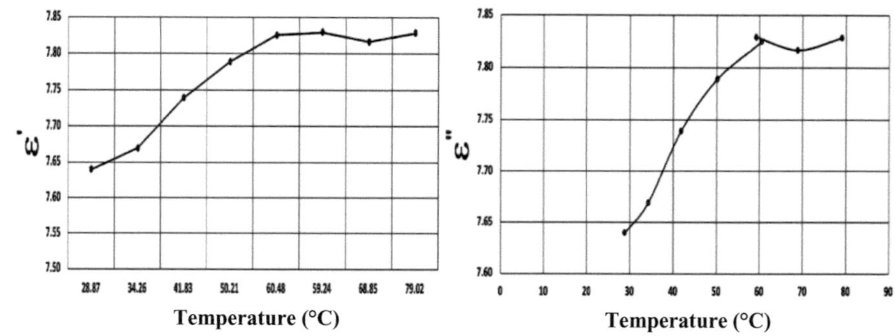

Fig. 5 Graph of the real (Ɛ′) and imaginary (Ɛ″) dielectric constant for PVDF 0.25 sample

3 Results and Discussion

3.1 *Dielectric Constant*

The dielectric constant is associated with specific materials such as those used in cable insulation, capacitors, and printed circuit board substrates. The imaginary dielectric constant (Ɛ″) is a complex number associated with dielectric losses and the real dielectric constant (Ɛ′) is a complex number that indicates the degree to which a material can be polarized. The greater the degree of polarization, the more responsive and ideal the PVDF film is for a biomedical sensor. Figure 5 displays the highest-performing pure PVDF film with a concentration of 0.25 g. The real and imaginary constants increased as the temperature increased. Thus, the increase in temperature is directly proportional to the increase in both dielectric constants.

3.2 *Figures of Merit*

Table 1 below displays the figures of merits for PVDF samples PVDF 0.025, PVDF 1.5, and PVDF 0.25. High current responsivity (F_I), high voltage responsivity (F_v), high detectivity (F_d), and energy harvesting (F_e) are displayed at temperatures at or around 40, 50, and 60 °C. High current responsivity (F_I) fundamentally is time-dependent temperature changes—the temperature difference divided by the difference in time. high voltage responsivity (F_v), high detectivity (F_d), and energy harvesting (F_e) [10, 13].

Figures of merit generally reflect an increase as the concentration of PVDF increases. This was observed mostly, as seen in Fig. 6, in the PVDF 0.15 and PVDF 0.25 samples at 40 °C. An anomaly was noted with the PVDF 0.025 film where its high detectivity (F_d) FOM was greater than the higher PVDF samples, PVDF 0.15

Table 1 Displays high current responsivity (F_I), high voltage responsivity (F_v), high detectivity (F_d), and energy harvesting (F_e) Figures of Merit (FOM)

Figures of merit (FOM)							
Sample ID	Dimethyl formamide solvent (mL)	PVDF Concen-tration (g)	Temp. (°C)	F_I = Ip/A(dT/dt)	F_v = P/ε′ (nC/cm² °C)	F_d = P/sqrt(ε″) (nC/cm² °C)	F_e = P²/ε′ (nC/cm² °C)²
				X10^{-2}	X10^{-3}	X10^{-2}	X10^{-5}
PVDF 0.025	5	0.025	40	1.762	2.173	17.62	3.829
			50	15.67	19.46	78.34	304.9
			60	34.50	43.18	345	1489
PVDF 1.5	5	0.15	37.59	0.8154	1.074	1.700	0.8759
			52.93	2.001	2.938	4.494	5.904
			65.79	10.83	19.13	23.63	207.2
PVDF 0.25	5	0.25	40	1.689	2.183	3.137	3.689
			50	2.267	2.910	4.007	6.598
			60	4.024	5.139	6.615	20.68

and PVDF 0.25. Additional research will help to identify the anomaly that led to the higher high detectivity (F_d) FOM for the smallest PVDF concentration.

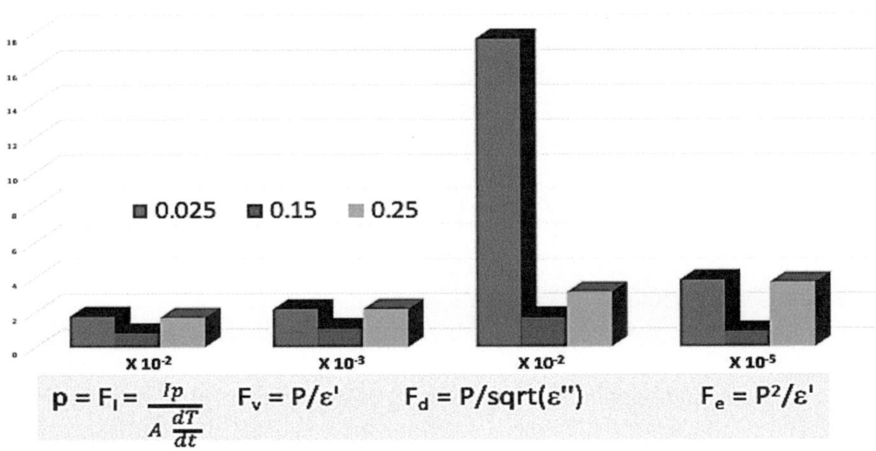

Fig. 6 Dielectric parametric figures of merit results at 40 °C

3.3 Spectrum of Applications

It is shown here that PVDF-based nanocomposites represent a versatile and high-performance class of materials that integrate the advantageous properties of PVDF with those of nanoscale fillers. These composites are extensively studied and utilized in various applications, particularly in energy harvesting and biomedical technologies, due to their enhanced piezoelectric, pyroelectric, and biocompatible properties.

Although an exhaustive discussion is beyond the scope of this publication, some of the energy harvesting applications include vibration energy, wearable energy, and hybrid sources of energy. In biomedical technologies, PVDF-based nanocomposites offer several advantages, including biocompatibility, flexibility, and the ability to respond to various stimuli. These properties make them suitable for a range of biomedical applications. Several applications that are considered include biosensors, flexible and wearable sensors, stimuli-responsive delivery, implantable devices and scaffolds for regenerative medicine. PVDF's flexibility and lightweight nature make it ideal for wearable energy harvesters that can convert energy from body movements or thermal gradients into power for small electronic devices, such as sensors or medical implants.

4 Conclusions and Future Directions

PVDF is a promising material for energy harvesting applications due to its excellent piezoelectric and pyroelectric properties, flexibility, and chemical resistance. The research reported here shows that PVDF-based nanocomposites have great potential for energy harvesters, such as in critical infrastructure monitoring [14], remote and off-grid sensors that entirely depend on mechanical vibration, thermal energy, or other sources of energy to be harvested to power these sensors, agriculture, and even defense applications. Biomedical applications that are being considered include the onset of essential tremors (ET) and Parkinson's disease [15], catheters, and monitors for infant sleep, prostate enlargement, and patient movement in hospitals [3, 6, 11, 16]. Ongoing research and development are likely to further enhance its performance and broaden its applications, potentially leading to more efficient and widely adopted energy harvesting solutions. Processing methods, such as electrospinning or stretching are used to align the PVDF molecules in the β-phase, thereby enhancing its piezoelectric properties. This approach is critical for developing high-performance PVDF nanogenerators [3, 11, 17].

While PVDF holds great promise, several challenges remain, viz. achieving and maintaining the β-phase of PVDF, which is crucial for its piezoelectric properties and requires precise processing techniques. Ensuring consistent quality and performance on a large scale is yet another challenge. Integrating PVDF energy harvesters into practical applications, such as in wearables or industrial sensors, requires further advancements in the scalability and durability of the materials. PVDF is a promising material for energy harvesting applications due to its excellent piezoelectric and pyroelectric properties, flexibility, and chemical resistance. Ongoing research and development are likely to further enhance its performance and broaden its applications, potentially leading to more efficient and widely adopted energy harvesting solutions. As these technologies mature, PVDF-based devices could play a significant role in powering the next generation of self-sustaining electronics and wearable technologies.

Acknowledgements The authors acknowledge NSF Award ID 2331969 CREST HBCU-RISE: Preparation and Characterization of Ceramics Embedded with Nanostructures for Specialized Applications and EPSCoR-GRSP award. This article is dedicated to one of the co-authors (Late Prof. Ashok Batra).

References

1. Vaseashta, Ashok, and Surik Khudaverdyan, ed. 2013. *Advanced Sensors for Safety and Security.* NATO Science for Peace and Security Series B: Physics and Biophysics. Dordrecht: Springer Netherlands. https://doi.org/10.1007/978-94-007-7003-4.
2. Vaseashta, A. 2006. Carbon Nanotubes Based Devices and Sensors. *Materials and Manufacturing Processes* 21: 710–716. https://doi.org/10.1080/10426910600613595.
3. Demir, Didem, Nimet Bolgen, and Ashok Vaseashta. 2023. Electrospun Nanofibers for Biomedical, Sensing, and Energy Harvesting Functions. *Polymers* 15: 4253. https://doi.org/10.3390/polym15214253.
4. Fukada, E. 2000. History and recent progress in piezoelectric polymers. *IEEE Transactions on Ultrasonics, Ferroelectrics and Frequency Control* 47: 1277–1290. https://doi.org/10.1109/58.883516.
5. Kawai, Heiji. 1969. The Piezoelectricity of Poly (vinylidene Fluoride). *Japanese Journal of Applied Physics* 8: 975. https://doi.org/10.1143/JJAP.8.975.
6. Sharvare Palwai, Ashok Batra, Sushma Kotru, and Ashok Vaseashta. 2022. Electrospun Polyvinylidene Fluoride Nanofiber Membrane-Based Flexible Capacitive Tactile Sensors for Biomedical Applications. *Surface Engineering and Applied Electrochemistry* 58: 194–201. https://doi.org/10.3103/S1068375522020089.
7. Sampson, James C, Sharvare Palwai, Ashok Batra, and Ashok Vaseashta. 2022. Dielectric Parametric Study in Electrospun PLZT/PVDF Nanocomposite Membranes. *Nanotechnology & Applications* 5. https://doi.org/10.33425/2639-9466.1032.
8. Bergman, J. G., J. H. McFee, and G. R. Crane. 1971. PYROELECTRICITY AND OPTICAL SECOND HARMONIC GENERATION IN POLYVINYLIDENE FLUORIDE FILMS. *Applied Physics Letters* 18: 203–205. https://doi.org/10.1063/1.1653624.
9. Nakamura, Ken'ichi, and Yasaku Wada. 1971. Piezoelectricity, pyroelectricity, and the electrostriction constant of poly(vinylidene fluoride). *Journal of Polymer Science Part A-2: Polymer Physics* 9: 161–173. https://doi.org/10.1002/pol.1971.160090111.
10. Batra, Ashok, James Sampson, Angela Davis, James Currie, and Ashok Vaseashta. 2023. Electrospun nanofibers doped with PVDF and PLZT nanoparticles for potential biomedical and energy harvesting applications. *Journal of Materials Science: Materials in Electronics* 34: 1654. https://doi.org/10.1007/s10854-023-11066-6.
11. Vaseashta, Ashok, and Ashok Batra. 2022. Electrospun Nanofibers of High-Performance Electret Polymers for Tactile Sensing and Wearable Electronics. In *Proceedings of the Sixth International Symposium on Dielectric Materials and Applications (ISyDMA'6)*, ed. Ashok Vaseashta,

Mohammed Essaid Achour, Mustapha Mabrouki, Didier Fasquelle, and Amina Tachafine, 31–52. Cham: Springer International Publishing. https://doi.org/10.1007/978-3-031-11397-0_3.

12. Wong, C. K., and F. G. Shin. 2005. Electrical conductivity enhanced dielectric and piezoelectric properties of ferroelectric 0-3 composites. *Journal of Applied Physics* 97: 064111. https://doi.org/10.1063/1.1862317.
13. Batra, A. K., M. E. Edwards, P. Guggilla, M. D. Aggarwal, and R. B. Lal. 2014. Pyroelectric Properties of PVDF:MWCNT Nanocomposite Film for Uncooled Infrared Detectors and Medical Applications. *Integrated Ferroelectrics* 158: 98–107. https://doi.org/10.1080/10584587.2014.957559.
14. Vaseashta, Ashok, Gor Gevorgyan, Doga Kavaz, Ognyan Ivanov, Mohammad Jawaid, and Dejan Vasović. 2021. Exposome, Biomonitoring, Assessment and Data Analytics to Quantify Universal Water Quality. In *Water Safety, Security and Sustainability*, ed. Ashok Vaseashta and Carmen Maftei, 67–114. Advanced Sciences and Technologies for Security Applications. Cham: Springer International Publishing. https://doi.org/10.1007/978-3-030-76008-3_4.
15. Batra, A. K., James Sampson, M. D. Aggarwal, Ashok Vaseashta, and Kunal Grover. 2021. Wearable Smart Sensor Ring to Monitor the Severity of Hand Tremor 253: 44–46.
16. Vaseashta, Ashok, Didem Demir, Burcu Sakım, Müge Aşık, and Nimet Bölgen. 2022. Hierarchical Integration of 3D Printing and Electrospinning of Nanofibers for Rapid Prototyping. In *Electrospun Nanofibers*, ed. Ashok Vaseashta and Nimet Bölgen, 631–655. Cham: Springer International Publishing. https://doi.org/10.1007/978-3-030-99958-2_22.
17. Vaseashta, Ashok, and Nimet Bölgen, ed. 2022. *Electrospun Nanofibers: Principles, Technology and Novel Applications*. Cham: Springer International Publishing. https://doi.org/10.1007/978-3-030-99958-2.

Dielectric Behavior and AC Electrical Conductivity of Composite Materials Based on Conducting Polymers

S. Barnoss, N. Aribou, A. J. Paleo, Mohammed Essaid Achour, and L. C. Costa

Abstract This work presents the electrical and dielectric properties of poly (methyl methacrylate) (PMMA) composites filled with conductive polypyrrole (PPy) particles in the frequency range from 600 Hz to 1 MHz, and in a temperature range from 290 to 380 K. Investigation of AC conductivity indicated a thermally activated conduction mechanism, explained by a correlated barrier hopping (CBH) model. Notable, the decrease in activation energy with increasing frequency observed was also discussed from the CBH model. The temperature dependence of the electrical and dielectric response has been analyzed for concentrations above the percolation threshold. The dielectric permittivity was analyzed using the Havriliak-Negami model taking into account the effect of electrode polarization. This analysis enabled us to describe quantitatively the experimental data, calculate the ohmic conductivity, and determine the parameters characterizing the contribution of ohmic conduction to the complex dielectric permittivity.

Keywords Conducting particles · Impedance spectroscopy · Havriliak-Negami model · Electrical conductivity · Dielectric polarization

1 Introduction

In recent years, there has been a surge of interest in the exploration of conducting polymers, igniting a fervent wave of research. These versatile composites have captured the attention of various industries due to their promising potential across a spectrum of

S. Barnoss (✉) · N. Aribou · M. E. Achour
Laboratory of Material Physics and Subatomic, Faculty of Sciences, Ibn-Tofail University, Kenitra, Morocco
e-mail: barnoss1702@gmail.com

A. J. Paleo
2C2T-Centre for Textile Science and Technology, University of Minho, Guimarães, Portugal

L. C. Costa
I3N and Physics Department, University of Aveiro, Aveiro, Portugal

A. Vaseashta et al. (eds.), *Proceedings of the Eighth International Symposium on Dielectric Materials and Applications (ISyDMA'8)*, Springer Proceedings in Physics 349, https://doi.org/10.1007/978-3-031-84809-4_19

applications [1–3]. Research endeavors in this domain are geared towards introducing novel materials primed for diverse technological applications, spanning from electronic devices to light-emitting diodes, electromagnetic shielding, antistatic coatings, and rechargeable batteries [4, 5]. The fusion of insulating polymers with conducting counterparts yields a polymeric hybrid endowed with unparalleled mechanical and electrical characteristics. Among these, polypyrrole (PPy) stands out as an essential representative of conducting polymers, boasting remarkable electrical conductivity, environmental resilience, and facile synthesis protocols. PPy demonstrates processability through doping with either inorganic or organic acids, a feature that enhances its utility across various applications. However, it's imperative to note that the electronic structure of doped conducting polymers is significantly influenced by an inherent disorder stemming from non-uniform doping. The charge carriers engendered during the doping process by conjugated polymeric chains typically manifest as polarons or bi-polarons, contributing to conductivity through mechanisms such as phonon-assisted hopping or tunneling [6].

Dielectric relaxation phenomena have been employed to investigate the transport mechanisms of various composite materials based on the PMMA matrix infused with conducting charges. For example, Logakis et al. [7] investigated the dielectric relaxation observed in PMMA/CNT percolating systems, aligning their findings with differential scanning calorimetry and dynamic-mechanical analysis. They uncovered a subtle interaction between the macromolecular chains and CNT particles, noting that the absence of crystallinity facilitates the attainment of high conductivity levels in these nanocomposites. Meanwhile, Maji et al. [8] explored the impact of temperature and frequency on the dipolar motion within PMMA/$Zn(NO_3)$ composites. Their findings suggest its suitability for various technical applications such as supercapacitors, high-speed computers, and gate dielectric materials for organic FETs. Similarly, Aribou et al. [9] investigated the electrical and dielectric properties of PMMA/PPy composites, revealing abnormal low-frequency dispersion in the response of these materials beyond the percolation threshold. This behavior is attributed to the hopping of charge carriers between localized states.

In this study, we investigated the dielectric response and AC conduction of PPy filled PMMA matrix at two concentrations above the percolation threshold. The AC conductivity of two samples was analyzed over a frequency range from 100 Hz to 1 MHz using the Jonscher power law, which allows for the specification of the appropriate electrical conduction mechanism using the exponent s in that law. Additionally, we applied the Havriliak-Negami model to interpret the parameters associated with the contribution of ohmic conduction to the complex dielectric permittivity.

2 Experimental

2.1 Preparation

The polypyrrole (PPy) powder was synthesized through the doping of intrinsic PPy with the tosylate (TS) anion. The degree of doping was determined using the XPS technique and was found to be approximately one tosylate ion TS^- for every four pyrrole monomers [10]. The average grain size of PPy falls within the range of 10–15μm. Samples were prepared by mixing the desired volume fraction, ϕ, of PPy with PMMA powder in various proportions. These mixtures were then pressed at 5,000 kg cm^{-2} and 423 K. Upon pressing, the samples were allowed to cool naturally to room temperature, resulting in solid disc-shaped samples suitable for measurements. All discs had a diameter of 13 mm and a thickness of 3–4 mm. The DC conductivities of PPy and the PMMA matrix are 54 and 3.10^{-5} S/m respectively [11], with densities of 1.20 and 1.14–1.20 g cm^{-3} respectively. The percolation threshold ϕ_c for this series of samples is approximately 3.85 wt%. Additionally, the glass transition temperature of PMMA polymer is $T_g \sim 388$ K [12].

2.2 Measurements

The complex permittivity function, denoted as $\varepsilon^*(\omega) = \varepsilon'(\omega) - j\varepsilon''(\omega)$, was determined using a Hewlett Packard Network Analyzer (Model 4192A). The relative permittivity, ε', and the loss factor, ε'', of the sample were computed from the complex admittance $Y^*(\omega) = G(\omega) + jB(\omega) = jC_0.\omega.\varepsilon^*$ of the equivalent circuit. This led to $\varepsilon' = (d/\varepsilon_0.A.\omega)B(\omega)$ and $\varepsilon'' = (d/\varepsilon_0.A.\omega)G(\omega)$, where A is the cross-sectional area of the sample, d is its thickness and ε_0 is the free space permittivity. The terms $B(\omega)$ and $G(\omega)$ are, respectively, the conductance and the susceptance of the samples. The measurements were conducted within the frequency range of 600 Hz to 1 MHz under isothermal conditions, spanning temperatures from 290 to 380 K. Then, we calculated the AC electrical conductivity $\sigma_{AC}(\omega)$ can be calculated using the following expression:

$$\sigma_{AC}(\omega) = (d/A\varepsilon_0)G(\omega) \quad (1)$$

3 Results and Discussion

3.1 AC Electrical Conductivity

The study of AC conductivity provides detailed insights into the time-dependent movements of charge, which lead to conductivity and dielectric relaxation phenomena. This analysis offers a well of insight into the distribution of electric field in the system and the field-induced perturbations. The frequency dependence of the AC conductivity at various temperatures for the PMMA/PPy composite is shown in Fig. 1. It is clear that the plots of the AC conductivity at various temperatures exhibit a plateau (frequency-independent) in the low-frequency region, named D1, and dispersion at higher frequencies, labeled as D2. In D1, the conductivity strongly depends on the temperature, while in D2, the observed increase in conductivity is dominated by frequency. This behavior is common not only in glassy and crystalline ionic solids but also in semiconducting solids [13]. The phenomenon of conductivity dispersion is generally analyzed using the universal Jonscher power law given as [14] $\sigma = \sigma_{DC} + A\omega^s$ was used in several studies, as seen, the crossover frequency ($\omega_c = 2\pi F_c$) at which the conductivity shows significant changes does not appear in this model. For this aim, an extended approximation model [15, 16] can be used to describe the electrical conductivity given as:

$$\sigma(\omega, T) = \sigma_{DC}(T)\left[1 + k\left(\frac{\omega}{\omega_c}\right)^{s(T)}\right] \quad (2)$$

where s(T) is a parameter, describing the degree of interaction between charge carriers and their environment, k is a constant and ω_c is the crossover frequency ($\omega_c = 2\pi F_c$). The solid lines in Fig. 1 show the curves found by modeling the Jonscher power law for the samples 6wt% and 8wt% of PMMA/PPy, obtained from the fit of Eq. (2). The

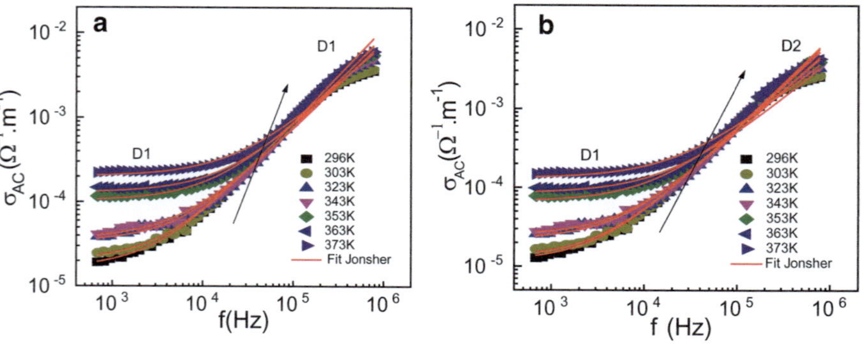

Fig. 1 The fitting and measurement curves of the AC electrical conductivity, according to the Jonscher power law for samples 6 wt% (**a**) and 8 wt% (**b**) of PMMA/PPy

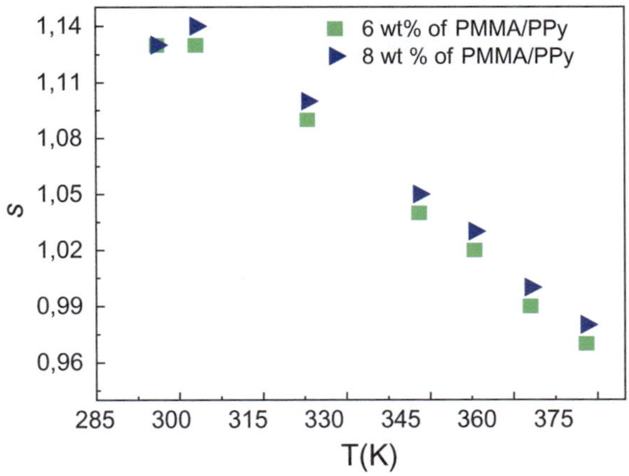

Fig. 2 Exponent s as a function of temperature, for the studied samples

value of the exponent s is calculated by fitting the curves of the conductivity plots (Fig. 1) at various temperatures.

Figure 2 shows the plot of the exponent *s* as a function of temperature which demonstrates that it has a decreasing trend with increasing temperature. This type of behavior has been previously explained by the Correlated Barrier Hopping (CBH) model also referred to as the Hopping Over the Barrier (HOB) mode [17–19]. According to this model, the thermal energy or thermal agitation increases with temperature, increasing contribution from the deep traps followed by the possibility of band-to-band transition. As a result of such transitions, the conduction process dominates, and the value of exponent s tends to zero since electronic conduction is independent of frequency. It is, therefore, evident from Fig. 2 that the AC conductivity depends significantly on the frequency in the low-temperature regime and as the temperature increases, the DC conductivity takes over [20, 21].

Figure 3 shows the Arrhenius plots of the AC conductivity at selected frequencies, i.e., 1, 21, 312, and 763 kHz. For all the frequencies, an Arrhenius behavior was observed in the AC conductivity which suggested a thermally activated hopping of charge carriers among different localized states. Activation energies were calculated (Table 1) at selected frequencies from the slope of the linear fit of the data in Fig. 3. The activation energy corresponding to 1 kHz was much less than the band gap which indicated an extrinsic type conduction in PPy particles at low frequencies. It can be seen from Table 1 that the activation energy decreases with increasing frequency which suggests that the dominant conduction mechanism is the hopping of charge carriers as per the CBH model [22, 23].

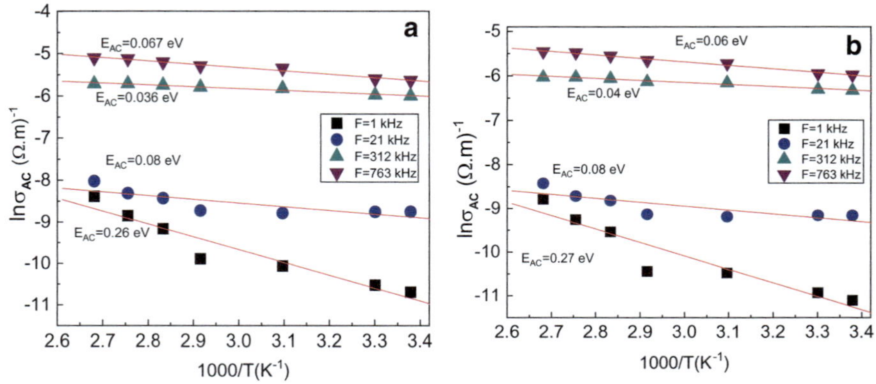

Fig. 3 Arrhenius plots of the AC conductivity at 1, 21, 312, and 763 kHz for samples 6 wt% (**a**) and 8 wt% (**b**) of PMMA/PPy

Table 1 Activation energies at selected frequencies (1, 21, 312, and 763 kHz)

Samples	E_{AC} (eV)			
	F = 1 kHz	F = 21 kHz	F = 312 kHz	F = 763 kHz
6 wt%	0.26 ± 0.03	0.07 ± 0.02	0.03 ± 0.003	0.06 ± 0.004
8 wt%	0.27 ± 0.04	0.08 ± 0.02	0.04 ± 0.003	0.06 ± 0.003

3.2 *Dielectric Relaxations Analysis*

The real and imaginary parts of the complex permittivity for the concentrations 6 wt% and 8wt% of PMMA/PPy as a function of frequency at various temperatures are shown in Fig. 4. The analysis of these figures yields the following results: (i) the dielectric constant ε', of these samples exhibits a frequency dependence. It can be observed that the ε' is higher at lower frequencies and increases with an increase in temperature. At low frequencies, the high value of the dielectric constant is due to the accumulation of free charges at the interface of the PPy and PMMA, as charges do not acquire enough energy to overcome the barrier hopping. However, at high frequencies, the permittivity has contributions from dipolar polarization due to the presence of permanent dipoles or dipolar effects produced by impurities [24]. The low-frequency area is said as extrinsic response of the material [25]. (ii) The frequency-dependent imaginary part, ε'', at different temperatures, shows an increase with the incorporation of PPy particles in the PMMA matrix. This increase is due to the electrically conductive nature of PPy. Additionally, the ε'' also exhibits significant dependence on the measuring temperatures. The increase in ε'' with an increase in temperature is brought on by the temperature-dependent increase in the DC contribution.

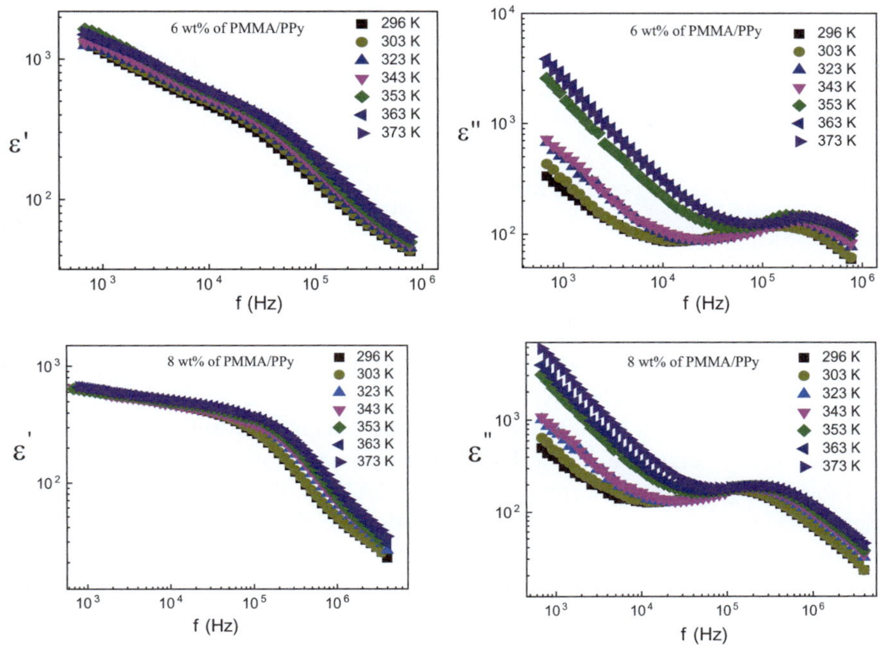

Fig. 4 Real and imaginary parts of the complex permittivity as a function of frequency, at different temperatures, for samples 6 wt% and 8 wt% of PMMA/PPy

3.3 *Havriliak-Negami Modeling*

The frequency dependence of the real and imaginary parts of the dielectric permittivity reveals that the relaxation processes in these composites are rather complex. To determine the parameters of this process, we analyzed the results using the Havriliak–Negami model, taking into consideration the ohmic conduction losses. In this model, the expression of the complex dielectric permittivity is given by Eq. (4) [26–28]:

$$\varepsilon^*(\omega) = \varepsilon_\infty + \frac{\varepsilon_s - \varepsilon_\infty}{(1 + (j\omega\tau)^\alpha)^\beta} + \frac{1}{j}\left(\frac{\sigma_{DC}}{\omega\varepsilon_0}\right)^n \tag{3}$$

where ε_s and ε_∞ are the dielectric constants at zero and infinite frequency respectively, τ is the relaxation time, and α and β are empirical exponents. σ_{DC} is the ohmic conductivity and n is a parameter characterizing the contribution of the ohmic conduction in the complex dielectric permittivity. The real, ε', and imaginary, ε'', parts of the complex permittivity in this model are defined as follows (Eqs. (5) and (6) [29]:

$$\varepsilon'(\omega) = \varepsilon_\infty + \frac{\Delta\varepsilon \cos\left(\beta \tan^{-1}\left[\frac{(\omega\tau)^\alpha \sin\left({\alpha\pi}/{2}\right)}{1+(\omega\tau)^\alpha \cos\left({\alpha\pi}/{2}\right)}\right]\right)}{\left[1 + (\omega\tau)^{2\alpha} + 2(\omega\tau)^\alpha \cos\left({\alpha\pi}/{2}\right)\right]^{\beta/2}}) \tag{4}$$

$$\varepsilon''(\omega) = \frac{\Delta\varepsilon \sin\left(\beta \tan^{-1}\left[\frac{(\omega\tau)^\alpha \sin\left({\alpha\pi}/{2}\right)}{1+(\omega\tau)^\alpha \cos\left({\alpha\pi}/{2}\right)}\right]\right)}{\left[1 + (\omega\tau)^{2\alpha} + 2(\omega\tau)^\alpha \cos\left({\alpha\pi}/{2}\right)\right]^{\beta/2}} + \left(\frac{\sigma_{DC}}{\omega\varepsilon_0}\right)^n \tag{5}$$

The initial values of τ, ε_∞, and $\Delta\varepsilon = \varepsilon_s - \varepsilon_\infty$ were estimated, respectively, from $f_{max} = 1/2\pi\tau$, ε' (1 MHz), and ε' (10 kHz). The initial values for the parameters α and β were chosen to be equal to 1.

Figure 5 shows that good agreement can be obtained between the Eqs. (5, 6) and the experimental data. The values of the obtained fitting parameters are given in Table 2, for the two concentrations of PMMA/PPy. This table reveals that σ_{DC} increases significantly with the PPy concentration. On the other hand, the parameter n varies between 0.6 and 1. The parameter β is close to 1, suggesting that the Cole–Cole model could potentially describe the variations of the dielectric permittivity with frequency, while α is less than 1. The relaxation time τ decreases with increasing temperature.

Up to now, we have not examined the behavior of ε', at low frequencies. A detailed analysis reveals that ε' is the sum of a term obtained from Eq. (4) and a second term (ε'_{sec}) having a linear behavior in the logarithmic representation, which accounts for the electrode polarization. In this case, the new equation for ε' becomes:

$$\varepsilon'_{sec}(\omega) = \varepsilon_\infty + \frac{\Delta\varepsilon \cos\left(\beta \tan^{-1}\left[\frac{(\omega\tau)^\alpha \sin\left({\alpha\pi}/{2}\right)}{1+(\omega\tau)^\alpha \cos\left({\alpha\pi}/{2}\right)}\right]\right)}{\left[1 + (\omega\tau)^{2\alpha} + 2(\omega\tau)^\alpha \cos\left({\alpha\pi}/{2}\right)\right]^{\beta/2}}) + \frac{A}{\omega^m} \tag{6}$$

where A is a parameter that quantifies the conductivity due to the space charges (localized charges), and m is a dimensionless exponent ($0 < m < 1$). For an ideal complex conductivity, $m = 1$. If $m < 1$, it implies that the polarization has a distribution of the carrier polarization mechanism [30].

As shown in Fig. 6, Eq. (7) can adequately describe the variations of ε' with frequency, in the explored range. The values of parameters A and m, obtained from the nonlinear fits, are summarized in Table 3. These parameters increase with temperature, suggesting a stronger and more frequency-dependent electrode polarization effect at higher temperatures. This can be attributed to enhanced charge carrier mobility, changes in the dielectric properties of the material, and thermal activation of charge carriers in the composite based on conducting polymers. This indicates that the term A/ω^m could be attributed to the effect of electrode polarization [31–33].

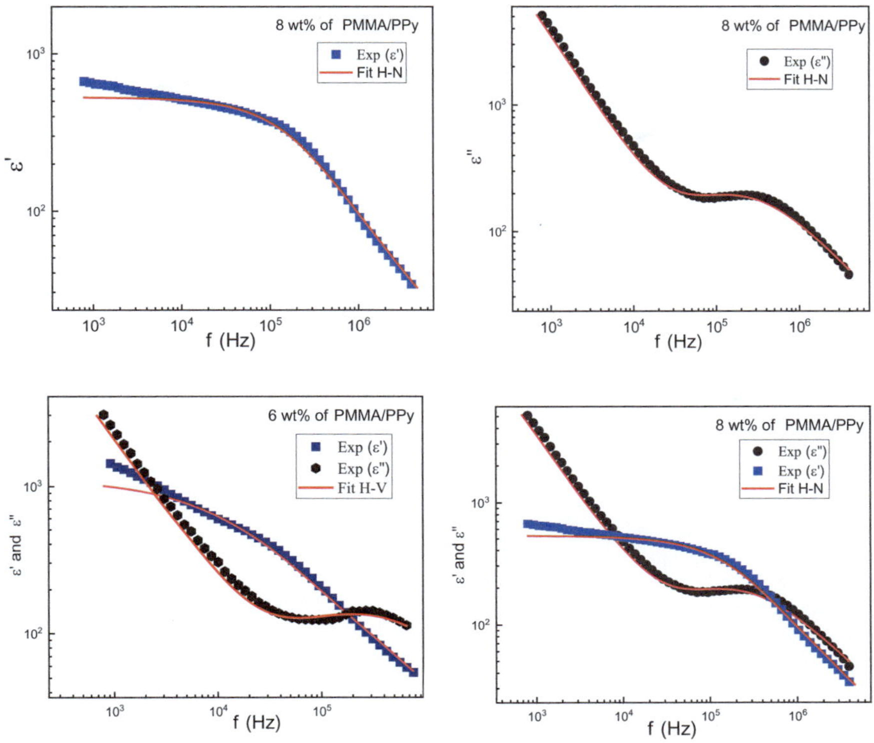

Fig. 5 Real and imaginary parts of the complex permittivity as a function of frequency for T = 373 K. Symbols are the experimental points and the solid lines are the fits for Eqs. (5, 6)

Table 2 Fitting parameters, obtained from the fits of the real and the imaginary parts of complex permittivity to the Havriliak–Negami model, for the two concentrations of PMMA/PPy

Samples	T (K)	α	β	τ (s)	σDC (Ω.m) -1	n
6 wt%	296	0.80	0.94	6.47 × 10–6	1.65 × 10–5	0.68
	343	0.82	0.98	4.05 × 10–6	2.46 × 10–5	0.80
	373	0.72	1.00	3.46 × 10–6	5.10 × 10–5	0.98
8 wt%	296	0.81	0.94	5.69 × 10–6	3.45 × 10–5	0.63
	343	0.79	1.00	3.68 × 10–6	3.79 × 10–5	0.83
	373	0.70	0.99	3.46 × 10–6	4.95 × 10–5	1.00

4 Conclusion

The frequency dependence of the AC conductivity for PMMA/PPy at different temperatures was analyzed using Jonscher's power law characteristic of disordered materials. The observed decrease in the frequency exponent, *s*, with increasing

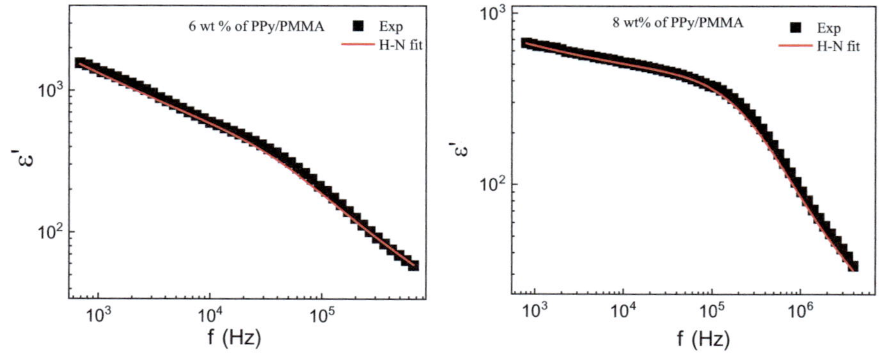

Fig. 6 Real part of the complex permittivity as a function of frequency, for T = 373 K. the solid lines are the fits for Eq. (7)

Table 3 Values of fitting parameters A and m derived from the fits of the real part of complex permittivity (ε′) for the two concentrations of PMMA/PPy

Samples	T (K)	A	M
6 wt%	296	2891	0.47
	343	3080	0.56
	373	50,741	0.69
8 wt%	296	26,702	0.43
	343	59,016	0.69
	373	77,445	0.86

temperature, suggests that the correlated barrier hopping (CBH) is the dominant transport mechanism in PPy. Furthermore, the dielectric permittivity fits well with the Havriliak-Negami model, which accounted for the effect of electrode polarization. From this model, this analysis enabled us to describe quantitatively the experimental data from which we extracted the ohmic conductivity and the parameters characterizing the contribution of the ohmic conduction in the complex dielectric permittivity. Additionally, conductivity increased with temperature and the effect of electrode polarization could explain the increase in the real part of dielectric permittivity, at low frequencies.

References

1. Bhadra, S., Khastgir, D., Singha, N. K. & Lee, J. H. Progress in preparation, processing and applications of polyaniline. *Prog. Polym. Sci.* **34**, 783–810 (2009).
2. Halik, M. *et al.* Relationship Between Molecular Structure and Electrical Performance of Oligothiophene Organic Thin Film Transistors - Halik - 2003 - Advanced Materials - Wiley Online Library. *Adv. Mater.* **15**, 917–922 (2003).

3. Soto-Oviedo, M. A., Araújo, O. A., Faez, R., Rezende, M. C. & De Paoli, M.-A. Antistatic coating and electromagnetic shielding properties of a hybrid material based on polyaniline/organoclay nanocomposite and EPDM rubber. *Synth. Met.* **156**, 1249–1255 (2006).
4. Yang, X., Zhao, T., Yu, Y. & Wei, Y. Synthesis of conductive polyaniline/epoxy resin composites: doping of the interpenetrating network. *Synth. Met.* **1–3**, 57–61 (2004).
5. De, S., De, A., Das, A. & De, S. K. Transport and dielectric properties of -zirconium phosphate–polyaniline composite. *Mater. Chem. Phys.* **91**, 477–483 (2005).
6. *Handbook of Conducting Polymers*. (M. Dekker, New York, 1998).
7. Logakis, E., Pandis, Ch., Pissis, P., Pionteck, J. & Pötschke, P. Highly conducting poly(methyl methacrylate)/carbon nanotubes composites: Investigation on their thermal, dynamic-mechanical, electrical and dielectric properties. *Compos. Sci. Technol.* **71**, 854–862 (2011).
8. Maji, P., Pande, P. P. & Choudhary, R. B. Effect of $Zn(NO_3)_2$ filler on the dielectric permittivity and electrical modulus of PMMA. *Bull. Mater. Sci.* **38**, 417–424 (2015).
9. Aribou, N. *et al.* Thermal and Spectral Dielectric Properties of Polypyrrole/Polymethylmethacrylate Composites. *Spectrosc. Lett.* **45**, 477–481 (2012).
10. Bouknaitir, I. *et al.* Electrical properties of conducting polymer composites: Experimental and modeling approaches. *Spectrosc. Lett.* **50**, 196–199 (2017).
11. Aribou, N., Nioua, Y., Hasnaoui, M. E., Achour, M. E. & Costa, L. C. Effect of filler/matrix interphase boundaries on the DC electrical conductivity of two- and three-dimensional composites. *J. Optoelectron. Adv. Mater.* **22**, 629–634 (2020).
12. Aribou, N., Barnoss, S., El Hasnaoui, M., Achour, M. E. & Costa, L. C. Structural, Electrical and Thermal Properties of Composites Based on Conducting Polymer. *Jordan J. Phys.* **13**, 157–163 (2020).
13. Tiwari, J. P. & Shahi, K. Super-linear frequency dependence of ac conductivity of disordered $Ag_2S–Sb_2S_3$ at cryogenic temperatures. *Philos. Mag.* **87**, 4475–4500 (2007).
14. Jonscher, A. K. & Frost, M. S. Weakly frequency-dependent electrical conductivity in a chalcogenide glass. *Thin Solid Films* **37**, 267–273 (1976).
15. Summerfield, S. Universal low-frequency behavior in the a.c. hopping conductivity of disordered systems. *Philos. Mag. B* **52**, 9–22 (1985).
16. Barnoss, S., Fengouchi, I. E., Graca, M. P. F., Hasnaoui, M. E. & Costa, L. C. Carbon Nanotubes Modified Oil-based Nanofluids: Electrical Conduction Mechanism Analysis. *Moroc. J. Chem.* **12**, 493–508 (2024).
17. Patel, H. K. & Martin, S. W. Fast ionic conduction in Na2S+B2S3 glasses: Compositional contributions to non-exponentiality in conductivity relaxations. *Solid State Ion.* **53–56**, 1148–1156 (1992).
18. Pike, G. E. ac Conductivity of Scandium Oxide and a New Hopping Model for Conductivity. *Phys. Rev. B* **6**, 1572–1580 (1972).
19. Okutan, M., Basaran, E., Bakan, H. I. & Yakuphanoglu, F. AC conductivity and dielectric properties of Co-doped TiO{sub 2}. *Phys. B Condens. Matter* **364**, (2005).
20. Chithambaram, V., Das, S. J. & Krishnan, S. Synthesis, optical and dielectric studies on novel semi organic nonlinear optical crystal by solution growth technique. *J. Alloys Compd.* **509**, 4543–4546 (2011).
21. Bharadwaja, S. S. N. & Krupanidhi, S. B. Alternating current electrical properties of antiferroelectric lead zirconate thin films by pulsed excimer laser ablation. *J. Appl. Phys.* **88**, 2072–2080 (2000).
22. Hayat, K., Rafiq, M. A., Durrani, S. K. & Hasan, M. M. Impedance spectroscopy and investigation of conduction mechanism in BaMnO3 nanorods. *Phys. B Condens. Matter* **406**, 309–314 (2011).
23. Sahay, P. P., Tewari, S., Nath, R. K., Jha, S. & Shamsuddin, M. Studies on ac response of zinc oxide pellets. *J. Mater. Sci.* **43**, 4534–4540 (2008).
24. Sakellis, I., Papathanassiou, A. N. & Grammatikakis, J. Effect of composition on the dielectric relaxation of zeolite-conducting polyaniline blends. *J. Appl. Phys.* **105**, 064109 (2009).

25. Lunkenheimer, P. *et al.* Origin of apparent colossal dielectric constants. *Phys. Rev. B* **66**, 052105 (2002).
26. Havriliak, S. & Negami, S. A complex plane analysis of α-dispersions in some polymer systems. *J. Polym. Sci. Part C Polym. Symp.* **14**, 99–117 (1966).
27. Havriliak, S. & Negami, S. A complex plane representation of dielectric and mechanical relaxation processes in some polymers. *Polymer* **8**, 161–210 (1967).
28. Signorelli, R., Ku, D. C., Kassakian, J. G. & Schindall, J. E. Electrochemical Double-Layer Capacitors Using Carbon Nanotube Electrode Structures. *Proc. IEEE* **97**, 1837–1847 (2009).
29. Samir, Z. *et al.* Temperature Effect on the Dielectric Response of Carbon Nanotubes Particles Filled Polyester Polymer Composites. in *Advanced Nanotechnologies for Detection and Defence against CBRN Agents* (eds. Petkov, P., Tsiulyanu, D., Popov, C. & Kulisch, W.) 261–272 (Springer Netherlands, Dordrecht, 2018). https://doi.org/10.1007/978-94-024-1298-7_25.
30. T. Prasit, T. Suwat, Y. Teerapon, M. Santi, Journal of Physics: Condensed Matter 20 (2008) 395227.
31. Garton, C. G. Dielectric loss in thin films of insulating liquids. *J. Inst. Electr. Eng. - Part III Commun. Eng. Proc. Wirel. Sect. Inst.* **88**, 23–40 (1941).
32. Bartnikas, R. Performance characteristics of dielectrics in the presence of space charge. *IEEE Trans. Dielectr. Electr. Insul.* **4**, 544–557 (1997).
33. Miyairi, K. Low-frequency dielectric response of polyethylene terephthalate (PET) films. *J. Phys. Appl. Phys.* **19**, 1973–1980 (1986).

Modeling Variation of Time, Temperature, and Composition on Thermal Treatment of Al-Si Alloys

Sara El Hamzi, Omar Ben Lenda, Nadia Dkhireche, Amina Ibnlfassi, and Elmadani Saad

Abstract The extraordinary expansion of the aluminum industry is due to the numerous beneficial typical design features of this valuable metal; hence, the unceasing demands for products made of aluminum or the integration of aluminum into their composition. To master and improve the quality and properties of the final products, the major industrial challenge lies in the possibility of controlling the morphology, and size of microstructures that reside within the molded pieces, as well as their defects; which is the fundamental reason for our interest in mastering the growth and germination of aluminum alloys, as well as the developing structures, during the solidification process. The modeling reveals a valuable aid in optimizing the formation of such heterogeneousness segregation cells, normally incompatible with industrial requirements. This work consists of studying the influence of homogenization, while systematically varying the three parameters: time, temperature, and composition of aluminum–silicon alloys. Usually, such alloys' microstructure and mechanical behavior are directly influenced by some parameters such as composition, cooling velocity, and homogenization process. One can then imagine the stake in understanding and mastering the morphology and evolution of the solidification interface.

Keywords Solidification · Homogenization · Aluminum silicone alloys · Modeling

S. El Hamzi (✉) · O. B. Lenda · E. Saad
Physical Chemistry of Processes and Materials, Hassan 1St University, Settat, Morocco
e-mail: s.elhamzi@uae.ac.ma

S. El Hamzi
High School of Tecnology, Abdelmalek Essaadi University, Tetouan, Morocco

O. B. Lenda
The National Higher School of Mining of Rabat (ENSMR), Rabat, Morocco

N. Dkhireche
Advanced Materials and Process Engineering, Ibn Tofail University, Kenitra, Morocco

A. Ibnlfassi
Environmental Sciences and Development, Hassan 1St University, Settat, Morocco

A. Vaseashta et al. (eds.), *Proceedings of the Eighth International Symposium on Dielectric Materials and Applications (ISyDMA'8)*, Springer Proceedings in Physics 349, https://doi.org/10.1007/978-3-031-84809-4_20

1 Introduction

The extraordinary growth that the aluminum industry has achieved is connected to all the positive characteristics of this metal, hence the demand for products made of aluminum or incorporating aluminum in their composition continues to grow year after year. It is used today for a wide range of applications; it is the material of mobility! It is used in all current means of transport (road vehicles, aeronautics, maritime transport, rail transport), in construction and public works; it is also used in packaging, electricity, and electronics, moreover, during its use phase, aluminum has many advantages in terms of sustainable development and is a permanent material! It is 100% recyclable, infinitely and without loss of its physical and chemical qualities.

Almost like those of all pure metals, the characteristics of pure aluminum are low, in particular its mechanical properties. Commercially pure aluminum has low mechanical properties in the annealed state; however, these properties can be considerably improved by work hardening, the addition of alloying elements, or heat treatment, as appropriate [1]. The very wide range of alloys allows you to find the one that corresponds to the constraints of use envisaged. The most notable effect of silicon in aluminum alloys is the improvement of forming (casting) characteristics. The addition of silicon to pure aluminum greatly increases the fluidity, resistance to hot cracking, and flow characteristics of the liquid metal. The composition most used in foundry aluminum alloys is that of the Al-Si family [2].

The objective of this work is to model the influence of heat treatment on the mechanical properties of Al-Si alloys. This essentially consists of first studying the influence of homogenization, while systematically varying the three parameters: time, temperature, and composition, and then studying the influence of the addition of silicon and the cooling speed on the different mechanical properties of the alloys in question. We then imagine how much is at stake to understand and master; the morphology and evolution of the solidification interface.

2 Material and Methods

Solidification modeling work essentially focuses on calculating the evolution of the thermal field in cast parts. This temperature field is obtained by solving the heat diffusion equation.

The entire domain consists of solid plus liquid, and the problem of heat diffusion with phase transformation can be expressed by the following formula [3]:

$$\mathrm{div}\left(\mathrm{K}\overrightarrow{\mathrm{grad}}\mathrm{T}\right) + \dot{\mathrm{Q}} = \rho \mathrm{c_p}.\frac{\partial \mathrm{T}}{\partial \mathrm{t}} \tag{1}$$

With:

K is the thermic conductivity;

ρ is the voluminal density;

c_p is the specific heat;

Q is the source term bound to the liberation of melting latent heat.

$$\dot{Q} = \rho L.\frac{\partial f_s}{\partial t} \tag{2}$$

With:

L is the melting latent heat;

f_s is the solid fraction.

The speed of a solution diffusion increases with temperature (network vibration) and expresses itself as follows [4]:

$$D_i = D_i^0.\exp\left(-\frac{Q_i}{RT}\right) \tag{3}$$

With:

D_i^0 is the pre-exponential term;

Q_i is the activation energy;

R is the perfect gas constant;

T is the temperature expressed in Kelvin.

Before any micro-structural investigation, it is necessary to perfectly study the balance diagram between Al-Si phases to know the phases susceptible to being encountered. Parts cast from aluminum alloys are mainly made of aluminum–silicon alloys [5]. Silicon increases the castability of aluminum and allows the metal to completely fill any desired cavities. Additionally, silicon prevents hot cracking and slightly increases the strength of the metal by forming hard silicon particles or fibers during solidification [5]. As shown in Fig. 1, the Al-Si binary system has a eutectic point of 12.6% Si (atomic %). The maximum solubility of silicon in aluminum is 1.5% (atomic %) at the eutectic temperature; on the other hand, only 0.016% (atomic %) of aluminum is soluble in silicon at 1190 °C [5].

Aluminum–silicon alloys can be classified into three main categories in Fig. 2 based on the eutectic point: from 2 to 9% Si, the alloy is in the hypoeutectic zone; for the eutectic alloys, they can contain from 10 to 13% Si; and those with more than 14% Si fall into the category of hypereutectic alloys [5, 6]. Due to the solubility limits of the alloying elements, we will limit the study of the Al-Si equilibrium diagram to areas of chemical composition between 0 and 12% by weight for the element Si, and temperatures of homogenization of: 600 °C which remains lower than the melting

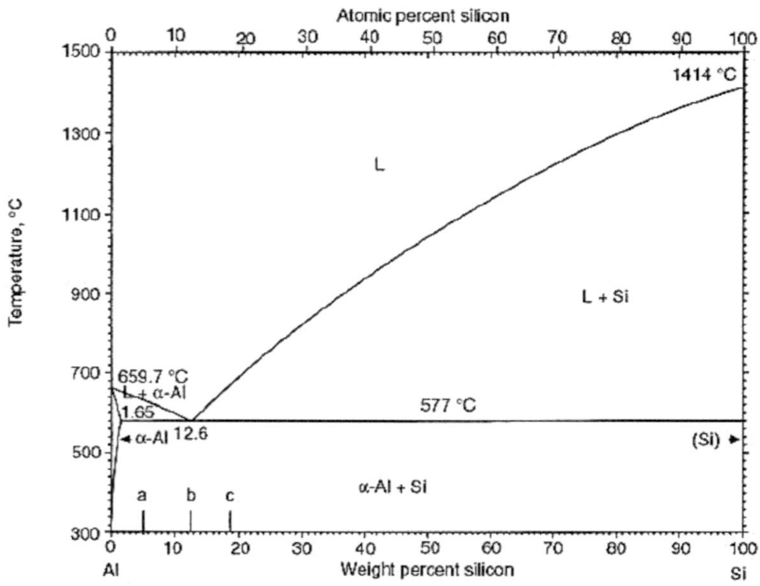

Fig. 1 Al-Si phase diagram [6]

temperature of aluminum, 400 and 200 °C which present temperatures below the eutectic level.

During our study, we assume that the geometric shape of our alloys is a cylindrical ingot of length 100 μm.

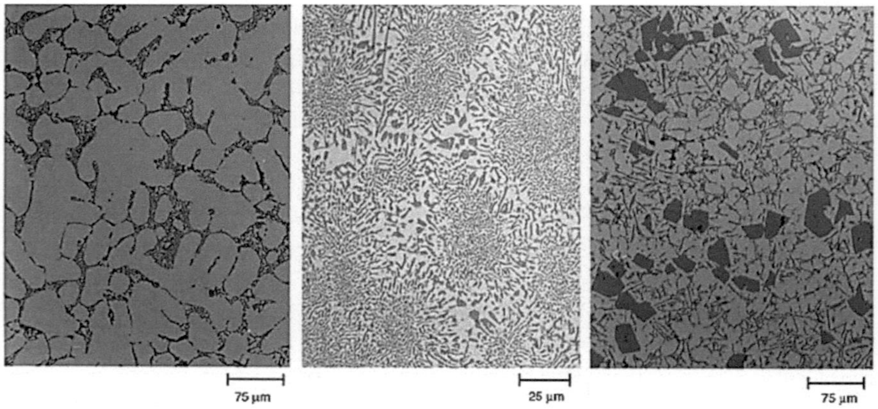

Fig. 2 Microstructure of Al-Si alloy; **a** hypoeutectic alloys, **b** eutectic alloys, **c** hypereutectic alloys [6]

3 Result and Interpretation

The vast majority of the thermomechanical properties of the alloy result from the type of microstructure obtained by the interaction of different physicochemical parameters during solidification. These parameters occur on the one hand from the physical conditions of solidification, and on the other hand, from the chemistry of the alloy (added alloying elements: Si in our case).

3.1 Influence of Homogenization

For the following study, we are interested in understanding the homogenization effect according to the three parameters: time, temperature, and composition, and seeing each one influence upon the silicon concentration according to the distance between the surface and the core of the ingot.

Influence of Time: Fig. 3 shows the evolution of the silicon concentration as a function of the distance between the surface and the core of the ingot over time, of the Al1%Si alloy at a temperature of 400 °C. We see that there is a mixture of the two metals with a tendency to form a homogeneous solid solution. Indeed, for prolonged maintenance over time; the Si concentration on the surface of the ingot increases slightly and goes from a minimum value of 0.13 (wt.%) at 3 h to 0.25 (wt.%) after 18 h, unlike the core of the ingot, the Si concentration decreases slightly between 3 and 18 h and goes from 0.45 to 0.28 (wt.%), however, we can also note that the Si concentration tends to be uniform at all points between the surface and the core of the ingot beyond 18 h. Qualitatively, it appears that the atoms of each element move over time from regions rich in this element towards poor regions which become enriched, more precisely under the influence of a concentration gradient which is the driving force of the diffusion phenomenon. It is therefore necessary to maintain sufficient homogenization time so that the diffusion of chemical species between the different phases can occur.

Influence of Temperature: For the three temperatures of maintenance: 200, 400, and 600 °C the silicon concentration evolutions according to the distance between the surface and the ingot core, of the Al1% Si alloy at the end of 3 h, are shown in Fig. 4. We observe that the Si concentration at 600 °C presents the highest concentration 0.99 (wt.%) and it remains constant, while at 400 °C, we see a progressive enrichment of Si on the surface 0.13 (wt.%) at the core of the ingot 0.45 (wt.%); these values tend towards equilibrium values.

For the temperature of 200 °C, the Si concentration does not seem to be affected because the speed of diffusion of the chemical species is considerably slower in the solid phase than in the liquid phase, the concentration remains zero and only begins to increase at beyond 41 microns: the variation is certainly small but also detectable.

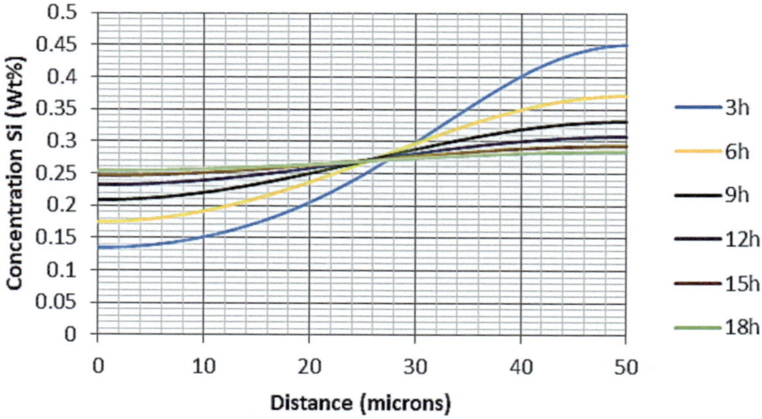

Fig. 3 Silicon concentration evolution according to the distance between the surface and the ingot core, shown as a function of time of the Al1%Si alloy (at T = 400 °C)

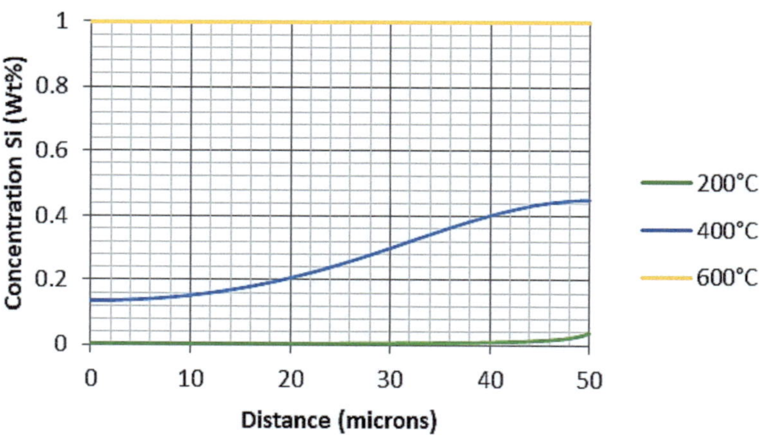

Fig. 4 Silicon concentration evolution according to the distance between the surface and the ingot core of the Al1%Si alloy at the temperatures of 200 °C, 400 °C et 600 °C

Furthermore, the comparison of the Si concentrations at all points of the ingot shows that the equilibrium concentration of Si is higher at 600 °C, which means that the diffusivity only becomes appreciable at high temperatures. We can therefore suggest that the diffusion of silicon in a solid solution is fast enough to reach the solubility limit from the first moments at high temperature.

Influence of Composition: To study influence, we have dealt with four different compositions: Al1%Si, Al2%Si, Al4%Si et Al12%Si. The silicon concentration evolution is represented in the following Fig. 5, which shows the evolution of the silicon concentration as a function of the distance between the surface and the core of the ingot, at the temperature 200 °C of the four alloys in question after 3 h. The curves of the Al1%Si, Al2%Si and Al4%Si alloys follow the same variation; we see a progressive enrichment of silicon on the surface at the heart of the ingot, while the Si concentration remains constant for the Al12%Si alloy containing more silicon.

The evolution of the silicon concentration as a function of the distance between the surface and the core of the ingot after 3 h, at the temperature of 400 °C of the Al1%Si, Al2%Si, Al4%Si and Al12%Si alloys is shown in Fig. 6.

There is a gradual increase in the Si concentration values for the Al1%Si, Al2%Si and Al4%Si alloys; in fact, on the surface of the ingot the alloys containing more

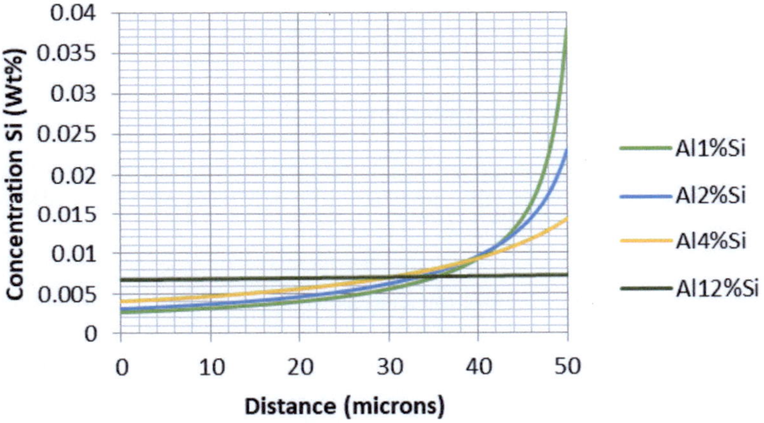

Fig. 5 Silicon concentration evolution according to the distance between the surface and ingot core at the temperature of 200 °C of the following alloys: Al1%Si, Al2%Si, Al4%Si and Al12%Si, and at the end of 3 h

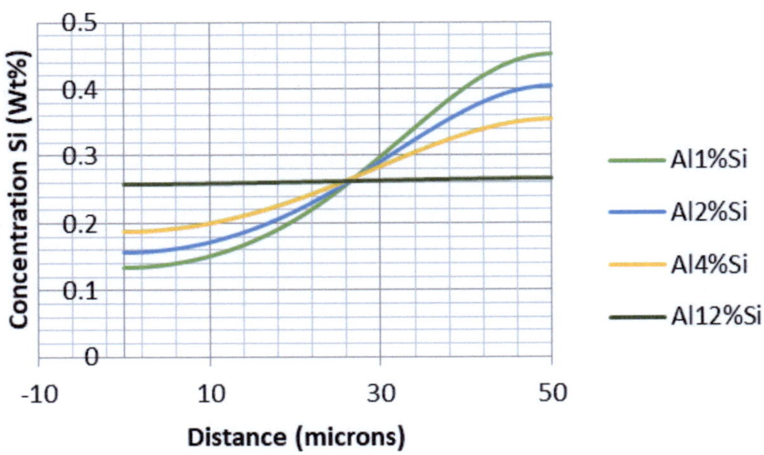

Fig. 6 Silicon concentration evolution according to the distance between the surface and ingot core at the temperature of 400 °C of the following alloys: Al1%Si, Al2%Si, Al4%Si and Al12%Si, and at the end of 3 h

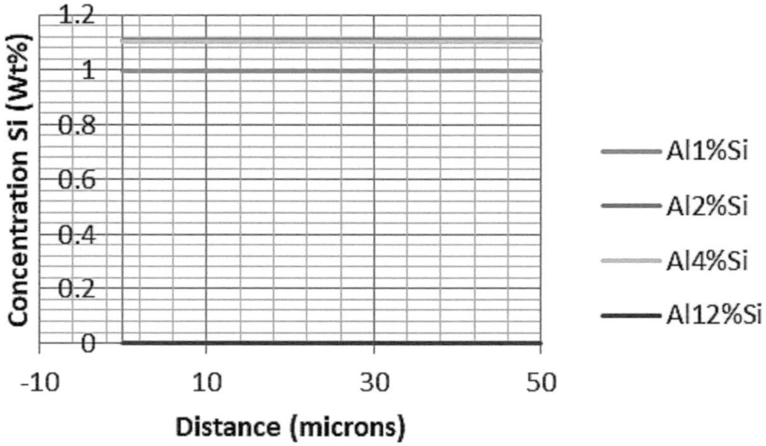

Fig. 7 Silicon concentration evolution according to the distance between the surface and ingot core at the temperature of 600 °C of the following alloys: Al1%Si, Al2%Si, Al4%Si and Al12%Si, and at the end of 3 h

silicon have the highest concentrations and as we approach the core, the concentrations of silicon in alloys containing less Si become the highest, however the Al12%S alloy containing more silicon keeps a constant value of the Si concentration at every point of the ingot.

Figure 7 describes the evolution after 3 h of silicon concentration as a function of the distance between the surface and the core of the ingot, at the temperature 600 °C of the alloys Al1%Si, Al2%Si, Al4%Si and Al12%Si. We can clearly observe that the concentration value remains constant for each alloy, in 0.99 (wt.%) for Al1%Si and in 1.10 (wt.%) for the Al2%Si and Al4%Si alloys; therefore, alloys containing more Si recognize the concentration stagnation at a higher value.

We conclude that with more Si, the diffusion of silicon in solid solution is sufficiently rapid at this temperature to reach the solubility limit after 3 h for the αAl + Si phase, with the exception of the Al12% alloy. If where we note the stagnation at 0 (wt.%), under these conditions the latter is no longer a solid solution but rather a liquid phase.

The influence of Si then results in the activation of the diffusion phenomenon, the driving force of the homogenization treatment which makes it possible to reduce the heterogeneity of composition due to the segregation phenomena which appear in the structure during solidification.

4 Conclusion

Based on this study, we conclude that it is necessary to keep the time of homogenization sufficiently large so that the diffusion of the chemical species between the various stages may occur. Under the influence of a gradient of concentration which is the driving force of the diffusion phenomenon, the atoms of each element move as a function of time under normal diffusion parameters. At high temperatures, the diffusion of silicon in a solid solution is rapid enough to reach the solubility limit from the first moments. The influence of silicon results in the activation of the diffusion phenomenon: the driving force of the homogenization treatment, which makes it possible to reduce the heterogeneity of composition due to the segregation phenomena that appear during the solidification. The addition of silicon affects the temperature as well as the transformation time. Si slows down the cooling process and it increases in the solidification time and therefore automatically on the microstructure of the alloy.

References

1. Baïlon, J.P., Dorlot, J.M., Des matériaux, 3e édition. Chapitre 11, Métaux et alliages non ferreux. ISBN-13: 978–2553007705 (2000).
2. Morin, S., Effect of magnesium, heat treatments, and porosity on the mechanical traction and fatigue properties of the alloy under pressure a380.1, the University of Quebec in Chicoutimi, (2002).
3. Adouane, S. Thermodynamic and morphological study of the growth of metals and alloys, Master's thesis in physics, University of BATNA, (2012).
4. Dehmas, M., Weisbecker, P., Geandier, G., Archambault, P., & Aeby-Gautier, E. (2005). Experimental study of phase transformations in 3003 aluminium alloys during heating by in situ high energy X-ray synchrotron radiation. Journal of Alloys and Compounds, 400, 116–124.
5. Moreau, C., Effects of alloy elements and quenching, during t4 heat treatments and artificial aging, on the microstructure and mechanical properties of type 413 aluminum-silicon alloys, the University of QUEBEC at CHICOUTIMI, (2004).
6. Aluminum-Silicon Casting Alloys: Atlas of Microfractographs, ASM International (2004).

Thermal Performance Dynamic Simulations of a Building Using Earth Bricks in Six Climatic Zones in Morocco

Mohammed Benfars, Youness Azalam, Abdelmounaim Alioui, Mourad Kaddiri, and Mustapha Mabrouki

Abstract As awareness of the environmental impacts of traditional building materials like cement and concrete—known for their significant greenhouse gas emissions and high energy requirements for heating and cooling—increases, there is growing interest in adopting eco-friendly and sustainable alternatives. This study evaluates the thermal performance of clay brick constructions across various climatic zones in Morocco, emphasizing their impact on indoor thermal comfort and energy efficiency. Using OpenStudio software, we analyzed annual indoor temperature variations and energy consumption for heating and cooling in six distinct cities: Rabat, Tangier, Fes, Ifrane, Benguerir, and Ouarzazate. Our results demonstrate that clay brick constructions significantly enhance thermal comfort. In Rabat, up to 46% of indoor temperatures remain between 20 and 26 °C, with a maximum reduction in indoor temperatures of 13.77 °C compared to outdoor temperatures. Similarly, Tangier shows 43% of indoor temperatures in the optimal range, with a maximum reduction of 12.69 °C. In regions with more extreme conditions, such as Fes and Ifrane, clay bricks achieve maximum reductions of 14.51 and 14.49 °C, respectively, while maintaining indoor fluctuations below 6 °C despite larger outdoor variations. Annual energy demands for heating and cooling vary across cities, with the highest energy loads observed in Ifrane (125.98 kWh/m^2) and the lowest in Rabat (69.47 kWh/m^2). These results indicate that clay brick constructions are particularly effective in moderate climates like Rabat and Tangier. For more extreme climates, such as Fes and Ouarzazate, additional measures such as reinforcing adobe bricks with natural fibers and integrating thermal insulation panels are recommended to further enhance energy efficiency and indoor comfort. This study highlights the effectiveness of clay brick constructions in improving indoor thermal stability and reducing energy consumption across

M. Benfars (✉) · Y. Azalam · A. Alioui · M. Kaddiri · M. Mabrouki
Laboratory of Industrial and Surface Engineering, Faculty of Science and Technologies, Sultan Moulay Slimane University, Beni Mellal, Morocco
e-mail: mohammed.fars@usms.ma

A. Alioui
Polydisciplinary Faculty, Sultan Moulay Slimane University, Beni Mellal, Morocco

A. Vaseashta et al. (eds.), *Proceedings of the Eighth International Symposium on Dielectric Materials and Applications (ISyDMA'8)*, Springer Proceedings in Physics 349, https://doi.org/10.1007/978-3-031-84809-4_21

different Moroccan climates. Future research should explore advanced construction techniques and materials to optimize thermal performance in varying climatic contexts.

Keywords Natural building material · Clay brick · Building energy efficiency · Dynamic thermal simulation

1 Introduction

Nowadays, the issue of climate change is crucial on a worldwide scale. It's important to note that over 40% of the energy produced worldwide is devoted to the building sector, contributing to over 30% of CO_2 emissions [1]. The widespread use of fired bricks in construction, despite their structural efficiency, poses significant environmental problems. Indeed, their manufacturing process requires a substantial amount of energy, ranging from 0.54–3.14 MJ/kg of material produced, emitting 70—282 g of CO_2 per kilogram [2]. Additionally, they generate a significant amount of waste during the construction and demolition phases, reaching nearly 14 million tons per year in Morocco [2]. Furthermore, with the projected global population rising from 7.7 billion in 2019 to 9.7 billion by 2050 [3], global housing is expected to double by 2050 [4]. In this context, there's a growing interest in sustainable building materials that offer improved thermal efficiency and have a minimal environmental impact [5]. Among these materials, unfired earth bricks present significant advantages, including their availability, recyclability, and ability to regulate indoor humidity while enhancing building thermal comfort [2]. Compared with concrete and fired clay bricks, these materials require little energy for their transformation process and have less impact on the environment. This research paper explores the potential use of unfired earth bricks in construction based on their thermal performance in the six distinct climatic zones of Morocco. By carefully examining the thermal properties of these bricks under varied climatic conditions, our objective is to assess their viability as a promising solution for reducing energy consumption and the environmental impact of the building sector.

2 Methodology

The main objective of adobe brick construction is to create a comfortable interior space by reducing the annual energy consumption related to heating and air conditioning. In this context, a series of thermal simulations was carried out in different climatic zones in Morocco using OpenStudio software. The thermal performance of a building varies depending on the climatic zone in which it is located; therefore, we used the division into six zones of Morocco according to the Moroccan Thermal

Regulation for Construction (RTCM) [6]: zone 1 (Rabat), zone 2 (Tangier), zone 3 (Fez), zone 4 (Ifrane), zone 5 (Benguerir), and zone 6 (Ouarzazate).

2.1 Case Study Description

The building studied is an individual, single-story construction, detached construction with a total surface area of 74 m^2 and a height of 3 m. It is divided into four distinct thermal zones comprising two bedrooms, a living room, and a bathroom, as illustrated in Figs. 1 and 2. Figure 1 provides an overview of the selected building. The geometric model was created using SketchUp.

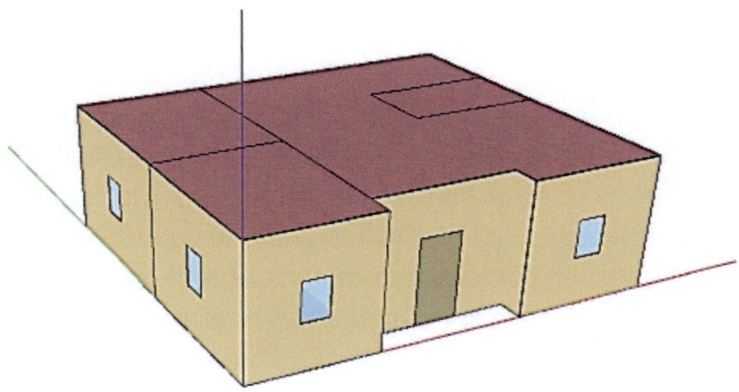

Fig. 1 Schematic of the building in the building energy model

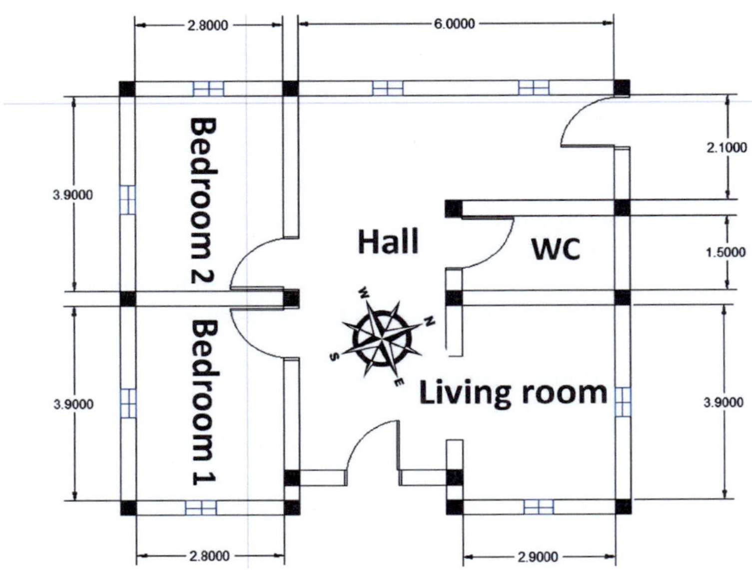

Fig. 2 Architectural plan

Table 1 Thermo-physical properties of the selected residential building envelope

	Building material	E (m)	λ (W/m.K)	ρ (kg/m^3)	C (J/kg.K)
External walls	Adobe brick [7]	0.30	0.57	1936	774

The following table presents the characteristics of the building envelope considered in this study (Table 1).

3 Result and Discussion

3.1 *Indoor Temperature*

For the building energy simulation exercise, annual hourly temperature values for each selected city are extracted using Meteonorm tool version number 8. Figure 3 illustrates the monthly variation of indoor air temperature under different climatic conditions in Morocco. These graphs provide valuable insights into information on how buildings react to various climatic conditions.

The results demonstrate that clay brick construction provides significant thermal comfort across the six climatic zones studied. In Rabat, up to 46% of indoor temperatures fall between 20 and 26 °C. During the summer, adobe brick constructions can reduce indoor air temperatures by up to 13.77 °C compared to outdoor temperatures. Additionally, while the maximum fluctuation in outdoor temperatures is 16.35 °C, indoor fluctuations do not exceed 5.37 °C.

In Tangier, 43% of indoor air temperatures are also between 20 and 26 °C. In summer, the maximum reduction in indoor air temperature compared to outdoor temperatures is 12.69 °C. Furthermore, despite a maximum outdoor temperature fluctuation of 12.69 °C, indoor fluctuations do not exceed 5.11 °C. In Fes, a maximum decrease of 14.51 °C in indoor air temperature compared to outdoor temperatures was observed during the summer, with a maximum indoor fluctuation of 5.58 °C, while the maximum outdoor fluctuation reaches 16.37 °C.

In Ifrane, the maximum reduction in indoor air temperature compared to outdoor temperatures is nearly identical to Fes, at 14.49 °C, with a similarly close indoor fluctuation of 5.49 °C. In Benguerir, 37% of indoor air temperatures fall within the 20–26 °C range. During summer, the maximum decrease in indoor air temperature compared to outdoor temperatures is 13.99 °C, while the maximum outdoor fluctuation reaches 18.87 °C and the maximum indoor fluctuation does not exceed 5.49 °C.

Finally, in Ouarzazate, clay brick construction achieves a maximum reduction in indoor air temperature of up to 14.76 °C compared to outdoor temperatures during the summer. The maximum indoor fluctuation does not exceed 6.41 °C, while the maximum outdoor fluctuation reaches 15.97 °C.

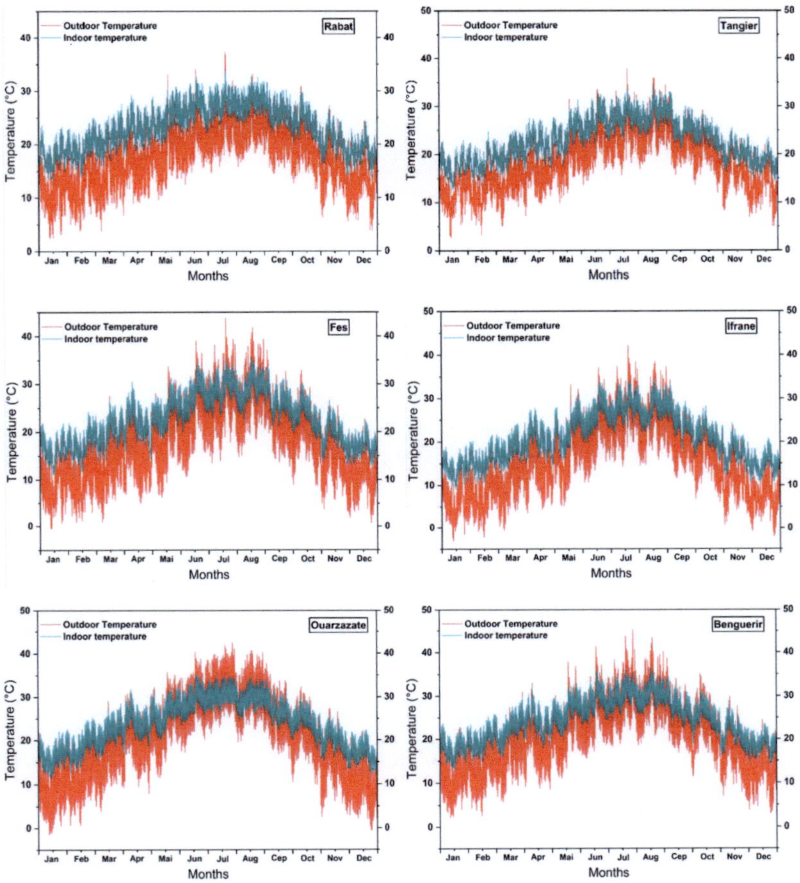

Fig. 3 Annual thermal needs in 6 climate zones

3.2 *Thermal Loads*

Comparing the annual energy consumption for heating and cooling a building in different cities of Morocco can provide valuable information on the building's energy efficiency under varying climatic conditions. According to the Moroccan standard RTCM [6], the setpoint temperatures are 20 and 26 °C for heating and cooling, respectively.

Figure 4 illustrates the annual energy demands for six distinct climatic zones in Morocco. The data reveal that the highest annual energy loads are observed in Ifrane, with a total of 125.98 kWh/m^2, followed by Ouarzazate (109.10 kWh/m^2), Fès (104.17 kWh/m^2), and Benguerir (97.34 kWh/m^2). Conversely, the lowest demands are noted in Rabat and Tangier, with values of 69.47 and 74.06 kWh/m^2, respectively. These results suggest that clay brick constructions exhibit exceptional thermal performance in Rabat and Tangier, comparable to those observed in other Moroccan cities.

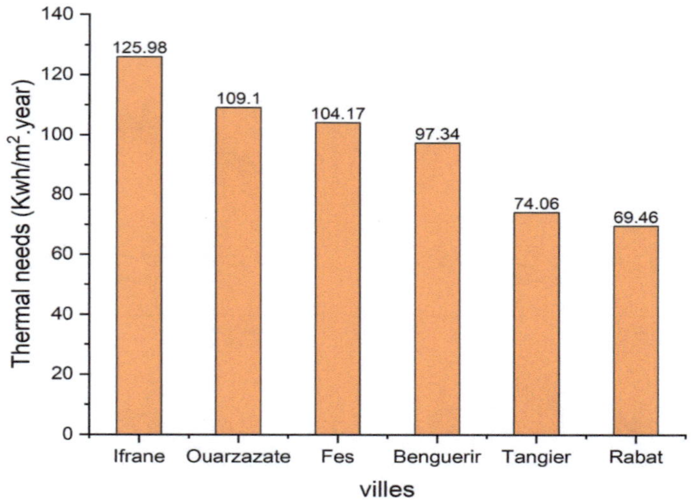

Fig. 4 Annual thermal loads for bedrooms in different cities of Morocco

To enhance thermal performance in the cities of Fès, Ifrane, Benguerir, and Ouarzazate, it is recommended to consider reinforcing adobe bricks with natural fibers and integrating thermal insulation panels.

4 Conclusion

This study provides a detailed examination of the thermal performance of clay brick constructions across various climatic zones in Morocco, highlighting their effectiveness in terms of indoor thermal comfort and energy efficiency. The results reveal that clay brick constructions offer a substantial improvement in indoor thermal stability and a notable reduction in temperature fluctuations compared to external climatic conditions.

In the cities of Rabat and Tangier, clay brick buildings manage to maintain up to 46 and 43%, respectively, of indoor temperatures within the optimal range of 20–26 °C. During the summer months, they allow a significant reduction in indoor temperatures compared with outdoor temperatures, reaching up to 13.77 °C in Rabat and 12.69 °C in Tangiers. In regions with more extreme climatic conditions, such as Fes, Ifrane, Benguerir, and Ouarzazate, clay brick constructions continue to demonstrate effective performance in the summer. The maximum reductions in indoor temperatures range from 12.69–14.76 °C compared to the outdoor temperatures. The data show that the maximum indoor temperature variations generally remain below 6 °C across all cities, despite significant fluctuations in outdoor temperatures ranging from 15.97 to 18.87 °C. In terms of energy consumption, clay brick constructions exhibit variable annual energy demands depending on the climatic zones, with the highest energy loads recorded in Ifrane (125.98 kWh/m^2) and the lowest in Rabat (69.47 kWh/m^2). The lower energy demands observed in Rabat and Tangier suggest that clay brick constructions are particularly effective in more moderate climates, while regions with extreme climatic conditions might benefit from additional solutions. It is recommended to consider reinforcing clay bricks with natural fibers and incorporating thermal insulation panels to enhance energy efficiency and indoor comfort in more demanding climatic contexts. In conclusion, this study provides valuable insights into the thermal performance of clay brick constructions across different Moroccan climates, optimizing both indoor comfort and energy consumption for heating and cooling. Future research should focus on improving construction techniques and materials to further maximize thermal efficiency in various climatic contexts.

References

1. "Directive 2010/31/EU of the European Parliament and of the Council of 19 May (2010) on the energy performance of buildings".
2. Bouchefra, I., Bichri, F. Z. EL, Chehouani, H., and Benhamou, B., "Mechanical and thermophysical properties of compressed earth brick reinforced by raw and treated doum fibers," Constr Build Mater, vol. 318, p. 126031, Feb. (2022), https://doi.org/10.1016/J.CONBUILDMAT.2021.126031.
3. H. L.-P. Societies and undefined 2020, "World population outlook: Explosion or implosion?," cairn-int.infoH Leridon Population Societies, 2020 cairn-int.info, Accessed: Mar. 26, 2024. [Online]. Available: https://www.cairn-int.info/journal-population-and-societies-2020-1-page-1.htm/journal-population-and-societies.htm.
4. Sokolowski, I.L., Maheshwari, A., and Malik, A., "Green buildings: a finance and policy blueprint for emerging markets," 2019, Accessed: Mar. 26, 2024. [Online]. Available: https://policycommons.net/artifacts/1271438/green-buildings/1854570/.
5. Wardi, F. Z. El, Ladouy, S., Atbir, A., and Khabbazi, A., "Study of the thermal and mechanical properties of local clay materials activated with quicklime, Sefrou (Morocco)," Mater Today Proc, vol. 58, pp. 1423–1430, Jan. 2022, https://doi.org/10.1016/J.MATPR.2022.02.348.
6. "Reglement Thermique de Construction au Maroc (RTCM) Agence Nationale pour le Développement des Energies Renouvelables et de l'Efficacité Energétique (ADEREE) (2014).", Accessed: Sep. 07, 2024. [Online]. Available: https://www.amee.ma/sites/default/files/inline-files/Reglement_thermique_de_construction_au_Maroc.pdf.
7. Alioui, A., Kaitouni, S.I., Azalam, Y., Al armouzi, N., Bendada, E.M., and Mabrouki, M., "Effect of straw fibers addition on hygrothermal and mechanical properties of carbon-free adobe bricks: From material to building scale in a semi-arid climate," Build Environ, vol. 255, p. 111380, (2024), https://doi.org/10.1016/J.BUILDENV.2024.111380.

Mechanical Performance Evaluation of Adobe Bricks Manufactured Using Different Clay Soils Extracted from Beni Mellal Region in Morocco

Youness Azalam, Abdelmounaim Alioui, Mohammed Benfars, Mustapha Mabrouki, and El Maati Bendada

Abstract This paper focuses on the comprehensive evaluation of the mechanical performance of adobe bricks fabricated from two distinct clay soils sourced from the Beni Mellal region. The study aims to compare the mechanical strengths of adobes crafted from these clay soils, shedding light on their structural suitability for construction practices. The research employs a dual-pronged approach. First, physical testing of adobe bricks is carried out to ascertain their compression and flexural strengths. The findings reveal that the highest recorded values for compression and flexural strengths are 14.57 and 0.5 MPa, respectively. This empirical data serves as a basis for evaluating the load-bearing capabilities of adobe bricks, an essential aspect of their feasibility as building components. By investigating these mechanical behaviors, the study provides insights into the structural response and durability of the adobe bricks, allowing for a comprehensive understanding of their performance under different conditions. This research paper offers a thorough examination of the mechanical properties of adobe bricks fabricated from clay soils in the Beni Mellal region. By comparing their mechanical strengths and interpreting the factors influencing their performance, the study contributes to an understanding of the suitability of earth materials in construction. With sustainability becoming a key focus in modern construction, the findings of this research offer valuable insights for architects, engineers, and policymakers.

Keywords Adobe bricks · Mechanical performances · Structural suitability

Y. Azalam (✉) · A. Alioui · M. Benfars · M. Mabrouki · E. M. Bendada
Laboratory of Industrial and Surface Engineering, Sultan Moulay Slimane University, Beni Mellal, Morocco
e-mail: youness.azalam@usms.ma

A. Vaseashta et al. (eds.), *Proceedings of the Eighth International Symposium on Dielectric Materials and Applications (ISyDMA'8)*, Springer Proceedings in Physics 349, https://doi.org/10.1007/978-3-031-84809-4_22

1 Introduction

The field of construction worldwide has undergone significant development in terms of the materials used and the techniques adopted. However, ordinary building materials, namely cement and steel, generate environmental concerns that increase the carbon footprint and energy consumption of this sector. In Morocco, for example, this sector is marked by an annual energy consumption representing approximately 33% of the country's total energy usage [1]. Additionally, the fast population expansion, with a growth rate of 4.27% from 2014 to 2017 and an expected rise of up to 30% by 2028 [2], highlights the need to reassess materials and construction methods.

Addressing these challenges has prompted significant research efforts to discover energy-efficient and eco-friendly methods, to reduce the CO2 emissions linked to the manufacturing of conventional construction materials such as cement and steel [3, 4]. Among these approaches, earthen construction stands out as a compelling choice, rooted in a history dating back nearly 10,000 years before the Common Era [5, 6]. Its current widespread use globally, providing shelter for approximately 30% of the world's population in earth-built structures [7–9], further underscores the impetus for continued research in this area.

This construction type, primarily utilizing clay as a raw material, provides notable advantages in terms of thermal comfort [10]. However, the mechanical performance of clay-based building materials is typically unsatisfactory and relies on the type of clay used, as well as potential reinforcement with natural fibers [11]. In this approach, numerous researchers have investigated the impact of incorporating natural fibers on the thermal and mechanical properties of clay-based adobes. They aim to enhance the bricks' strength through the tensile strength of the fibers. However, it's worth noting that alternative methods exist to reinforce and stabilize earthen bricks, such as adding artificial materials like cement or quicklime. These methods are intended to mitigate the susceptibility of this construction type to mechanical stresses.

Another aspect of the assessment of earthen construction is the study of the soil used as a raw material, the soil used directly impacts the mechanical performance of the brick, several scientific articles have studied the performance of adobes using different types of clay since each country has a distinct soil type, identification tests such as particle size, Atterberg limits, and methylene blue value are important factors influencing the mechanical performance of adobe. Internationally recognized standards, such as ASTM and DIN, outline the essential characteristics required for soils in earthen constructions. These standards specify that the soil used must contain a clay content exceeding 40%. Additionally, the grain size distribution should exhibit a uniformity of grain sizes ranging from fines to larger soil elements. Therefore, this study aims to assess the mechanical properties of two soil types extracted from the Beni Mellal region and their suitability for earth construction, specifically in the adobe technique. Additionally, it seeks to establish correlations between the physical characteristics of the raw materials and their mechanical performance.

2 Materials and Methodology

2.1 Materials

For raw materials, we use two types of soils (red clay and black clay) that were extracted from two villages called Ouled Mbarek and Mghila in the north-central part of Morocco Fig. 1. These types of soils were chosen for their abundance and the amount of clayey material contained, which gives them potential use as a construction material.

As a description of the soils studied, the first sample Fig. 2a is reddish-brown soil with various sizes of grains, and the second sample Fig. 2b is black clay with an important amount of fine elements and medium-sized aggregates.

2.1.1 Physical Properties of the Studied Soils

In order to characterize the two samples of soils used in this study, several identification tests were carried out. The Atterberg limits, such as the plastic limit (PL), liquidity limit (LL), and plasticity index, were measured according to NM ISO 17892-4.

As depicted in Table 1, the results of the Atterberg limits measured for the two samples of soils measured according to NM ISO 17892-4 4 [12] align with the earthquake-resistant regulations of Moroccan earthen construction. The liquidity limit has a value of 30% for the red soil and 32% for the black soil, while the plastic limit and plasticity index have values of 16 and 14%, respectively.

Additionally, the particle size analysis of the soils shown in Fig. 3 shows an important amount of clay material, around 40% for the red soil and 48% for the black soil. These results confirm the potential use of both soils as suitable materials for earth construction technique [13].

2.1.2 Sample Preparation

Following the removal of visible organic elements such as leaves and plant roots, the sols were sieved manually using a mesh opening of 1 cm. This step was conducted to remove large aggregates and ensure the uniformity of the raw material before adobe brick production. The approach adhered to traditional methods observed in southern Morocco, which have proven to be highly effective [14].

Subsequently, the two types of soil are blended with water and mixed by hand to have a workable paste that will be used after the manufacture of adobe bricks, the paste is then left to ferment for 48 h. This fermentation process leads to a more consistent paste, which is essential for producing adobe bricks of uniform quality and strength.

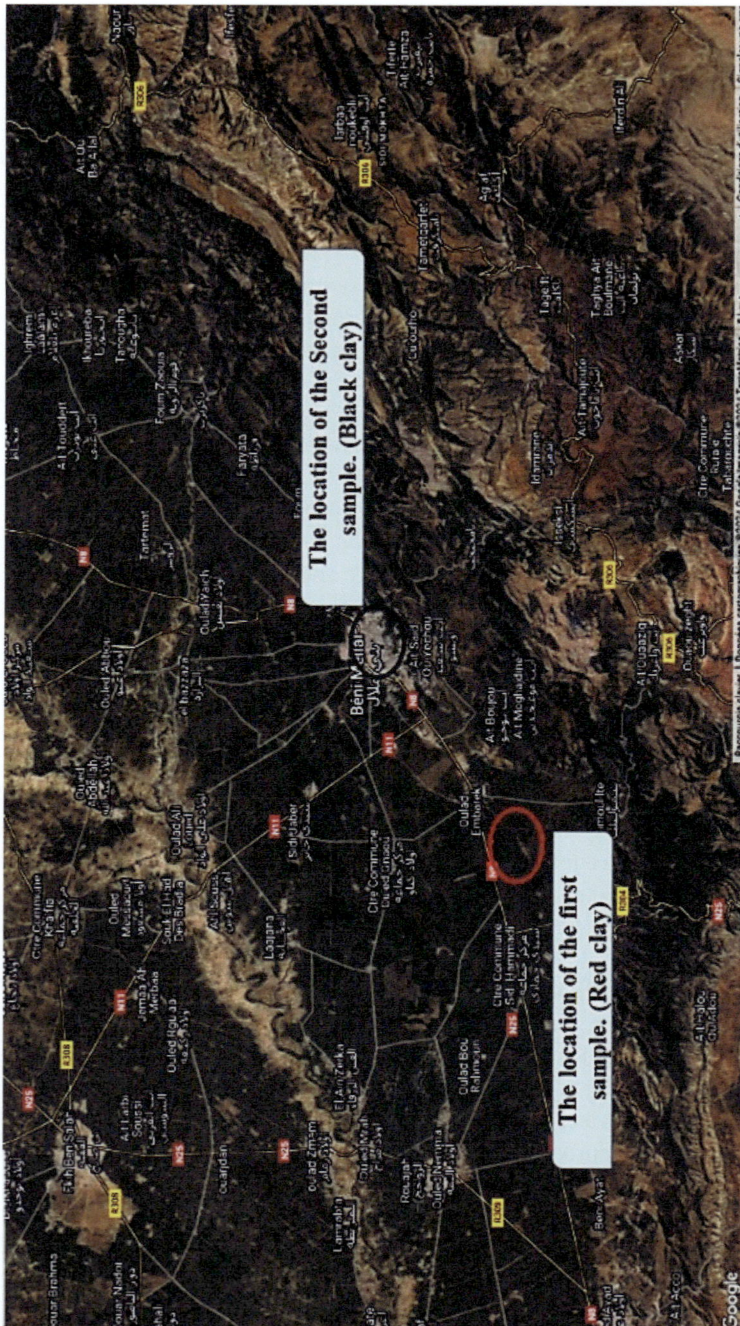

Fig. 1 Locations of the soil samples extracted

Fig. 2 The two samples of clay (Red and Black) used in the study

Table 1 Physical properties of the soils studied

Characteristics of the studied soil	Value	
	Red soil	Black soil
Water content (%) (W)	2.0	2.14
Liquidity limit (%) (LL)	30	32
Plasticity limit (%) (PL)	16	19
Plasticity index (IP = LL-PL)	14	13
Methylene blue value (VBS)	2.67	2.86

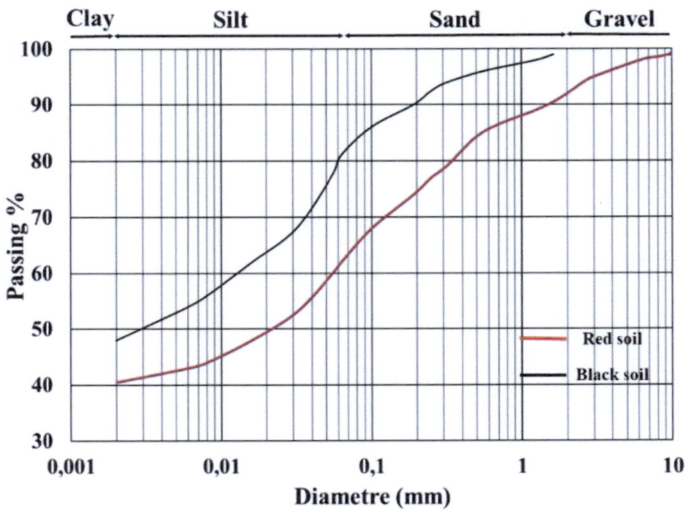

Fig. 3 Particle size distribution of soils studied

Fig. 4 Preparation of adobe brick samples

Subsequently, the fermented clay paste is carefully poured into rectangular molds with dimensions of 30*15*10 cm3. Manual compaction is performed to ensure the mixture becomes compact and solid. The molds are then carefully shaped and smoothed by hand using a trowel, resulting in a sleek and refined surface.

The adobe bricks are then left to air-dry for four weeks at an ambient temperature of 25 ± 5 °C. The drying process continues until the mass of the bricks stabilizes at a certain value. Once the bricks are thoroughly dried and polished, they are ready for testing (Fig. 4).

3 Experimental Results

3.1 Bulk Density Test

After manufacturing three samples of each soil type, red soil and black soil, they were dried in the open air for 28 days. Table 2 shows the measurement of the apparent density of the adobe samples.

Table 2 Bulk density measured of adobe samples

	Adobe sample	Bulk density (kg/m3)	Average Bulk density (kg/m3)
Red soil	Sample 1	1994.5	1963
	Sample 2	1928.5	
	Sample 3	1966	
Black soil	Sample 1	1972	1963.5
	Sample 2	1985	
	Sample 3	1933.5	

3.2 Compressive and Flexural Test

The compressive strength of the adobe bricks was carried out using a UTEST compression machine. The compression tests were conducted according to Moroccan standards NM 10.1–042 and XP P 13–90, with an average crushing speed of 7.5 kN/s. Furthermore, three-point bending tests were executed to determine the bending strength of the samples. The controlled displacement speed for these tests was maintained at 51 mm/min, in accordance with the Moroccan standard NM EN 772-6-2015.

As shown in Table 3 and Fig. 5, the results found (force applied and compressive strength) of the adobes manufactured based on red earth have reached a maximum value of 190.2 KN for the compression load and 14.72 MPa for compressive strength. These values are significantly higher than those for adobes manufactured from black earth, which are 79.18 KN and 6.12 MPa, respectively. This difference can be attributed to the granulometric nature of the red earth, as illustrated in the figure. The red earth exhibits a homogeneity of the granular skeleton, with a better balance between fine particles (clay) and coarse particles (sand and gravel). This optimal distribution has enhanced the bonding between particles, resulting in higher compressive strength. In contrast, the black earth contains a high proportion of fines, leading to the formation of damaging cracks during compression tests. This, in turn, reduces its compressive strength and causes the adobe to break. The values that are found either for red or black soil exceed the threshold of resistance prescribed in various international standards [13, 15]

After determining the compressive strength of adobes manufactured from both soil types, a bending test was conducted on three samples of each type to measure an average flexural strength. This test was performed in accordance with the Moroccan standard NM EN 772-6 using a three-point bending setup. In this setup, the sample is positioned in the crushing machine with its longitudinal axis perpendicular to the rollers, while an axial force is applied until the adobe sample fractures.

Table 4 and Fig. 6 present the measured results, indicating a lower bending strength compared to values reported in the literature. This reduced strength is attributed to the absence of fibers or stabilizers in the crushed adobe samples. When comparing the two soil types, it is evident that red earth performs better even in bending, reaching a maximum value of 0.5202 MPa, whereas black earth does not exceed 0.3630 MPa.

The presence of a high proportion of fines in black earth significantly impacts its resistance to bending and compression. This influence becomes evident once the

Table 3 Compressive force and strength of the adobe bricks tested

Reference	Force applied (KN)				Compressive strength (MPa)			
	Trial 1	Trial 2	Trial 3	Average	Trial 1	Trial 2	Trial 3	Average
Red soil	182.52	190.2	185.30	186	14.44	14.72	14.53	14.56
Black soil	76.2	72.34	79.18	75.90	6.08	6.02	6.12	6.25

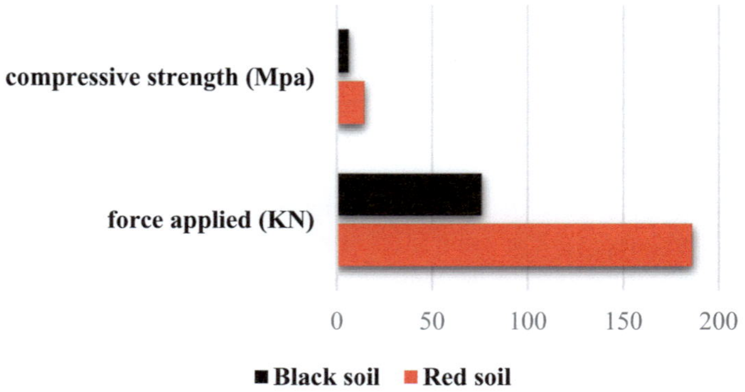

Fig. 5 Compressive strength results of adobe samples

Table 4 Flexural strength of the adobe bricks tested

Reference	Flexural strength (MPa)			
	Trial 1	Trial 2	Trial 3	Average
Red soil	0.4822	0.4620	0.5202	0.4881
Black soil	0.2903	0.3630	0.3385	0.3306

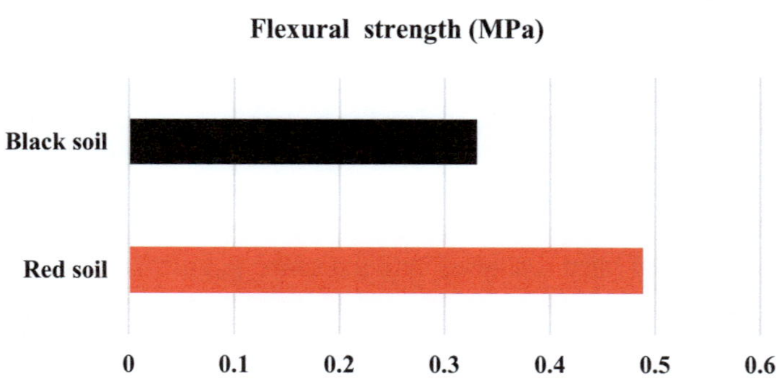

Fig.6 Flexural strength of adobes tested

samples are dried, as numerous cracks appear on the surface of adobes made from black soil, in contrast to those made from red soil.

4 Conclusion

This study compares clay bricks made from two types of soil (red and black) sourced from the Beni Mellal region in the north-central of Morocco, evaluating their mechanical performance, particularly focusing on a characteristic typically viewed as a drawback of this construction technique. The results indicate significant compressive strength values exceeding 14 MPa. However, these bricks have reached a value of 1.96 kg/cm3 which surpasses the limit of 1.75 kg/cm3 prescribed for lightweight bricks by the Moroccan standard NM 10.1009–2014, making them heavy but high-performing.

The comparison between adobe bricks made from red soil and those made from black soil revealed a significant difference in mechanical resistance. The red soil bricks exhibited about 58% higher compression resistance and 32% higher bending resistance compared to the black soil bricks. This difference is attributed to the distinct grain sizes of the two soil types. The black soil has a higher proportion of fine particles, while the red soil has a more continuous and homogeneous particle distribution. This characteristic of the red soil promotes better adhesion between grains, resulting in a more compact block that can withstand greater loads.

References

1. AMEE | Efficacité énergétique dans le bâtiment. Accessed: Jul. 25, (2023). [Online]. Available: https://www.amee.ma/fr/expertise/batiment.
2. "World Bank: Kingdom of Morocco systematic country... - Google Scholar."
3. "The properties of earth as a building material," Building with Earth, pp. 19–35, Nov. (2007), https://doi.org/10.1007/3-7643-7873-5_2/HTML.
4. M. S. Zami and A. Lee, "Economic benefits of contemporary earth construction in low-cost urban housing - State-of-the-art review," Journal of Building Appraisal, vol. 5 (3), pp. 259–271, Dec. (2010), https://doi.org/10.1057/JBA.2009.32.
5. A. Sánchez, H. Varum, T. Martins, and J. Fernández, "Mechanical properties of adobe masonry for the rehabilitation of buildings," Construction and Building Materials, vol. 333. Elsevier Ltd, May 23, (2022). https://doi.org/10.1016/j.conbuildmat.2022.127330.
6. H. Schroeder, "Sustainable building with earth," Sustainable Building with Earth, pp. 1–560, Oct. (2015), https://doi.org/10.1007/978-3-319-19491-2/COVER.
7. O. Khtou, I. Aalil, M. Aboussaleh, and F. Z. EL Wardi, "Mechanical analysis of fiber reinforced adobe," Civil Engineering and Architecture, vol. 9, no. 7, pp. 2160–2168, Dec. (2021), https://doi.org/10.13189/cea.2021.090705.
8. E. Avrami, H. Guillaud, and M. Hardy, "Terra Literature Review: An Overview of Research in Earthen Architecture Conservation."
9. H. Houben, H. G.-(No Title), and undefined 1994, "Earth construction: a comprehensive guide," Intermediate Technology Publications, (1994).
10. K. Rashid, E. U. Haq, M. S. Kamran, N. Munir, A. Shahid, and I. Hanif, "Experimental and finite element analysis on thermal conductivity of burnt clay bricks reinforced with fibers," Constr Build Mater, vol. 221, pp. 190–199, Oct. (2019), https://doi.org/10.1016/j.conbuildmat.2019.06.055.
11. C. Babé, D. K. Kidmo, A. Tom, R. R. N. Mvondo, B. Kola, and N. Djongyang, "Effect of neem (Azadirachta Indica) fibers on mechanical, thermal and durability properties of adobe bricks," Energy Reports, vol. 7, pp. 686–698, Nov. (2021), https://doi.org/10.1016/j.egyr.2021.07.085.

12. "NM ISO 17892-4 | IMANOR." Accessed: Aug. 08, (2023). [Online]. Available: https://www.imanor.gov.ma/Norme/nm-iso-17892-4/.
13. D. U. M. ROYAUME, "Règlement parasismique des constructions en terre," RPCTerre, Ministère de l'interieur, Ministère de l'equipment et du transport, Ministère de l'habitat, de l'urbanisme et de la politique de la ville, Rabat, (2001).
14. J. M. L. Osorio, A. M. Lozano, and U. M. Codes, "Rammed-earth construction in southern Morocco: A reappraisal of the technology," Rammed Earth Conservation, pp. 169–174, Jan. (2012), https://doi.org/10.1201/B15164-32/RAMMED-EARTH-CONSTRUCTION-SOUTHERN-MOROCCO-REAPPRAISAL-TECHNOLOGY-L.
15. N. M. Code, "14.7. 4 NMAC (2015) New Mexico Earthen Building Materials Code," New Mexico Administrative Code.

Incorporating Recycled Paper into Eco-Friendly Clay Bricks Using Sustainable Manufacturing to Enhance Thermal Properties of Composite Material

Abdelmounaim Alioui, Mohammed Benfars, Youness Azalam, Samir Idrissi Kaitouni, El Maati Bendada, and Mustapha Mabrouki

Abstract The environmental problems caused by the use of conventional building materials, such as cement, and their rapid depletion, make it necessary to develop environmentally friendly building materials made from renewable resources. Inadequate management of waste office paper in landfill sites leads to air and groundwater pollution. In this context, the exploitation of wastepaper as a renewable resource, with its use as a reinforcement in the manufacture of unfired adobe bricks, appears to be a solution that can resolve these environmental problems in a more environmentally friendly way. In this study, waste office paper was used as an additive in adobe clay bricks, with weight proportions of 0, 2, 4, and 10%, to enhance their thermal performance. Then, by using a heat flux meter (HFM), the thermal properties of the prepared samples were experimentally analyzed. The results showed promising potential for the future application of this composite material in building construction, to improve thermal comfort and reduce greenhouse gas emissions. Furthermore, a standard building model was simulated using the clay-waste paper envelope in the semi-arid climate of Morocco to analyze its thermal behaviors and determine its ability to provide optimal comfort by minimizing excessive summer heat and preserving warmth during winter without artificial heating and cooling systems.

Keywords Composite material · Adobe bricks · Waste paper · Thermal comfort · Thermophysical properties · Building construction

A. Alioui (✉) · M. Benfars · Y. Azalam · E. M. Bendada · M. Mabrouki
Laboratory of Industrial and Surface Engineering, Faculty of Science and Technologies, Sultan Moulay Slimane University, Beni Mellal, Morocco
e-mail: abdelmounaim.aliouifpb@usms.ac.ma

A. Alioui · E. M. Bendada
Polydisciplinary Faculty, Sultan Moulay Slimane University, Beni Mellal, Morocco

S. I. Kaitouni
Materials, Environment and Sustainable Development (MEDD), Faculty of Sciences and Techniques of Tangier, Abdelmalek Essaadi University, Tangier, Morocco

A. Vaseashta et al. (eds.), *Proceedings of the Eighth International Symposium on Dielectric Materials and Applications (ISyDMA'8)*, Springer Proceedings in Physics 349, https://doi.org/10.1007/978-3-031-84809-4_23

1 Introduction

The construction industry is one of the most important factors in socio-economic development. However, it is also acknowledged as a significant source of adverse environmental impacts. One of the main concerns relates to traditional building materials such as cement, which notably contribute to consuming energy, emitting greenhouse gases, and depleting natural resources. Globally, the construction sector accounts for around 40% of total energy consumption, leading to 33% of CO_2 emissions [1, 2]. Faced with these pressing environmental issues, the pursuit of sustainable and ecologically sensitive solutions within the construction field has become an imperative necessity. In the regions of North Africa, distinguished by a dry summer climate, the regulation of indoor climate is based on traditional and vernacular architectural methods. These methods involve the use of materials such as clay, straw, natural fibers, and materials with high thermal inertia [3–6]. A range of strategies, integrated into the concept of ecological construction, are devised to minimize heat transfer from the external environment to the building interiors. Construction decisions guided by environmental awareness prioritize environmental preservation through the utilization of local resources and the design of buildings that consider climatic limitations. Consequently, ensuring the ideal construction of building walls becomes imperative for improving thermal comfort, especially given that conventional walls fail to meet the necessary thermal standards for efficient construction materials [7, 8]. Given the limited economic capacity of much of the North African population, it would be prudent to opt for cost-effective building materials, such as reinforced adobe bricks, for constructing suitable walls. The use of earth-based materials has significant advantages, including ecology, sustainability, recyclability, comfort, and socio-economic aspects. It reduces the energy required for extraction and transportation, thereby contributing to the development of a roadmap for zero-energy buildings. Today, the exploration of soil stabilizers like waste and natural by-products represents a new area of research. They can serve as building materials for bioconstruction, helping to decrease raw material consumption [9, 10]. The main aim of this study is to use wastepaper as a reinforcement to improve the thermophysical performance of adobe bricks, making them more insulating. This also aims to reduce energy consumption and provide thermal comfort without recourse to heating and cooling systems, by using the clay and waste office paper envelope in a semi-arid climate where climatic conditions are very difficult.

2 Materials and Experimental Methods

2.1 Materials

The clay used in this study was extracted from the Beni Mellal region, located in southern Morocco. The office wastepaper is a natural material composed mainly of cellulose fibers, the most common and renewable natural polymer on the planet. Office wastepaper is collected in municipal solid waste, and then finely shredded into small pieces. These fragments can then be used as a potential reinforcement to improve the performance of green bricks.

2.2 Sample Preparation

The method used in this study to manufacture the samples is the traditional Adobe production method used in southern Morocco. After the preparation of the raw material (sieving of earth and grinding of paper) (Fig. 1-a). An amount of 19% water was added to the dry weight of the soil to prepare the clay-water paste. Then, the mixture was kneaded for 30 min to obtain a homogeneous dough. This paste was left to ferment naturally for three days, this process being essential in traditional adobe manufacturing to strengthen the strength and stability of the material over time. Then, quantities of wastepaper of 0, 2, 4, and 10% were added to the prepared paste and mixed for 15 min to obtain a convenient consistency (Fig. 1b). Finally, an iron mold of dimensions $200 \times 180 \times 40$ mm^3 was used to shape the samples. All samples were allowed to air dry naturally for 28 days (Fig. 1c).

Fig. 1 Steps for preparing adobe samples

2.3 Experimental Methods

In this study, the Lambda HFM 446 flowmeter (NETZSCH) shown in Fig. 2 was used to determine the thermal conductivity and specific thermal capacity of adobe samples. The HFM 446 operates according to the thermal flow meter method, following ASTM C518 and DIN EN 12667 [11, 12]. The experimental procedure involves sandwiching a sample between two heating plates to establish a controlled temperature gradient. The heat flow through the sample is then quantified using precisely calibrated heat flow sensors. The thickness of the sample is measured, and once the thermal equilibrium is reached, Fourier's law is applied to determine the thermal conductivity of the sample according to Eq. (1):

$$\lambda = \frac{Q \times L}{\Delta T} \tag{1}$$

We can easily calculate the thermal diffusivity of the samples from the determined values of the density, the thermal conductivity, and the heat capacity of each sample using the following the Eq. (2):

$$a = \frac{\lambda}{\rho \times C_p} \tag{2}$$

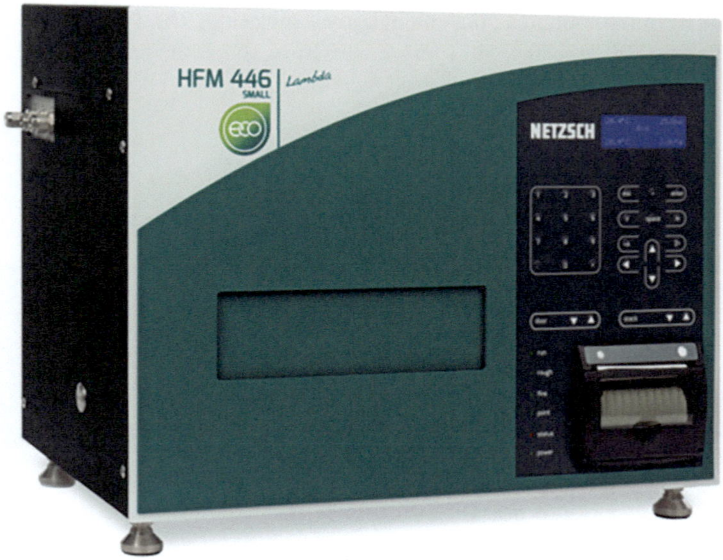

Fig. 2 Heat flow meter (HFM 446) METZCH used for investigations

Table 1 Density (ρ), thermal conductivity (λ), heat capacity (Cp), and thermal diffusivity (a) of the composite adobe samples (clay—wastepaper)

Samples	ρ (Kg/m^3)	λ (W/m K)	Cp (KJ/kg.K)	a × 10^{-7} (m^2/s)
B0	1936	0.57	0.774	3.8
B2	1703	0.40	0.752	3.12
B4	1614	0.31	0.743	2.58
B10	1390	0.19	0.686	1.99

3 Results

In this study, to determine the thermophysical properties of the samples studied, three tests were carried out for each sample, and the average value of each characteristic was presented in Table 1. From the results, we can see a reduction in density due to the addition of wastepaper, from 1936 kg/m^3 for the reference brick to 1390 kg/m^3 for bricks containing 10% wastepaper, i.e., a reduction of 28.2%. This reduction affects the thermal properties of the samples. Furthermore, the results show that thermal conductivity decreases as the percentage of wastepaper in the clay mass increases. This demonstrates that wastepaper has a significant effect on reducing thermal conductivity, with a rate of 66.66% for the composite material (clay + 10% wastepaper), which improves the thermal insulation performance of the bricks.

4 Thermal Dynamic Simulation

To carry out the dynamic thermal simulation, Trnsys software was used during a typical meteorological year (TMY) in the semi-arid climate of Morocco to improve the energy efficiency of the clay-paper composite material. The building studied is located in the town of Benguerir city, in southern Morocco, facing east, and has a total surface area of 74 m2 and a ceiling height of 3 m. The building is a single-story detached construction, divided into 5 thermal zones, as shown in Figs. 3 and 4.

Figure 5 shows the results of dynamic thermal simulation carried out on two types of envelopes: pure clay (B0) and clay with 10% paper waste (B10). It illustrates indoor temperature variations for both types of building envelopes as a function of ambient outdoor temperature throughout the year. It can be seen that thermal resistance values vary from $R_{B0} = 0.7$ m^2.K/W for the pure clay envelope to $R_{B10} =$ 2.105 m^2.K/W for the same wall thickness (e = 40 cm). It can be deduced that the addition of wastepaper to adobe bricks results in a clear reduction in heat transfer through the building's walls. This composite material reduces fluctuations in interior temperature. Analysis of the simulation results shows that this clay-paper waste composite material improves thermal comfort inside the building, despite the difficult external conditions, which are very hot in summer and very cold in winter. During the summer period, when the maximum temperature exceeds 46 °C, the building's

Fig. 3 The simulated real building

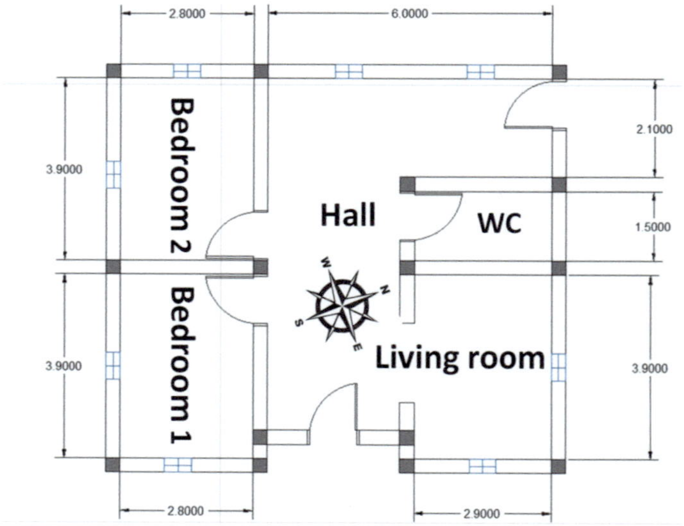

Fig. 4 The 2D plan of the case study building with the thermal zones

interior temperatures drop to 34 °C for the B10 envelope and 38 °C for the B0 envelope. The results indicate that adding paper waste to the walls of earthen buildings can enhance their thermal performance and improve passive heating and cooling by controlling indoor temperature variations.

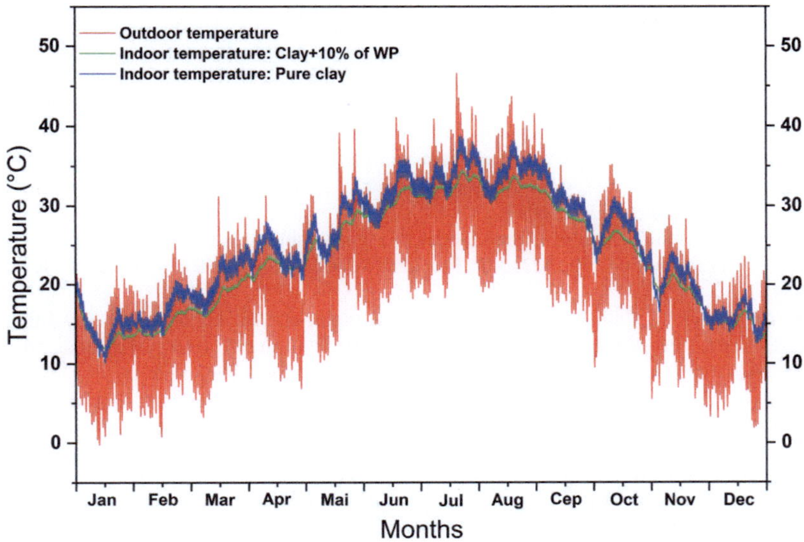

Fig. 5 Hourly temperature variations inside and outside the structure in 2022

5 Conclusion

The incorporation of wastepaper as reinforcement in adobe bricks offers promising prospects for improving the thermal performance and durability of building materials. This study focused on the experimental evaluation of the thermophysical properties of this new composite based on clay and wastepaper. The results show that the addition of office wastepaper significantly reduces the material's thermal conductivity and diffusivity. These modifications improve the thermal comfort of buildings constructed with this material without the need for additional heating or cooling systems. Additionally, the use of this innovative composite contributes to the reduction of greenhouse gas emissions by lowering energy consumption, thus offering an ecological and economical solution for sustainable construction. The results of dynamic thermal simulations confirm the energy efficiency of this material, highlighting the potential of wastepaper to play a crucial role in the development of environmentally friendly and energy-efficient construction techniques.

References

1. Ú. Cárdenas-Mamani, R. Kahhat, and I. Vázquez-Rowe, 'District-level analysis for household-related energy consumption and greenhouse gas emissions: A case study in Lima, Peru', Sustain Cities Soc, vol. 77, p. 103572, Feb. (2022), https://doi.org/10.1016/J.SCS.2021.103572.

2. M. Wakil, H. El Mghari, S. Idrissi Kaitouni, and R. El Amraoui, 'Thermal energy performance of compressed earth building in two different cities in Moroccan semi-arid climate', Energy and Built Environment, Jun. (2023), https://doi.org/10.1016/J.ENBENV.2023.06.008.
3. Y. Bahammou, M. Kouhila, Z. Tagnamas, H. Lamsyehe, A. Lamharrar, and A. Idlimam, 'Hygroscopic behavior of water absorbed by capillarity and stabilization of a bio-composite building material: Clay reinforced with Chamarrops humilis fibers', International Communications in Heat and Mass Transfer, vol. 135, p. 106077, Jun. (2022), https://doi.org/10.1016/J.ICHEATMASSTRANSFER.2022.106077.
4. S. H. Sameh, 'Promoting earth architecture as a sustainable construction technique in Egypt', J Clean Prod, vol. 65, pp. 362–373, Feb. (2014), https://doi.org/10.1016/J.JCLEPRO.2013.08.046.
5. K. El Azhary, Y. Chihab, M. Mansour, N. Laaroussi, and M. Garoum, 'Energy Efficiency and Thermal Properties of the Composite Material Clay-straw', Energy Procedia, vol. 141, pp. 160–164, Dec. (2017), https://doi.org/10.1016/J.EGYPRO.2017.11.030.
6. A. Alioui, S. Idrissi Kaitouni, Y. Azalam, N. Al armouzi, E. M. Bendada, and M. Mabrouki, 'Effect of straw fibers addition on hygrothermal and mechanical properties of carbon-free adobe bricks: From material to building scale in a semi-arid climate', Build Environ, vol. 255, p. 111380, May (2024), https://doi.org/10.1016/J.BUILDENV.2024.111380.
7. S. Charoenkit and S. Yiemwattana, 'Living walls and their contribution to improved thermal comfort and carbon emission reduction: A review', Build Environ, vol. 105, pp. 82–94, Aug. (2016), https://doi.org/10.1016/J.BUILDENV.2016.05.031.
8. L. Shashua-Bar, I. X. Tsiros, and M. Hoffman, 'Passive cooling design options to ameliorate thermal comfort in urban streets of a Mediterranean climate (Athens) under hot summer conditions', Build Environ, vol. 57, pp. 110–119, Nov. (2012), https://doi.org/10.1016/J.BUILDENV.2012.04.019.
9. A. E. Losini, A. C. Grillet, M. Bellotto, M. Woloszyn, and G. Dotelli, 'Natural additives and biopolymers for raw earth construction stabilization – a review', Constr Build Mater, vol. 304, p. 124507, Oct. (2021), https://doi.org/10.1016/J.CONBUILDMAT.2021.124507.
10. A. Mellaikhafi, A. Tilioua, and A. Benallel, 'Thermal performance assessment of a wall built with earth-based adobes and reinforced with pinnate leaves fibers', Mater Today Proc, vol. 58, pp. 1535–1540, Jan. (2022), https://doi.org/10.1016/J.MATPR.2022.03.296.
11. A. Standard, 'C518; Standard Test Method for Steady-State Thermal Transmission Properties by Means of the Heat Flow Meter Apparatus', ASTM International: West Conshohocken, PA, USA, (2021).
12. E. N. DIN, '12667, Thermal Performance of Building Materials and Products-Determination of Thermal Resistance using Guarded Hot Plate and HEAT flow Meter Methods-Products of High and Medium Thermal Resistance', *German version EN*, vol. 12667, (2001).

Machine Learning-Driven Optimization of Green Hydrogen Production in Porous Electrode Alkaline Electrolysis Cells

Mohamed-Amine Babay, Mustapha Adar, Souad Touairi, Ahmed Chebak, and Mustapha Mabrouki

Abstract This study develops a detailed mathematical model for alkaline water electrolyzer cells to simulate and analyze their performance, incorporating factors such as gas evolution, dissolution, bubble formation, and charge transport through a two-phase Euler-Euler method approach. Validated against experimental data at various current densities, the model assesses the effects of potassium hydroxide concentration, separator porosity, and electrolyte flow on flow dynamics and bubble behavior. Gas bubbles formation at electrodes decreases ionic conductivity and available surface area for reactions, increasing overpotential. The study also highlights the use of neural networks and ensemble tree models for predicting hydrogen production rates, achieving an average R-squared value of 0.98, indicating a strong prediction accuracy and potential for process optimization.

Keywords Alkaline water electrolysis · Green hydrogen · Bubble dispersion · ML

1 Introduction

Hydrogen, positioned primarily as an energy transporter rather than a primary resource, presents itself as a crucial contender amid challenges arising from depleting traditional energy supplies. As the need for alternatives that can be stored and transported, akin to hydrocarbons, becomes imperative, the focus on energy storage gains paramount importance. This need is particularly emphasized in the integration of renewable sources into the electrical grid due to their intermittent nature. Alkaline water electrolysis emerges as a commercially viable technology for hydrogen

M.-A. Babay (✉) · M. Adar · S. Touairi · M. Mabrouki
Laboratory of Industrial and Surface Engineering, Faculty of Science and Technology, Sultan Moulay Slimane University, Beni Mellal, Morocco
e-mail: mdamine.babay@gmail.com

A. Chebak
Green Tech Institute (GTI), Mohammed VI Polytechnic University, Ben Guerir, Morocco

A. Vaseashta et al. (eds.), *Proceedings of the Eighth International Symposium on Dielectric Materials and Applications (ISyDMA'8)*, Springer Proceedings in Physics 349, https://doi.org/10.1007/978-3-031-84809-4_24

production, offering durability, cost-effectiveness, and technical maturity over proton exchange membrane electrolysis.

Recent technological advancements have significantly heightened the efficiency of alkaline water electrolysis (AWE). Factors such as activation losses, ohmic losses, surface reductions, and transport resistance caused by bubbles play a crucial role in influencing the process's efficiency. Researchers have explored innovative solutions, such as anion exchange membranes and zero-gap cell designs, to minimize electrolyte resistance. The production of hydrogen and oxygen during AWE induces various flow patterns of two-phase electrolytes, with bubble production influencing the transfer of electroactive substances1].

Understanding the behavior of bubbles is pivotal, evident in numerous studies exploring their impact on AWE cell performance. Investigations utilizing computational fluid dynamics (CFD) models, such as Euler-Euler and Euler–Lagrange, have predicted or simulated the influence of bubbles on efficiency. Significant research efforts by Huang, Guan, and Eigeldinger contribute to the synthesis and characterization of materials for electrolysis, highlighting the importance of strain effects and electrolyte velocity in bubble movement [2, 3].

Studies by Reigel, Hine and Murakami, Abdelouahed, Boissonneau, El-Askary, Mandin, Ziegler, and Dahlkild provide valuable insights into bubble dynamics, current distribution, gas evolution, mass transfer, velocity, and bubble distribution in liquid electrolytes [3–6]. The utilization of CFD models in electrolysis systems deepens our comprehension of phenomena, providing opportunities to optimize AWE technologies for applications like hydrogen fuel production and metal electroplating. This study aims to bridge research gaps by employing Artificial Neural Networks, an ensemble tree model, for predicting hydrogen production through water electrolysis. With a focus on alkaline water electrolysis, the proposed ANN models offer a promising avenue for optimizing this process in hybrid solutions, contributing significantly to the field's advancement.

2 Computational Modeling Approach

2.1 Model Considerations

The model simplifies the complex dynamics of alkaline water electrolysis by treating the gas–liquid mixture as a Newtonian fluid, without turbulence, and represents gases as non-deforming, non-coalescing bubbles in a laminar flow. It assumes incompressibility and isotropy in the gas and electrolyte phases, with electrodes modeled as porous media characterized by their permeability and porosity. Operating at a constant temperature, the cell efficiently produces hydrogen and oxygen gases at opposite electrodes, with the mixed effluent exiting from the top, as illustrated in the provided Fig. 1.

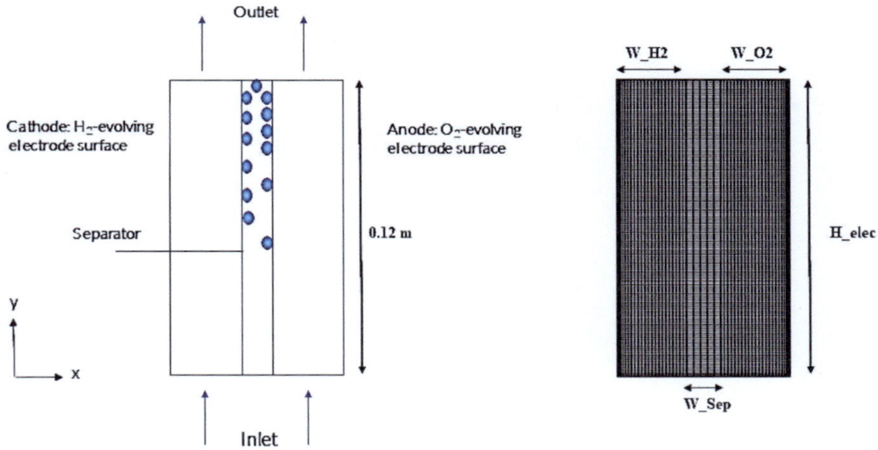

Fig. 1 The geometry of the domain

The operational dynamics of the cell involve the inflow of electrolytes from the lower section. A vital ion-conducting separator, also known as a diaphragm, segregates the electrolyte into distinct compartments. On the two vertical electrode surfaces, hydrogen and oxygen undergo evolution. The resulting gas and electrolyte mixtures elegantly exit the cell from its upper part, completing the intricate process.

2.2 Electrochemical Reactions

	Equations [7–9]
Mass conservation	$\frac{\partial}{\partial t}(\varepsilon\rho_m) + \nabla\cdot(\rho_m\vec{v}_m) = 0$
Momentum conservation	$\frac{1}{\varepsilon}\left[\frac{\partial}{\partial t}(\rho_m\vec{v}_m) + \frac{1}{\varepsilon}\nabla(\rho_m\vec{v}_m\vec{v}_m)\right] = -\nabla p + \nabla\left[\rho_m\nu_m\left(\nabla\vec{v}_m + \nabla\vec{v}_m^T\right)\right]$ $+\rho_m\vec{g} + \vec{F} + \nabla\left(\sum_{k=1}^{n} s^k\rho^k\vec{v}_{dr}^k\vec{v}_{dr}^k\right) + S_u$
Momentum source	$S_u = -\left(\frac{\rho_m\nu_m}{K}\right)\vec{v}_m$

(continued)

(continued)

	Equations [7–9]
Species conservation	$\varepsilon \frac{\partial C_i^k}{\partial t} + \nabla \cdot \left(\gamma_i C_i^k \vec{v}_m\right) =$ $\nabla \cdot \left[\sum_{k=1}^{n} D_i^{k,eff} \nabla\left(s^k C_i^k\right)\right] - \nabla \cdot \left[\sum_k \frac{mf_i^k}{MW_i j^k}\right] + S_i$ Advection correction factor: $\gamma_i = \frac{\rho_m\left(\sum_k^n \lambda^k mf_i^k\right)}{\sum_{kk}^n j^k \rho^k Y_i^k}$ Fickian diffusion flux: $D_i^{k,rff} = \varepsilon^r\left(s^k\right)^{r-1} D_i^k$ Consumption/production rate of species: $S_i = -\left(\frac{t_i}{z_i} + \frac{s_i}{n}\right)\frac{j}{F}$

2.3 *Dynamics of Porous Elements in AWE: A Two-Phase Model Analysis*

The two-phase mixture model is fundamental in investigating porous components within AWE cells. This model treats the gas and liquid electrolyte blend as a unified two-phase system, addressing crucial physical and chemical phenomena occurring in porous electrodes. Based on the principles of Newtonian fluids, it assumes laminar flow and constant temperature, while representing hydrogen and oxygen as dispersed gas bubbles. By incorporating key parameters such as activation energies, it offers a comprehensive understanding of how porous components behave in AWE cells. This study presents a two-phase mixture model employing the inhomogeneous Euler-Euler approach, focusing on shared pressure and utilizing mixture velocity and diffusive mass flux to depict multiphase flow. The transient isothermal model, guided by conservation equations, succinctly describes mass behavior within the mixture. In the realm of alkaline water electrolysis, electrodes produce hydrogen and oxygen gases through the application of electrical current to a solution, typically containing potassium hydroxide (KOH). Various measures, like expanding electrode surface area, controlling electrolyte flow, or optimizing operational parameters, are explored to enhance efficiency. Some electrolysis systems incorporate features for gathering and reutilizing produced gases, reducing bubble dispersion, and further refining process efficiency. Table 1 presents a compilation of equations utilized in this investigation.

Table 1 Conservation equation of bubble formation

	Equations ([7])
Bubble dispersion	$\vec{F}_k = \vec{F}_D + \vec{F}_L + \vec{F}_{BD}$ $\vec{F}_k = \underbrace{-\frac{3}{4}\phi_d \rho \frac{C_D}{d_b}\lvert U_{slip}\rvert U_{slip}}_{Dragforce} \underbrace{- \phi_d \rho C_L \lvert U_{slip}\rvert rot\left(\vec{V}_l\right)}_{Liftforce} \underbrace{- \phi_d \rho \frac{K_g}{d_b}\lvert U_{slip}\rvert \vec{\nabla}\phi_d}_{Bubbledispersionforce}$

3 Results and Discussion

Figure 1 displays the mesh layout of the Electrolyzer cell, which includes a porous separator positioned between two electrodes with elongated channels. To enhance computational efficiency, geometric symmetry was utilized to minimize the computational domain. This research utilized a two-phase model to simulate a 12 cm-long cell (H_elec). In Fig. 2, polarization curves obtained from experimental data and simulations are compared, acquired under a constant flow rate of 0.1 m/s. The findings demonstrate a significant correspondence between the model's predictions of overpotentials and the actual performance of components concerning current density. It was anticipated that hydrogen and oxygen bubbles would be present in the aqueous KOH electrolyte due to oversaturation, influenced by factors such as Henry's constant and gas–liquid interfacial energy.

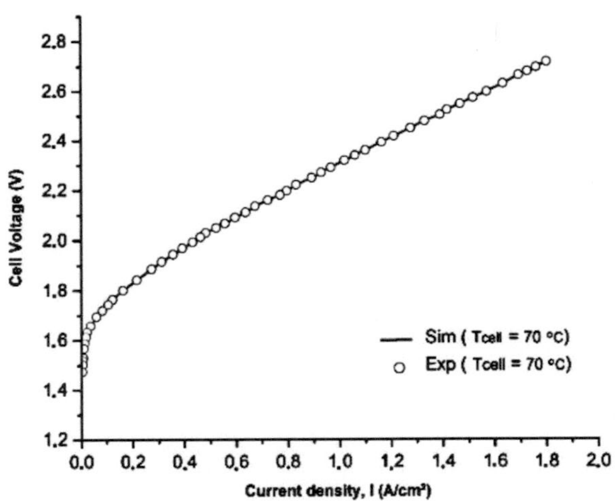

Fig. 2 Simulated and experimentally measured I-V curves at 70 °C

3.1 Influence of Velocity Magnitude

This research explores how varying KOH feed concentrations and separator porosities impact velocity magnitude in the continuous phase of electrolysis cells. By analyzing solutions with KOH concentrations of 500/1500/3000 mol/m^3 and different porosities, the study aims to discern the effects on flow dynamics. Velocity distribution is scrutinized at three points along the channel: the entrance, midpoint, and near the electrodes' ends, across each KOH and porosity setting. Findings, depicted in Fig. 3, reveal the nuanced relationship between KOH concentration, porosity, and flow patterns. Notably, increases in voltage, temperature, and reactant concentration accelerate reaction rates and the velocity of ions and gas bubbles. The study also observes that gas bubble size, specifically 50 μm bubbles, affects ascent speed, with smaller bubbles moving faster.

Figure 3 showcases the effects of KOH concentration and separator porosity on fluid dynamics within the system. At a low porosity setting, increasing the KOH concentration from 500 to 1500 mol/m^3 significantly boosts the peak velocity, suggesting more vigorous fluid movement. However, a further increase to 3000 mol/m^3 does not alter the flow much, hinting at a threshold beyond which higher concentrations don't enhance fluid motion. With a medium porosity ($\varepsilon = 0.4$), the trend in velocity magnitude remains similar but at a lower scale, indicating that a more porous separator leads to lower peak velocities and steadier flow. The highest porosity settings showed the least velocity, marking a pronounced slowdown in fluid movement and a more even flow distribution. The comparison highlights the profound impact of KOH concentration and separator porosity on how quickly fluids move, with these variables influencing the fluid's viscosity, the separator's effectiveness, and overall flow dynamics in meaningful ways.

3.2 Predicting Hydrogen Production Rates

Electrolysis, a technique that utilizes an electric current to split water molecules into hydrogen and oxygen, presents a promising avenue for hydrogen generation. This process, facilitated by electrolytes like potassium hydroxide, is affected by several factors such as electrolyte concentration, separator porosity, temperature, bubble size, cell voltage, and fluid velocity. To comprehend and model the intricate connections among these variables and hydrogen production rates, researchers utilize statistical and machine learning approaches, including neural networks inspired by the human brain. In this investigation, a MATLAB code incorporating the Neural Network Toolbox is employed to forecast hydrogen production rates based on diverse input parameters, thereby aiding in the enhancement of electrolysis efficiency [10].

During the data collection and preprocessing stage, hydrogen production rates were collected from an alkaline water electrolysis procedure conducted under

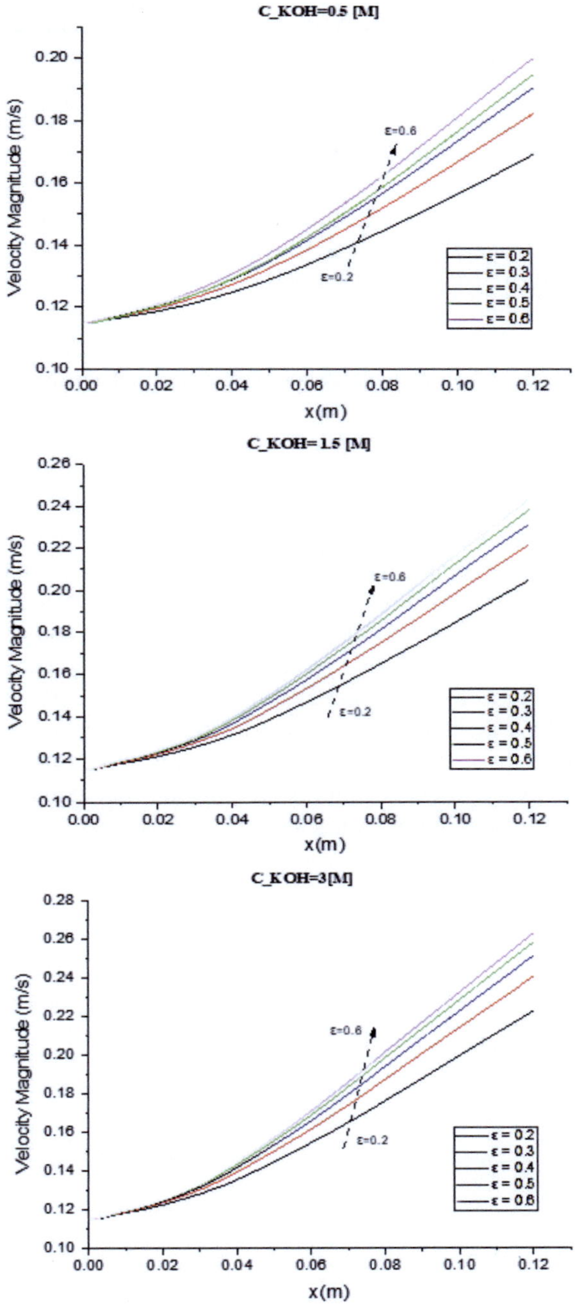

Fig. 3 An exploration was conducted into how velocity magnitude correlates with different separator porosities across various concentrations of potassium hydroxide (KOH)

different conditions. Various parameters including electrolyte concentration, separator porosity, temperature, bubble size, cell voltage, and fluid velocity were documented for each condition, forming a matrix where rows represented different conditions and columns represented parameters. To format the data for input into the neural network, a structure containing input parameters and target outputs was established. Subsequently, the target outputs matrix was reshaped into a column vector. Next, we utilized it to train and evaluate a neural network with a single hidden layer. The training process utilized 70% of the collected data, while 15% each was allocated for validation and testing purposes. The Levenberg–Marquardt algorithm was employed for training the network. The network's performance was assessed by computing the mean squared error and monitoring the number of epochs required for convergence. The trained neural network exhibited accurate predictions, achieving an MSE of 0.016 and converging after 200 epochs. Visual examination, depicted in Fig. 4, revealed a strong correlation between predicted outputs and target outputs, with a coefficient of 0.978, affirming the network's reliability.

The research identified the most effective input parameters for improving hydrogen production in alkaline water electrolysis. According to the ensemble tree model findings, the optimal conditions entail a moderate concentration of potassium hydroxide (c_KOH = 0.35), a moderate separator porosity (sep_porosity = 0.6), a high cell voltage (cell_voltage = 1.4), and a relatively small bubble diameter (bubble_diameter = 0.0013). These refined parameters resulted in the highest predicted hydrogen production rate of 24.42. The ensemble tree model exhibited a 30% explanation of variance in hydrogen production rates, with an average R-squared value of 0.3020, underscoring the significance of potassium hydroxide concentration on hydrogen production. Despite demonstrating moderate fit, further research incorporating additional variables and a larger dataset may enhance model accuracy and

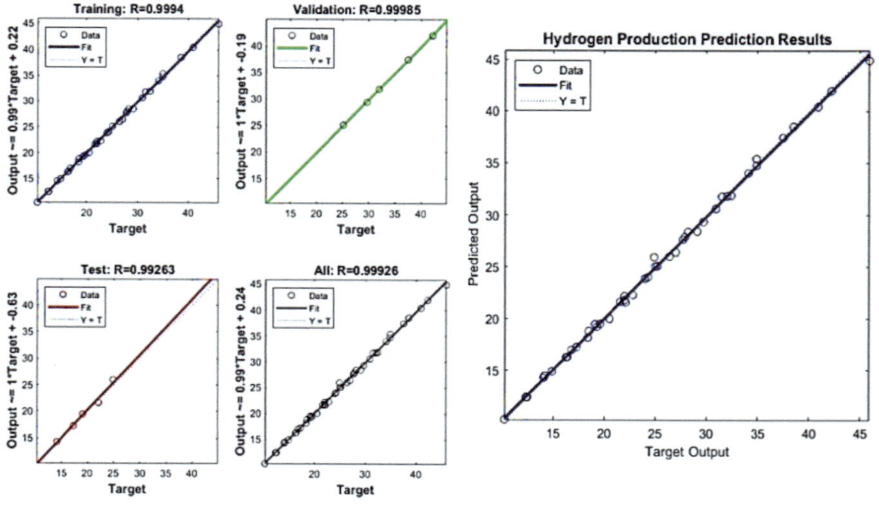

Fig. 4 Performance of the ANN model to predict HP

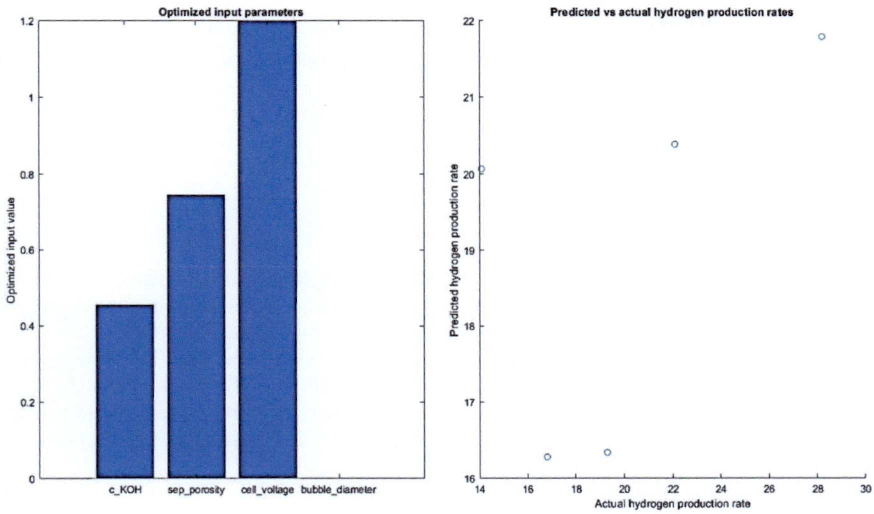

Fig. 5 Predicting hydrogen rates with the optimized input value

predictability. Figure 5 revealed slight variations in the optimized input parameters across different cross-validation folds. Nonetheless, the average optimized values remained steady, with c_KOH = 0.35, sep_porosity = 0.50, cell_voltage = 1.5, and bubble diameter = 2.40×10^{-3}. These findings suggest that the ideal conditions for hydrogen production involve a moderate concentration of potassium hydroxide, a moderate separator porosity, a high cell voltage, and a relatively small bubble diameter.

4 Conclusion

A 2D electrochemistry-transport model was devised for zero-gap alkaline water electrolysis cells, employing an inhomogeneous Euler-Euler mixture approach. The model depicted hydrogen and oxygen volume fraction contours, revealing notable overpotentials attributed to bubble nucleation and two-phase flow. The hydrogen fraction predominated due to lower oversaturation, notably impacting the negative electrode. The study proposed a new bubble transfer description, which aligned well with experimental findings and improved numerical convergence. However, challenges persist in achieving optimal grid resolution. Further exploration is necessary to grasp the influence of current density and bubble diameter on output parameters. Additionally, when comparing two neural network models, the second showed superior performance in terms of Mean Squared Error (MSE), suggesting efficient learning of underlying data patterns. While the hydrogen production model exhibited some predictive capability, additional research is warranted to identify additional factors affecting accuracy.

References

1. Babay M-A, Adar M, Chebak A, Mabrouki M (2024) Exploring the sustainability of serpentine flow-field fuel cell, straight channel PEM fuel cells high temperature through numerical analysis. Energy Nexus 14:100283. https://doi.org/10.1016/j.nexus.2024.100283.
2. Eigeldinger J, Vogt H (2000) Bubble coverage of gas-evolving electrodes in a flowing electrolyte. Electrochim Acta 45:4449–4456. https://doi.org/10.1016/S0013-4686(00)00513-2.
3. Hine F, Murakami K (1980) Bubble Effects on the Solution IR Drop in a Vertical Electrolyzer Under Free and Forced Convection. J Electrochem Soc 127:292. https://doi.org/10.1149/1.2129658.
4. Ziegler D, Evans JW (1986) Mathematical Modeling of Electrolyte Circulation in Cells with Planar Vertical Electrodes: II . Electrowinning Cells. J Electrochem Soc 133:567–576. https://doi.org/10.1149/1.2108620.
5. Boissonneau P, Byrne P (2000) An experimental investigation of bubble-induced free convection in a small electrochemical cell. J Appl Electrochem 30:767–775. https://doi.org/10.1023/A:1004034807331.
6. El-Askary WA, Sakr IM, Ibrahim KA, Balabel A (2015) Hydrodynamics characteristics of hydrogen evolution process through electrolysis: Numerical and experimental studies. Energy 90:722–737. https://doi.org/10.1016/j.energy.2015.07.108.
7. Babay MA, Adar M, Chebak A, Mabrouki M (2023) Dynamics of Gas Generation in Porous Electrode Alkaline Electrolysis Cells: An Investigation and Optimization Using Machine Learning. Energies 2023, Vol 16, Page 5365 16:5365. https://doi.org/10.3390/EN16145365.
8. Mohamed-Amine Babay, Mustapha Adar, Souad Touairi, et al (2024) Numerical Simulation and Thermal Analysis of Pressurized Hydrogen Vehicle Cylinders: Impact of Geometry and Phase Change Materials. J Adv Res Fluid Mech Therm Sci 117:71–90. https://doi.org/10.37934/arfmts.117.2.7190.
9. Babay M-A, Adar M, Nouri R, et al (2024) Integrated Thermodynamic Analysis and Channel Variation Effects on Solid Oxide Electrolysis for Efficient Hydrogen Generation. Procedia Comput Sci 236:152–159. https://doi.org/10.1016/j.procs.2024.05.016.
10. Babay M-A, Adar M, Chebak A, Mabrouki M (2025) Forecasting green hydrogen production: an assessment of renewable energy systems using deep learning and statistical methods. Fuel 381:133496. https://doi.org/10.1016/j.fuel.2024.133496.

Development and Implementation of a Blind Source Separation Algorithm to Extract the Fetal Electrocardiographic Signal in Real-Time

Mohcin Mekhfioui, Aziz Benahmed, Ahmed Chebak, Rachid Elgouri, and Laamari Hlou

Abstract In this paper, we present an effective solution for analyzing and extracting electrocardiogram (ECG) signals from the abdomen and thorax of pregnant women, with the primary aim of isolating fetal ECG (fECG) and maternal ECG (mECG) signals. To solve problems such as the low amplitude of the fetal ECG, the various noise sources during signal acquisition, and R-wave overlap, we developed a new method based on blind source separation techniques. This method uses independent component analysis algorithms to accurately detect and extract fECG and mECG signals from abdominal and thorax data. To validate our solution, we carried out experiments using a reliable database specifically designed to evaluate fECG extraction algorithms. In addition, to demonstrate real-time applicability, we implemented our method on an embedded board connected to electronic modules of the pregnant woman's body temperature. This configuration also includes the transmission of fetal and maternal data to physicians for diagnosis of fetal and maternal conditions. The accuracy of our method in isolating fECG and mECG signals under difficult conditions is demonstrated by our results, as well as by the calculation of heart rates (fBPM and mBPM). This method offers good potential for improving fetal monitoring and maternal health care during childbirth.

Keywords ECG · Fetal ECG · Blind separation source · Embedded system · Real time implementation

M. Mekhfioui (✉) · A. Chebak
Green Tech Institute (GTI), Mohammed VI Polytechnic University, Benguerir, Morocco
e-mail: mohcin.mekhfioui@um6p.ma

M. Mekhfioui · L. Hlou
Faculty of Science, University Ibn Tofail, Kenitra, Morocco

A. Benahmed
ERSC Team, Mohammadia Engineering School, Mohammed V University, Rabat, Morocco

R. Elgouri
Laboratory of Electrical Engineering and Telecommunications Systems, ENSA, Ibn Tofail University, Kenitra, Morocco

A. Vaseashta et al. (eds.), *Proceedings of the Eighth International Symposium on Dielectric Materials and Applications (ISyDMA'8)*, Springer Proceedings in Physics 349, https://doi.org/10.1007/978-3-031-84809-4_25

1 Introduction

The electrocardiogram (ECG) is a technique for recording the electrical activity generated by the heart. The ECG allows clinicians to analyze a patient's heart status and do a more detailed study or diagnosis. The ECG signal is typically measured using electrodes implanted on the skin. In the event of a pregnant woman, obtaining a fetal ECG is difficult since direct contact with the fetus is dangerous [1].

Furthermore, ECG recordings taken from the surface of a maternal belly contain a wide range of bioelectrical potentials, including maternal cardiac activity, fetal cardiac activity, maternal muscle activity, fetal activity, noise, and so on. These signal discrepancies exacerbate the challenge of recreating the prenatal ECG [2]. Several investigations have been undertaken on this subject using various signal separation approaches [3].

Blind source separation, specifically Independent Component Analysis (ICA) [4], is the most widely published and commonly used non-adaptive approach for obtaining the fECG signal. This method assumes that the components are statistically independent and uses as many electrodes on the maternal abdomen as there are uncorrelated signal sources. As a result, when separating fetal and maternal components from the abdominal signal, ICA requires at least two electrodes. It is not required to utilize too many electrodes because each one generates its own noise. ICA preprocessing includes centering, which transforms the vector into a zero-mean variable, and whitening, which produces a new vector with uncorrelated components and unit variances. Several ICA-based approaches have been developed, including FastICA [5], Single-Channel ICA (SCICA), and Joint Approximate Diagonalization of Eigenmatrices (JADE) [6]. In the extraction of fetal R peaks that overlap with the mQRS complex, ICA has an accuracy of around 97.47%.

Our application aims to separate fetal cardiac activity (fECG) from maternal cardiac activity (mECG) in real time. The aim is to extract specific components of the ECG signal for effective fetal monitoring during pregnancy and delivery. Separation will be performed using the method developed in the previous paper [5] using blind source separation algorithms and the DUE Arduino board [7], This paper focuses on the implementation of an algorithm combining several source separation methods on a low-cost Arduino board. Based on the calculation of the signal-to-interference ratio (SIR) of each method, the best-performing method will be used to estimate the output signals. The index of this method will be displayed on LEDs connected to the on-board board.

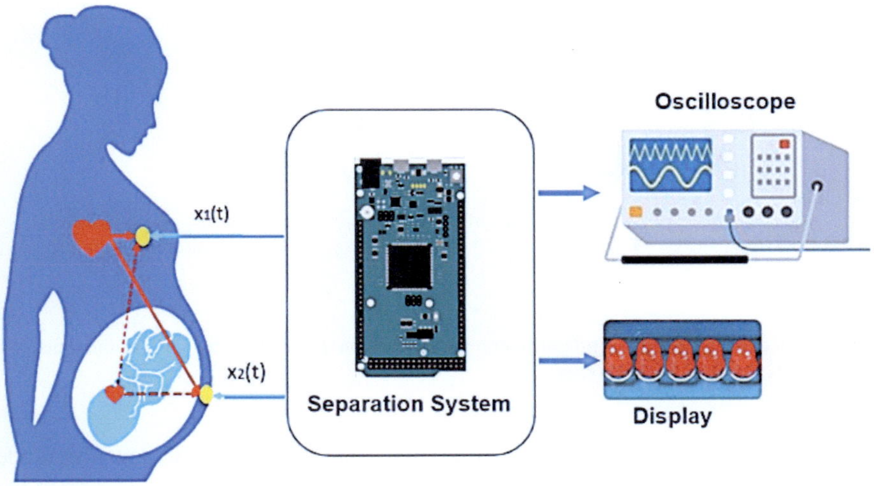

Fig. 1 Architecture of our system

2 Theory and Methods

2.1 System Architecture

The paper presents a low-cost technique to extract ECG signals from pregnant women and their fetuses' using signals collected from the woman's thorax, which is an excellent indicator of fetal health while minimizing the hazards associated with it. The solution developed consists of multiple steps and systems (Fig. 1).

- **The acquisition system** is made up of two electrodes, one on the abdomen and the other on the thorax of pregnant women. The signals recorded by these electrodes are transmitted in real time to a separation and processing system.
- **Separation and processing system** block includes a high-performance, low-cost on-board board that extracts the ECG signal from the fetus and its mother using blind source separation techniques and displays the results on the oscilloscope.

2.2 ECG Signal of a Pregnant Woman

The signal obtained from the abdomen of a pregnant woman during an electrocardiogram (ECG) is a complex mixture of many important components. The mother's ECG (mECG), which shows the electrical activity of the mother's heart, is the first component of this indication. With its complex QRS and P, T, and R waves, this part of the signal resembles a typical ECG indication. The mECG is superimposed on the fetal ECG (fECG) [8].

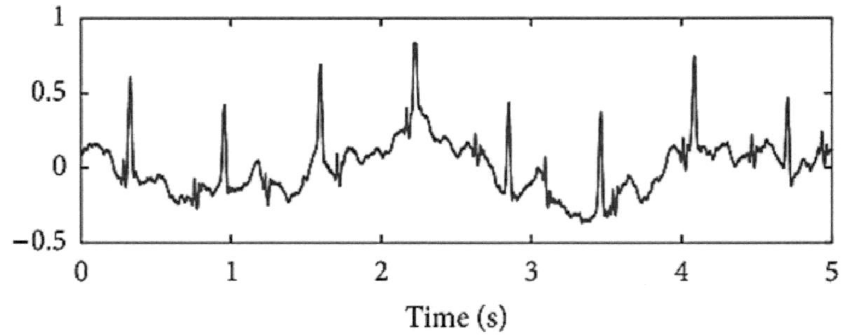

Fig. 2 Electrocardiogram of a pregnant woman. *Please note* F: indicates the QRS of the fetus, and M: indicates the QRS of the mother

The fetal ECG is generally more difficult to determine due to its low amplitude and the interference caused by the mECG. However, isolating the fECG from the general signal is essential for fetal monitoring, as it provides valuable information on fetal heart rate and rhythm. In addition, the abdominal ECG includes other signals resulting from uterine movements and muscle contractions. These additional signals, often called noise, can make extracting the fECG signal even more difficult. This calls for more sophisticated signal processing methods to separate these data without altering the ECG components (Fig. 2).

2.3 Blind Source Separation

Blind source separation (BSS) is a technique that aims to recover a set of original signals from a set of mixed signals, without knowing how the signals were mixed or what their characteristics are. BSS can be applied to various types of data, such as audio, images, or biomedical signals. BSS is also known as source separation or signal separation [9].

One of BSS's main challenges is finding a suitable criterion to measure the independence of the recovered signals. Different methods have been proposed for this purpose, such as principal component analysis (PCA), independent component analysis (ICA), or non-negative matrix factorization (NMF). These methods use different assumptions and mathematical tools to find a transformation that maximizes the independence of the signals.

The principle of the BSS can be depicted in Fig. 3, It is to find in the ideal case the matrix W of size $n \times m$ which provides the output vector:

$$\boldsymbol{y}(\boldsymbol{k}) = \boldsymbol{W}\boldsymbol{x}(\boldsymbol{k}) = \boldsymbol{W}\boldsymbol{H}\boldsymbol{s}(\boldsymbol{k}) \approx \boldsymbol{s}(\boldsymbol{k}) \tag{1}$$

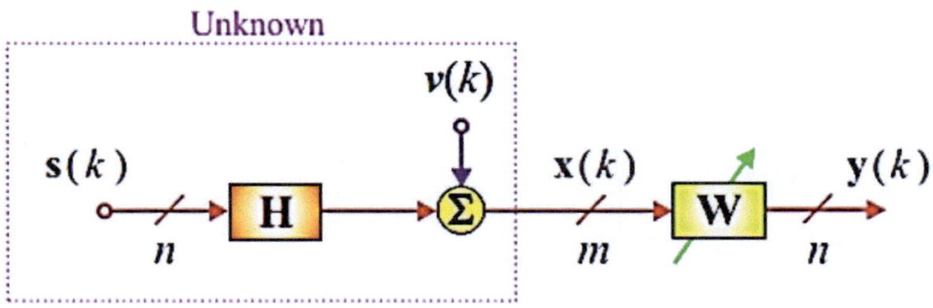

Fig. 3 Principle of the separation of sources [5]

The estimated sources are obtained from the vector **s(k)** and their related projections for the various sensors are determined from the estimated mixture array: $H = W^{-1}$ [5].

3 Implementation Result

3.1 Database

The data used to validate this method are extracted from the FECGSYNDB database [10], which is a reliable source for testing fetal ECG extraction algorithms. The FECGSYNDB database contains ECG recordings from a pregnant woman using two electrodes, one placed on the abdomen and the other on the chest. The sampling duration is four seconds at a sampling frequency of 250 Hz, resulting in a signal of 1000 samples. The database utilized for the algorithm is presented in Fig. 4. As seen in the samples, the data contains six beats of maternal ECG (mECG) and eleven beats of fetal ECG (fECG), where the mECG has a much higher amplitude than the fECG.

The fECG signal, if obtained from the abdomen area, will show very little fluctuation because of the distances between each electrode and the fetal heart's core. This quality is lost when working with the mECG because the signal will vary greatly based on where the electrodes are placed on the abdomen. These properties allow the mECG to be distorted by combining signals from several electrodes while maintaining the fECG with very slight modifications. For reasons of ease of use and electrode placement, it is therefore not necessary to use more than two electrodes for this technique. The separation procedure will be made easier by the combination of the two signals, which will produce a signal with a high power ratio between the fECG and the mECG.

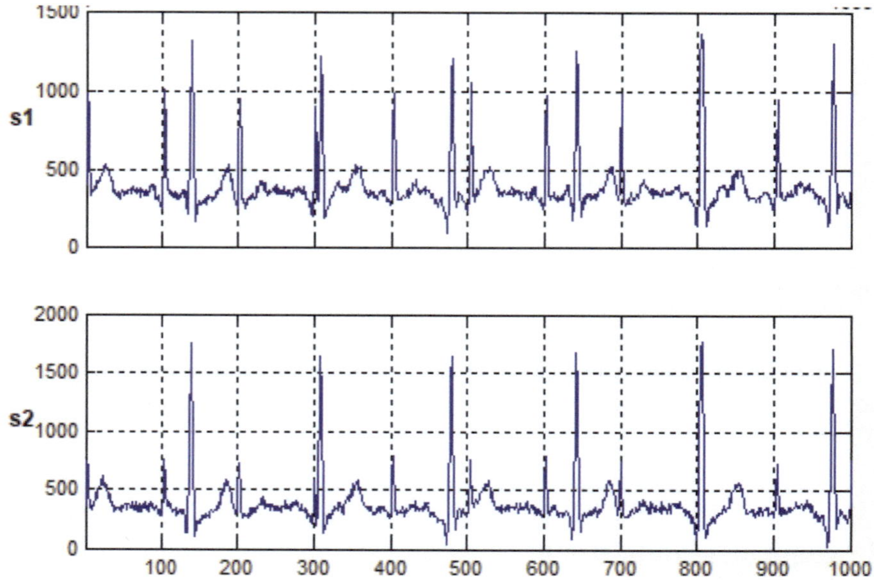

Fig. 4 ECG of a pregnant woman

3.2 Hardware Implementation

In this section, our separation method is implemented on an embedded board, ECG signals are captured by electrodes placed on the pregnant woman's abdomen and thorax, then sent to the Arduino board via the analog pins, to eliminate cable noise, signals are sent from Matlab using the "from workspace" block in "external mode" for simulation. The figure shows the original signals captured on the maternal abdomen and thorax (×1 and × 2), together with the estimated signals (y1 and y2). Signal y1 represents the fetal ECG extracted from signals × 1 and × 2 using the proposed algorithm, while signal y2 represents the ECG signal from the pregnant woman. The proposed method can extract a clean fetal ECG signal in real time without the loss of R waves. From these results, we can calculate several parameters such as heart rate and duration of each wave (Figs. 5 and 6).

On the four LEDs connected to the system, we can see that the index displayed is 4 (0100 in binary). This index corresponds to the SYM-WHITE algorithm in the implemented method table [5]. The results of the real-life implementation demonstrate superior performance in extracting fECG and mECG signals in real-time, without the need for a computer or separation software.

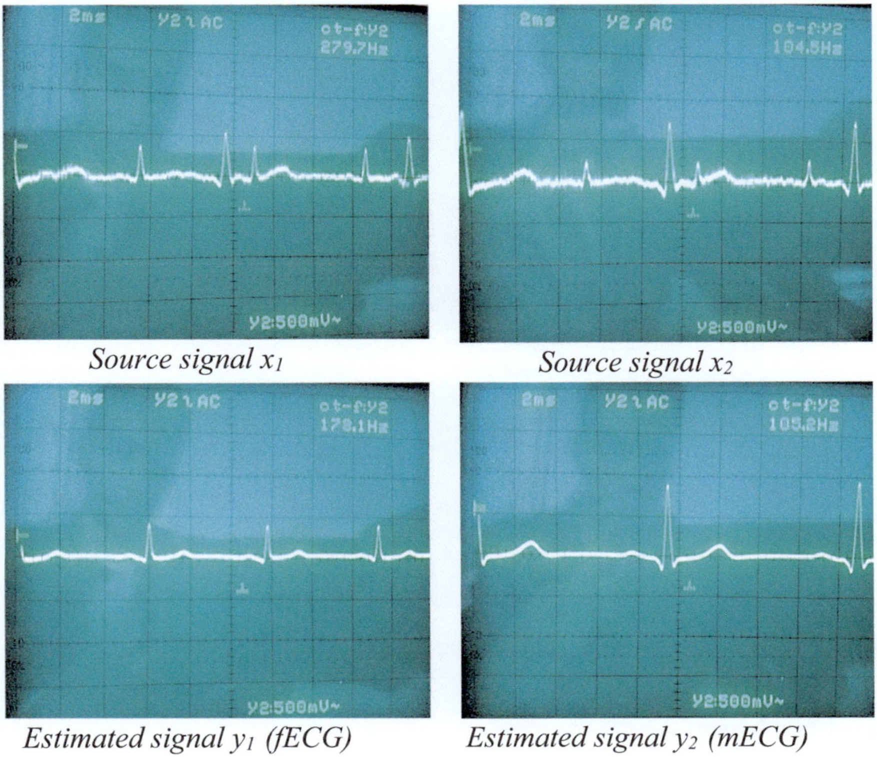

Fig. 5 Real-time extraction result

Fig. 6 The index of the most efficient separation algorithm

4 Conclusion

In this paper, we presented a new approach for real-time extraction of fetal and maternal ECG signals using blind source separation techniques. By implementing several independent component analysis methods on a low-cost embedded system, we efficiently isolated the fetal ECG from complex signals recorded from the abdomen and thorax of pregnant women. The method demonstrated high accuracy and reliability, validated by experiments with the FECGSYNDB database. Our approach has significant potential for improving fetal surveillance and maternal health care, providing a practical and efficient solution for clinical environments without depending on significant hardware or computing resources. Future work will focus on optimizing the algorithm for various clinical environments and exploring other applications in maternal and fetal health monitoring.

References

1. Raj, A., Brablik, J., Kahankova, R. et al. Nature-inspired method for noninvasive fetal ECG extraction. Sci Rep 12, 20159 (2022).
2. S. Mirza, K. Bhole and P. Singh, "Fetal ECG Extraction and QRS Detection using Independent Component Analysis," 2020 16th IEEE International Colloquium on Signal Processing & Its Applications (CSPA), Langkawi, Malaysia, 2020, pp. 157-161.

3. Cao S, Xiao H, Gong G, Fang W, Chen C. Morphology extraction of fetal ECG using temporal CNN-based nonlinear adaptive noise canceling. PLoS One. (2022) Dec 15;17(12):e0278917.
4. Tharwat, A. (2021). Independent component analysis: An introduction. Applied Computing and Informatics, 17(2), 222–249.
5. Mekhfioui, M.; Elgouri, R.; Satif, A.; Hlou, L. Real-time implementation of a new efficient algorithm for source separation using Matlab & Arduino due. Int. J. Sci. Technol. Res. (2020), 9, 531–535.
6. Khan JB, Jan T, Khalil RA, Saeed N, Almutiry M. An Efficient Multistage Approach for Blind Source Separation of Noisy Convolutive Speech Mixture. Applied Sciences. 2021; 11(13):5968.
7. Mekhfioui, M.; Elgouri, R.; Satif, A.; Moumouh, M.; Hlou, L. Implementation of the least mean square algorithm using Arduino & Simulink. Int. J. Sci. Technol. Res. (2020), 9, 664–667.
8. Sarafan, S.; Le, T.; Lau, M.P.H.; Hameed, A.; Ghirmai, T.; Cao, H. Fetal Electrocardiogram Extraction from the Mother's Abdominal Signal Using the Ensemble Kalman Filter. Sensors (2022), 22, 2788.
9. Benahmed, A.; Mekhfioui, M.; Hamzaoui, E.M.; Guennoun, Z. FPGA Implementation of Adaptive Zero-Tracking Algorithm For Real-Time Doa Estimation Using A Linear Antenna Array. ASEAN Eng. J.(2024), 14, 207–212.
10. Behar, J.; Andreotti, F.; Zaunseder, S.; Li, Q.; Oster, J.; Clifford, G.D. An ECG model for simulating maternal-fetal activity mixtures on abdominal ECG recordings. Physiol. Meas. (2014), 35, 1537–1550.

Energy Recovery from Beam Vibrations-Based Piezoelectric Material

Salam Khrissi, Salma Kaotar Hnawi, Yassine Tabbai, and Houda Lifi

Abstract The work presented in this article focuses on the recovery of vibratory energy using a piezoelectric beam. The device implemented in our laboratory is quite far from industrial reality, but it has the advantage of being easily reproducible and inexpensive. In the first step, we will model the functioning of the piezoelectric beam using an electromechanical model resulting from experimental tests. After validation on a basic recovery system (diode bridge) studied both from an experimental point of view and by simulation, the model is used on a more powerful recovery system (SSHI).

Keywords Piezoelectric beam · Energy recovery · SSHI control · Piezoelectric · Energy harvester

---

S. Khrissi · H. Lifi
Laboratory of Materials, Processes, Environment and Quality, National School of Applied Sciences, Cadi Ayyad University, Safi, Morocco

S. Khrissi
Laboratory Spectrometry of Materials and Archaeomaterials (LASMAR), Faculty of Science, Moulay Ismail University, Meknes, Morocco

S. K. Hnawi (✉)
School of Electrical Engineering, Xi'an University of Technology, Xi'an 710048, China
e-mail: hnawi.salma@gmail.com

Molecular Chemistry Laboratory, Coordination Chemistry and Catalysis Unit, Faculty of Sciences Semlalia, Cadi Ayyad University, Marrakech, Morocco

Y. Tabbai
Laboratory of Engineering Sciences for Energy, National School of Applied Sciences of El Jadida, Chouaib Doukkali University, El Jadida, Morocco

H. Lifi
Laboratory of Innovation in Science, Technology, and Modeling (LISTM), Faculty of Science, Chouaib Doukkali University, El Jadida, Morocco

A. Vaseashta et al. (eds.), *Proceedings of the Eighth International Symposium on Dielectric Materials and Applications (ISyDMA'8)*, Springer Proceedings in Physics 349, https://doi.org/10.1007/978-3-031-84809-4_26

1 Introduction

Low-power communication devices are increasingly used for home automation or health monitoring of people [1, 2]. The recovery of energy from ambient vibrations [3–6] is an alternative to the use of batteries or to increase the lifetime of the systems. We present here the possibilities offered by a PZT-5H piezoelectric recuperator (a piezoelectric membrane fixed on a vibrating beam) to power a low-power wireless communication module. This module is the basic building block of a wireless sensor network. The piezoelectric device is presented in the first part with its excitation and measurement system. The second part describes the experiments necessary for the modeling and explains the resulting model. The energy recovery is then implemented in the third part with a comparison between the experimental results and those from simulations for the simplest device. The last part analyzes the amount of energy needed to power the communication module and compares it with the energy recovered. The converters to be inserted between the storage and the load are also discussed in this part.

2 Experimental Setup

For cost and availability, the recuperator consists of a vibrating beam (metal reglet) on which a piezoelectric membrane is glued (buzzer-piezo: Kingstate KPSG-100). The ruler is fixed to a frame. In practice, the end of the beam can be excited regularly (by the wind [4]) or episodically (the end of the ruler is moved manually). The excitation can also come from the movement of the frame (for example, a bicycle [5] or a motor [6]). The frame is fixed for our tests, and an electromagnet powered by an LM12 audio amplifier is used to vibrate the beam (Fig. 1). A function generator sets the frequency and the level of excitation. A single-axis accelerometer (ICP ref 333B32) and its conditioner (Anicia CEI) deliver a voltage image of the acceleration (Fig. 2). The accelerometer also serves as a seismic mass [7–10].

2.1 *Modeling*

The deformation of a PZT-5H piezoelectric membrane induces a polarization of the piezoelectric material, and a charge transfer proportional to the strain:

$$q = \alpha . u \tag{1}$$

Classically, in the first approach, the beam and PZT-5H piezoelectric membrane device are modeled near the resonance by a current source i_{PZ} in parallel with a capacitor C (Fig. 3).

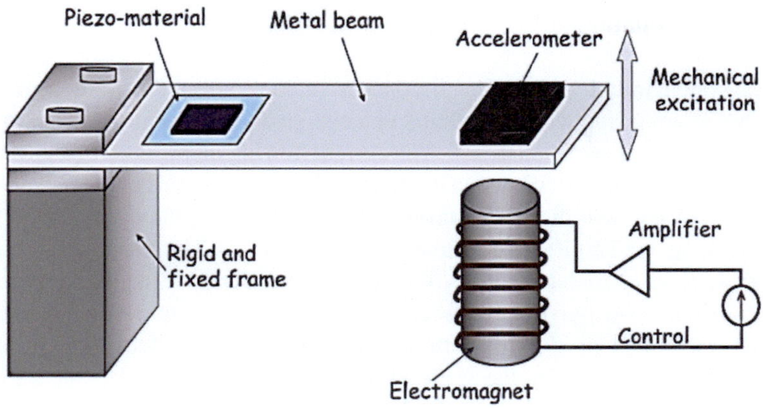

Fig. 1 Excitation and measurement equipment

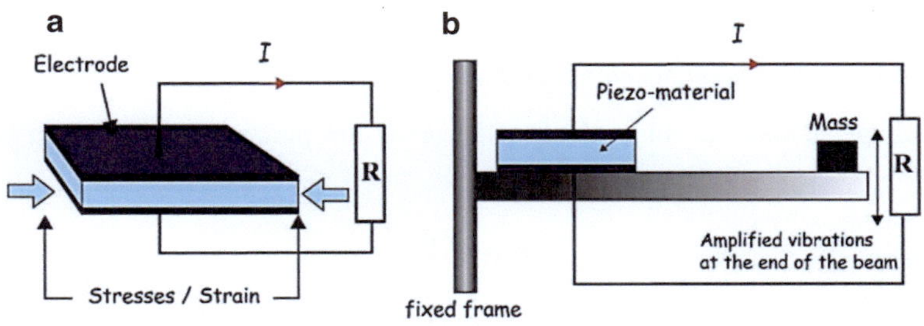

Fig. 2 **a** a piezoelectric converter, and **b** a piezoelectric cantilever

Fig. 3 Electrical model at resonance

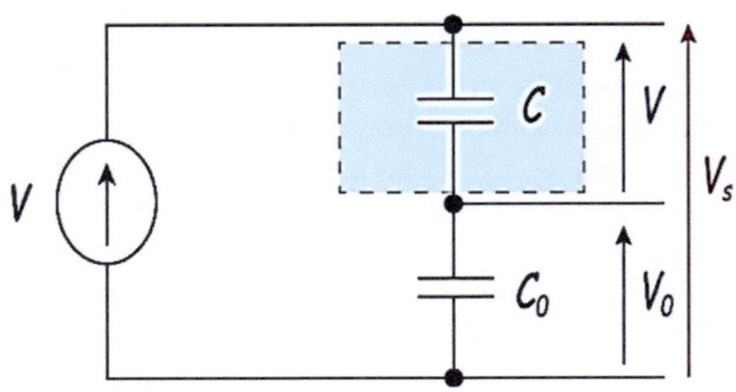

2.1.1 Determination of Resonance Frequency

To determine the resonance frequency of the beam. A constant stress is applied to the beam, and then this stress is released. The test allows determining the frequency of the damped oscillations, which will be confused with the natural frequency because of the low damping rate. Let us say: freq. = 18Hz.

2.1.2 Determination of C

With the system initially at rest, the piezo is disconnected from the oscilloscope; then a constant strain is applied. This results in a charge displacement to capacitor C; as the strain persists, $u = c^{te}$ and $\dot{u} = 0$. The current flowing into C is zero, V_{PZ} remains constant. At $t = t_A$, the oscilloscope is connected to the terminals of the piezoelectric (the capacitance C will discharge through the impedance of the oscilloscope (R_{OSC} = 1 MΩ, C_{OSC} = 12 pF) with a time constant R_{OSC}. C. The visualization of this discharge allows us to determine C (the influence of COSC is negligible here). We find C = 70 nF ([4–6] gives C between 65 and 85 nF). It is interesting to carry out the same test with an oscilloscope attenuator probe to see the effect of the resistance of the probe and to confirm the value of the capacitance by another measurement (R_{OSC} = 10 MΩ, C_{OSC} = 15pF for the N2862B).

2.1.3 Determination of A

The first method uses an accelerometer

The voltage across the internal capacitance of the piezoelectric is given by the equation:

$$q\prime = C.V_{PZ} \tag{2}$$

When the piezoelectric is not connected to a circuit, the displaced charge q and the charge q′ of C merge, resulting in the following:

$$a.u = C.V_{PZ} \tag{3}$$

To determine *a*, the image of the deformation, as noted by *u*, must be known. Measuring *u* with precision is difficult with the equipment present in the laboratory (rulers), but we can access u with the accelerometer mounted at the end of the ruler.

Exciting the piezoelectric actuator in a sinusoidal manner and assuming a linear and sinusoidal (for *u* and *Vpz*) regime to the pulsation ω, the preceding relationship becomes effective for the values: from a practical point of view, the assumption of linearity is not always completely verified (the acceleration signal may contain harmonics). We will therefore make either a direct measurement or the FFT of V_{PZ} and V_{acc} to determine the RMS values of the fundamentals. We recall that the measurement in dBv is related to the RMS value by a series of tests conducted around the resonance between 14 and 24 Hz leads to a value of,

$$\alpha = 1,45.10^{-4}\mathrm{C/m} \tag{4}$$

with a standard deviation of 5% from the mean.

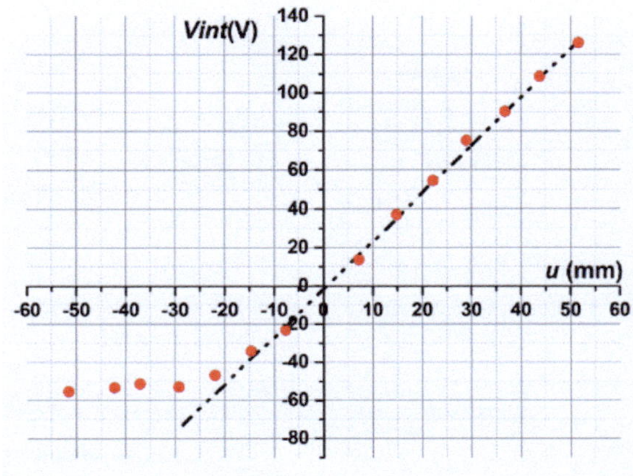

Fig. 4 Electrical model at resonance

The second method, without an accelerometer

We can also use a static method. First, we make sure that VPZ is zero and that the probe is disconnected. We then deform the beam by noting the deviation u. (rq1: *u* is counted positively downwards. The deformation induces a charge transfer and thus a voltage across the piezoelectric membrane. The probe is reconnected and a similar pattern to the one shown in Fig. 3 is obtained. The initial value, V_{init}, of the V_{PZ} voltage decay is recorded. The results are presented in Fig. 4.

We find saturation in the behavior of the upwardly curved piezoelectric membrane, but in the region of linearity: $V_{init} = 1{,}75.10^3 \times u$. Using Eq. 8 from Eq. 3 at $t = t_A{}^+$ we find $\alpha = 1{,}23.10^{-4}$ C/m. The results are in good agreement with 3.

$$q = a.u = C.V_{init} \tag{5}$$

2.2 *Energy Recovery*

2.2.1 Elementary Structure

The objective of this part is to design a system for the recovery and storage of the charges displaced during the movements of the beam and the piezoelectric membrane to supply the data logger. Given the alternating nature of the voltage produced by the transducer, the simplest solution to implement is a single-phase rectifier with capacitive filtering.

We cable the assembly presented in Fig. 5, replacing the load (data logger) with a voltage probe (the load resistance is then the impedance of the probe, i.e., 10 M). The V_{PZ} voltage is read using a high-impedance isolation probe and sent to the

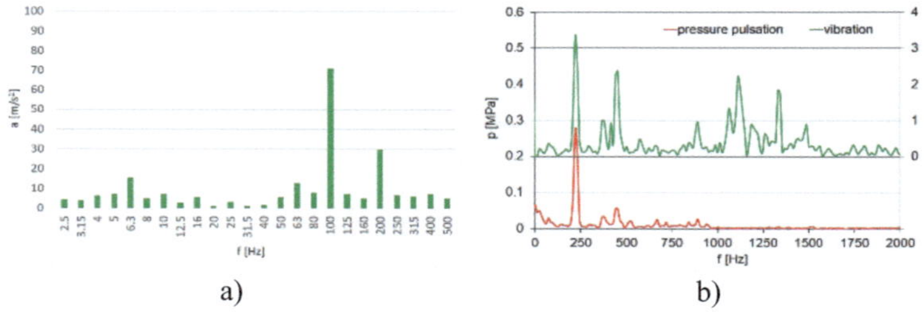

Fig. 5 Elementary Assembly (after [11])

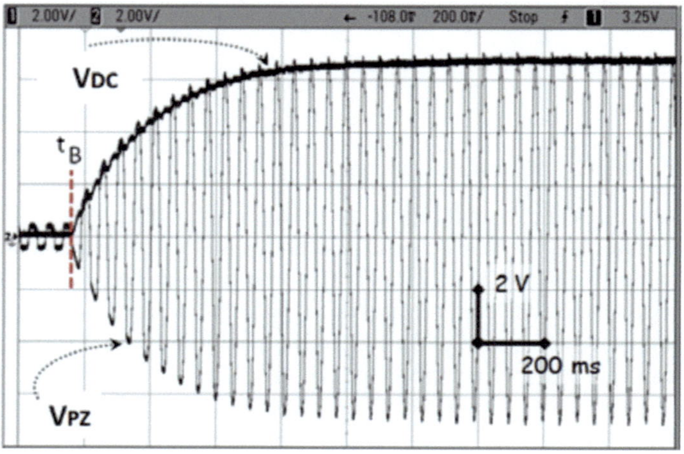

Fig. 6 Evolution of V_{PZ} (t) and V_{DC} (t)

oscilloscope. At the same time, the VDC voltage is measured (see Fig. 6). At $t = t_B$, the switch K initially closed is opened.

2.2.2 Validation of the Model

This part allows us to validate the model of the piezoelectric element obtained previously. In the sinusoidal regime, the measurement of the maximum acceleration allows one to go back to the maximum speed and, with the coefficient of deformation α, to determine the maximum value of the current delivered by the equivalent current source (see Fig. 3). During a test corresponding to the charge of a capacitor $C_L = 1\mu F$, we noted, in addition to $V_{PZ}(t)$ and $V_{DC}(t)$: $Vacc_{max} = 3{,}44V$ and $f = 18Hz$, that is,

$$u_{\max} = k\frac{Vacc_{\max}}{2\pi f} \tag{6}$$

Fig. 7 Comparison simulation/measurement for $V_{DC}\ (t)$ and $V_{PZ}\ (t)$

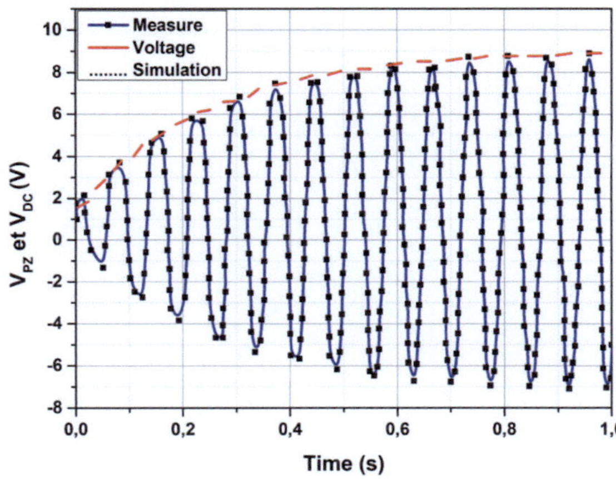

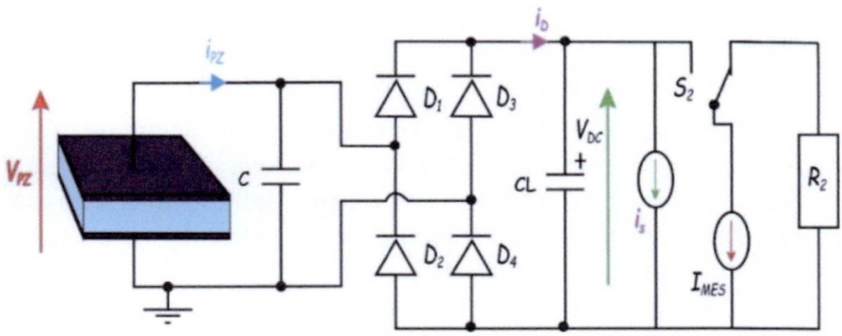

Fig. 8 Model of the complete device (recuperator, storage, load)

We find $\dot{u}_{max}$=0.372 ms^{-1}. By deriving Eq. 1, we obtain:

$$i_{PZ_{\max}} = \alpha.u_{\max} \tag{7}$$

$i_{pzmax} = 53.6\mu A$. This value of current comes (with C = 70nF) to complete the model used in the PSIM software. We can thus simulate the circuit (Fig. 8) and compare it with the experiment (Fig. 7).

The results obtained here confirm the validity of the electrical model of the piezoelectric, both for the value of the capacitance C and for the electromechanical constant α.

2.2.3 First Use of the Model

This model is then used in PSIM by adding the recovery device. At first, the switch S2 remains in the position given in Fig. 8. We only simulate a rectifier delivering a

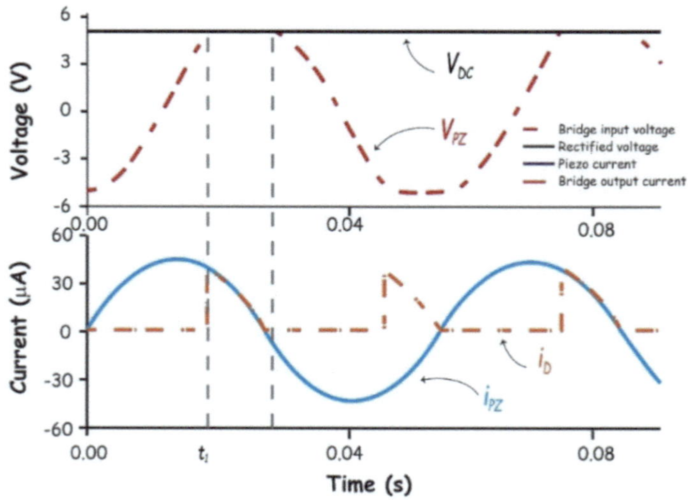

Fig. 9 Voltage and current at the input and output of the bridge

constant DC current source IS (changing IS changes the consumption of the load). To understand the operation and calculate the recoverable power in a steady state, we observe the input and output currents and voltages of the rectifier with capacitive filtering (Fig. 9). This part is important for the understanding the system because the currents are difficult to measure in practice.

Assuming that the value of CL is sufficient for V_{DC} to remain constant and assuming perfect diodes, the steady state calculations can be conducted analytically as presented below. From $t = 0\ t_o\ t = t_1$ the diode bridge does not conduct, the current i_{pz} charges the capacity of the piezoelectric, and allows the voltage at its terminals to rise from $-V_{DC}\ to + V_{DC}$ in a time:

$$t_1 = \frac{1}{\omega} Ar\cos\left(1 - \frac{2.C.\omega.V_{DC}}{i_{PZ-\max}}\right) \tag{8}$$

At the output of the rectifier, the energy E transferred between time t_1 and *T/2* is expressed as follows:

$$E = \int_{t1}^{T/2} .V_{DC}(t).i_D(t)dt = \frac{V_{DC}.i_{PZ-\max}}{\omega}(1 - \cos\omega t_1) \tag{9}$$

The average power supplied at the output of the bridge is:

$$\langle P_{PZ} \rangle = \frac{E}{T/2} \tag{10}$$

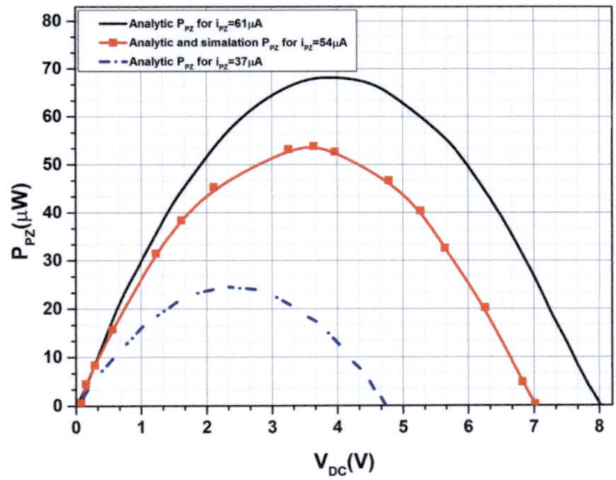

Fig. 10 Power supplied by the PPZ recuperator VDC

By injecting the time t1 into the previous equation, we find:

$$\langle P_{PZ} \rangle = \frac{2}{\pi} V_{DC}.i_{PZ\_\max} \left(1 - \frac{C\omega.V_{DC}}{i_{PZ\_\max}} \right) \tag{11}$$

It can be seen that the power supplied depends on the voltage of the DC stage and the current $I_{pz\text{-}max}$ (which is a function of the maximum strain). This expression allows us to plot the power recovered as a function of the voltage of the DC stage. In Fig. 10, we present the evolution of the power supplied to the load as a function of the VDC voltage for different currents (and thus different levels of excitation of the beam).

They use the simulation tool (for $i_{pz\text{-}max} = 53\mu A$ and $C_L = 10\mu F$) and by modifying the current consumed by the load (I_s), they read P and V_{dc} and obtain the points in o in Fig. 11. The analytical calculation and the simulations give concordant results (blue curve and points o). The simulation allows us to understand in a more intuitive way than an analytical expression that the power supplied (and thus dissipated in the load) is directly linked to the level of the rectified voltage. The problem of impedance matching is addressed here by simulation, but it can also be studied practically by varying the load resistance R_C and calculating the power PC supplied in a steady state from the V_{DC} measurement.

$$P_C = V_{DC}^2 / R_C \tag{12}$$

2.2.4 Power Supply for a Data Logger

The recovery system is now completed with a current consumption model of the data loader (because to a first approximation, the currents vary little with the supply

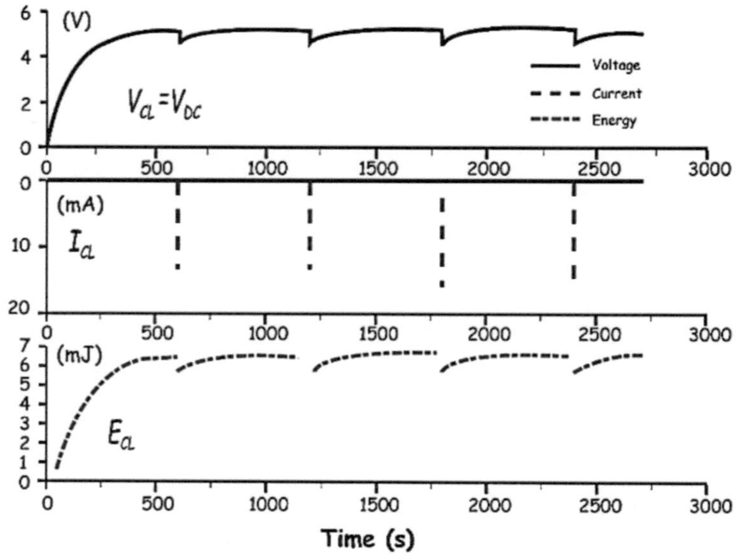

Fig. 11 Evolution of the CL capacitor parameters: voltage, current, and energy

voltage). The data logger (Jennic 5148) has a continuous so-called standby consumption of IS = 7μA at 5V, and a pulse consumption of IMES when it performs a measurement and sends data. This consumption is *16mA* under 5V during TMES = 12 ms which is an EMES energy of 937μJ. The measurements and transmissions are performed every $T_{CYCLE} = 10$ min. These consumptions are modeled respectively in Fig. 9 by the DC current source IS and the source IMES which are connected to the storage capacitor only during 12 ms each cycle by opening the switch S2. The piezoelectric recuperator must therefore provide an average power over a measurement cycle of:

$$\langle P\rangle = V_{DC}\cdot I_s + \frac{E_{MES}}{T_{CYCLE}} \tag{13}$$

Assuming that CL absorbs the consumption peaks without any noticeable voltage drop, we return to the case simulated in §4.3 Fig. 11 with $\langle P_{PZ}\rangle = \langle \mathrm{P}\rangle = 36{,}5\,\mu W$ and therefore $i_{pz\text{-}max} = 53\mu A$ et $V_{DC} = 5V$. To guarantee an almost constant bus voltage VDC, we dimension the CL capacitor so that VDC changes from Vmax = 5V to $V_{min} = 4.5V$ during the short measurement and transmission phase.

$$E_{MES} = \frac{1}{2}C_L\left(V_{\max}^2 - V_{\min}^2\right) \tag{14}$$

Which gives a CL value close to 400 μF.

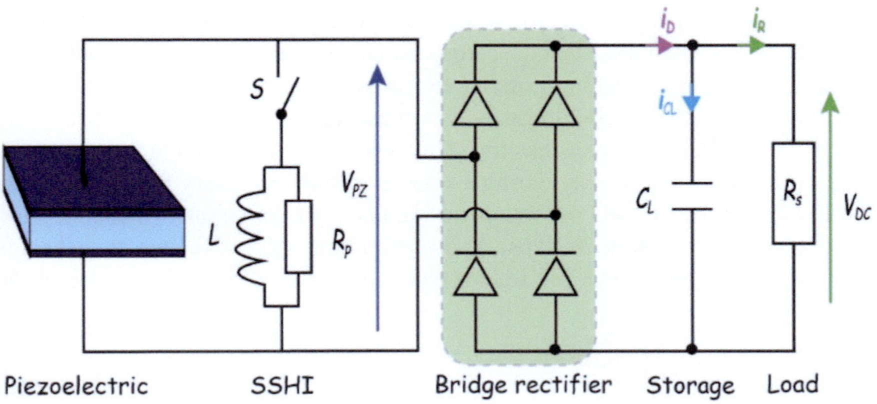

Fig. 12 Diagram of the parallel SSHI

The dimensioning of the system is now complete, one validates by simulation the electric supply of the datalogger by the piezoelectric (associated with the bridge diode). These simulations are relatively heavy (we simulate a total duration of 30 min, taking into account the rectified signals at 36 Hz). It is important to adjust the number of points per period (not more than 5), i.e., a calculation step of 5 ms.

We observe the operation of the system in Fig. 11. We can see that the voltage at the terminals of the capacitor is within the expected range. The current pulses (red curve: measurement transmission) do not cause the voltage of the capacitor CL to fall too much.

The device is correctly dimensioned for the application but (see Fig. 12) it does not function at the point of operation where the recoverable power is maximum (this point is located in our conditions at a voltage of approximately 3.5 V). Indeed, and as we saw previously, the recoverable power depends very strongly on the voltage VDC.

The recovery is not optimal, because, as we can see in Fig. 11, the current output of the rectifier is almost null during a great part of the period. Indeed, at each half-wave, the voltage at the terminals of the piezoelectric must be inverted, and the dynamics of this inversion is proportional to the excitation frequency of the piezoelectric. To improve the energy recovery, it is possible, on the one hand, to add to this device a mini converter to adapt to the impedance of the load seen by the recovery device, and on the other hand, to use another control structure, as we discuss in the following paragraph with an SSHI structure.

2.2.5 Structure SSHI

This part presents a more efficient control than the structure previously studied. There are several SSHI (Synchronized Switch Harvesting on Inductor) devices [7, 12–15]. The principle is globally the same for all of them. It consists of inverting the voltage at the terminals of the piezoelectric when the i_{PZ} current passes through zero. In this way, the i_{PZ} current and the voltage across the piezoelectric are in phase (unlike the simple diode bridge). It is thus possible to recover more power. We propose in this work to study a parallel SSHI structure as presented in Fig. 12. An inductor L is inserted in parallel to the piezoelectric. This inductor is connected to the rest of the circuit (via the switch S) only during a very short time necessary for the inversion of the voltage across C, which corresponds, in first approximation, to the half period of the LC resonant circuit. We chose an inversion time of $t_{inv} = 0.1$ ms (to make this time insignificant in comparison to the beam's resonance period). The value of the inductance is therefore:

$$L = \frac{t_{inv}^2}{\pi^2.C} = 14.4mH \tag{15}$$

This device is quite complex to set up (switch control), so we use simulation to compare it to the capacitive filtering rectifier. For the simulation to be realistic, the inductance is not perfect; we add to it a resistor Rp in parallel, which takes into account the global losses. We arbitrarily choose Rp = 3kΩ. The difficulty of the simulation is correctly controlling the switch. To do this, we must detect the passage through 0 of the current i_{PZ} to command the closing. The opening is done after the time t_{inv}. We propose the following model:

The comparator and edge detector will detect the current zero crossings. A monostable allows setting the switch closing time to t_{inv}. The device is then simulated using the same parameters as those used with the diode bridge, i.e., CL = 10μF and i_{PZ}-max = 53μA. In the end, the power delivered by each of the systems can be compared as a function of the current drawn by the load. The results are shown in Figs. 13 and 14.

As for the diode bridge, we observe that the recoverable power (as well as the DC bus voltage) evolves according to the current consumed. The SSHI device allows much more power to be recovered. If we consider that between the SSHI and the

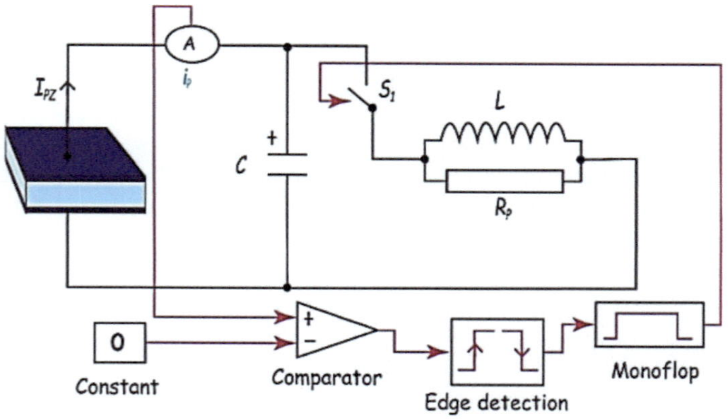

Fig. 13 SSHI structure–switch control

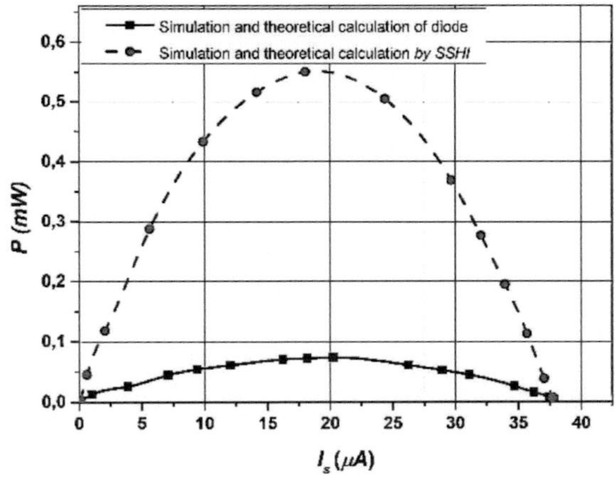

Fig. 14 Recoverable power with a SSHI device and with a simple rectifier with capacitive filtering

datalogger we have a lossless system that allows impedance matching (i.e., voltage leveling), then we can work at the point where the recoverable power is highest (370 μW for the SSHI and 46 μW for the diode bridge alone). The time to harvest the energy needed to turn on the datalogger (507 mJ) would therefore be:

- 1350 s or 22.5 min for the SSHI device
- 10,870 or 3 h for the diode bridge

The advantage of SSHI is visible here. This device remains complex to implement, and the simulations presented use ideal characteristics (except for the losses in the coil). It would be necessary to take into account the consumption of the MOS switch as well as that of its control to be closer to reality.

3 Conclusion

The objectives aimed at were reached: to interest the students, to incite them to find by themselves the manipulations to be carried out to model the device, to reflect on the validity of the measurements obtained, and to make the link between experimentation, modeling, and simulation on an original system. The students also asked themselves questions related to the problems of impedance matching to optimize the power transfer and finally, they were able to have an initiation to efficient adapted controls for energy recovery. The large size of the experimental recovery device can be criticized but is justified for a device that will be used by many hands. The dimensions of the ruler also allow the accelerometer to be easily fixed. Commercially available piezoelectric beams (MIDE) or much smaller complete devices also exist and are commercially available. They can be used to show the possible miniaturization of such devices.

References

1. Lifi, H. Ennawaoui, C. Hajjaji, H. Touhtouh, S. Benjelloun, M. et Azim, A, Elaboration, Characterization and Thermal Shock Sensor Application of Pyroelectric Ceramics PMN–xPT, Sens. Lett., sept. (2017).
2. Yessari, M. Lifi, H. Hajjaji, H. Rguiti, H. Benjelloun, M. et Arbaoui, A., Elaboration and characterization of PMN–xPT solid ceramics used in a numerical simulation of a piezoelectric transducer for a loudspeaker application, Opt. Quantum Electron., vol. 5, n° 48, p. 1-12, (2016).
3. Lifi, H., Sensors and energy harvesters based on (1–x)PMN-xPT piezoelectric ceramics, Eur. Phys. J. Appl. Phys., vol. 88, n° 1, p. 10901, Oct. (2019).
4. Khrissi, S. Haddad, M. Bejjit, L. Lyazidi, S,A. Amraoui, M. E. et Falguères, C. « Raman and XRD characterization of Moroccan Marbles », IOP Conf. Ser. Mater. Sci. Eng., vol. 186, p. 012028, mars (2017).
5. Khrissi, S. Bejjit, L. Haddad, S. Falguères, C. Ait Lyazidi, S. et El Amraoui, M. « Study of marbles from Middle Atlas (Morocco): elemental, mineralogical and structural analysis », Mater. Sci. Eng. Conf. Ser., vol. 353, n° 1, p. 012013, may (2018).
6. Khrissi, S. Lifi, H. Lifi, M. Nossir, N. Tabbai, Y. et Hnawi, S. K., Piezoelectric Vibration Energy Harvesters with Distinct Interdigitated Electrodes were Used for Toxic Gas Detection and in a Numerical Simulation for a Glucose Sensor Application. Int. J. Sens. Wirel. Commun. Control, vol. 12, p. 1-1, (2022).
7. Innocenti, B. « Biomechanics: a fundamental tool with a long history (and even longer future!) », Muscles Ligaments Tendons J., vol. 7, n° 4, p. 491-492, avr. (2018).
8. Ravi, K. V. P. et Rohit, R. M., A Wrapper Based Feature Selection Using Grey Wolf Optimization for Botnet Attack Detection, Int. J. Sens. Wirel. Commun. Control, vol. 11, n° 9, p. 951-956, Oct. (2021).
9. Chilakala, S. Ashutosh, S. et Vunnava, A., Improved Detection Performance of Energy Detection Based Spectrum Sensing in Cognitive Radio Networks, Int. J. Sens. Wirel. Commun. Control, vol. 11, n° 9, p. 957-962, Oct. (2021).
10. Ahmed, A. Khaled, M. N. et Korany, R. M., Enhancement for Content Caching and Delivery with Energy Consumption Awareness in Wireless Networks, Int. J. Sens. Wirel. Commun. Control, vol. 11, n° 9, p. 963-975, Oct. (2021).
11. Lifi, H. Alaoui-Belghiti, A. Lifi, M. Khrissi, S. Nossir, N. Tabbai Y. et Benjelloun, M., Mechanical Energy Harvesting System from the Human Arm Movement for Continuous Blood Pressure Measurement Int. J. Sens. Wirel. Commun. Control, vol. 12, p. 1-1, (2022).
12. Abala, I. « Density, Viscosity, Refractive Index, and Related Thermophysical Properties of Dibutyl Ether +2-Butanol + Cyclohexane Ternary Systems », J. Chem. Eng. Data, vol. 67, no 12, p. 3532-3542, déc. (2022).
13. Lifi, M. Rubio-Pérez, G. Lifi, H. Muñoz-Rujas, N. Aguilar, F. et Alaoui, F. E. M. Excess properties of mixtures containing 2-(2-ethoxyethoxy)ethanol, methylcyclohexane, cyclohexane and 1-hexene as fuel representative, J. Chem. Thermodyn., vol. 177, p. 106933, févr. (2023).
14. Lifi, H., Investigation of Complex Impedance and Modulus Properties of the Relaxor PMN-XPT Ceramics, in Proceedings of the Sixth International Symposium on Dielectric Materials and Applications (ISyDMA'6), Cham, p. 109-117, (2022).
15. Khrissi, S., Mechanical Energy Harvesting from Human Arm Using a Piezoelectric Ceramic, in Proceedings of the Sixth International Symposium on Dielectric Materials and Applications (ISyDMA'6), Cham, p. 185-194, (2022).

Correction to: Proceedings of the Eighth International Symposium on Dielectric Materials and Applications (ISyDMA'8)

Ashok Vaseashta, Mohammed Essaid Achour, Mustapha Mabrouki, Amina Tachafine, and Mustapha Aitali

Correction to:
A. Vaseashta et al. (eds.), *Proceedings of the Eighth International Symposium on Dielectric Materials and Applications (ISyDMA'8)*, Springer Proceedings in Physics 349, https://doi.org/10.1007/978-3-031-84809-4

In the original version of the book, the book was not included in the series "Springer Proceedings in Physics". This has now been rectified and the Series information has been updated".

The updated version of this book can be found at
https://doi.org/10.1007/978-3-031-84809-4

A. Vaseashta et al. (eds.), *Proceedings of the Eighth International Symposium on Dielectric Materials and Applications (ISyDMA'8)*, Springer Proceedings in Physics 349, https://doi.org/10.1007/978-3-031-84809-4_27

Author Index

A. Vaseashta et al. (eds.), *Proceedings of the Eighth International Symposium on Dielectric Materials and Applications (ISyDMA'8)*, Springer Proceedings in Physics 349, https://doi.org/10.1007/978-3-031-84809-4

Subject Index

A. Vaseashta et al. (eds.), *Proceedings of the Eighth International Symposium on Dielectric Materials and Applications (ISyDMA'8)*, Springer Proceedings in Physics 349, https://doi.org/10.1007/978-3-031-84809-4

MIX
Papier aus verantwortungsvollen Quellen
Paper from responsible sources
FSC® C105338

If you have any concerns about our products,
you can contact us on
ProductSafety@springernature.com

In case Publisher is established outside the EU,
the EU authorized representative is:
Springer Nature Customer Service Center GmbH
Europaplatz 3, 69115 Heidelberg, Germany

Printed by Libri Plureos GmbH
in Hamburg, Germany